Groundwater Contaminants and their Migration

Geological Society Special Publications

Series Editor A. J. FLEET

GEOLOGICAL SOCIETY SPECIAL PUBLICATION NO. 128

Groundwater Contaminants and their Migration

EDITED BY

J. MATHER

Royal Holloway and Bedford New College
University of London
UK

D. BANKS

Geological Survey of Norway
Norway

S. DUMPLETON

British Geological Survey
Keyworth
UK

M. FERMOR

Environmental Simulations Ltd
UK

1998

Published by

The Geological Society

London

THE GEOLOGICAL SOCIETY

The Society was founded in 1807 as The Geological Society of London and is the oldest geological society in the world. It received its Royal Charter in 1825 for the purpose of 'investigating the mineral structure of the Earth'. The Society is Britain's national society for geology with a membership of around 8000. It has countrywide coverage and approximately 1000 members reside overseas. The Society is responsible for all aspects of the geological sciences including professional matters. The Society has its own publishing house, which produces the Society's international journals, books and maps, and which acts as the European distributor for publications of the American Association of Petroleum Geologists, SEPM and the Geological Society of America.

Fellowship is open to those holding a recognized honours degree in geology or cognate subject and who have at least two years' relevant postgraduate experience, or who have not less than six years' relevant experience in geology or a cognate subject. A Fellow who has not less than five years' relevant postgraduate experience in the practice of geology may apply for validation and, subject to approval, may be able to use the designatory letters C Geol (Chartered Geologist).

Further information about the Society is available from the Membership Manager, The Geological Society, Burlington House, Piccadilly, London W1V 0JU, UK. The Society is a Registered Charity, No. 210161.

Published by The Geological Society from:
The Geological Society Publishing House
Unit 7, Brassmill Enterprise Centre
Brassmill Lane
Bath BA1 3JN
UK
(*Orders*: Tel. 01225 445046
         Fax 01225 442836)

First published 1998

The publishers make no representation, express or implied, with regard to the accuracy of the information contained in this book and cannot accept any legal responsibility for any errors or omissions that may be made.

British Library Cataloguing in Publication Data
A catalogue record for this book is available from the British Library.

ISBN 1-897799-95-0
ISSN 0305-8719

Typeset and printed by Alden Group, Oxford, UK.

Distributors

USA
AAPG Bookstore
PO Box 979
Tulsa
OK 74101-0979
USA
(*Orders*: Tel. (918) 584-2555
         Fax (918) 560-2652)

Australia
Australian Mineral Foundation
63 Conyngham Street
Glenside
South Australia 5065
Australia
(*Orders*: Tel. (08) 379-0444
         Fax (08) 379-4634)

India
Affiliated East-West Press PVT Ltd
G-1/16 Ansari Road
New Delhi 110 002
India
(*Orders:* Tel. (11) 327-9113
         Fax (11) 326-0538)

Japan
Kanda Book Trading Co.
Tanikawa Building
3-2 Kanda Surugadai
Chiyoda-Ku
Tokyo 101
Japan
(*Orders*: Tel. (03) 3255-3497
         Fax (03) 3255-3495)

Contents

Preface   vii

1. Groundwater pollution: the UK legal framework   **1**

R. C. Harris: Protection of groundwater quality in the UK: present controls and future issues   3

K. Mylrea: Recent UK legal developments relating to pollution of water resources   15

R. P. Ashley: Foreseeability of environmental hazards: the implications of the Cambridge Water Company case   23

2. New perspectives on groundwater contaminants and their migration   **27**

W. G. Burgess, J. Dottridge & R. M. Symington: Methyl tertiary butyl ether (MTBE): a groundwater contaminant of growing concern   29

G. M. Williams, P. J. Hooker, D. J. Noy & C. A. M. Ross: Mechanisms for ^{85}Sr migration through glacial sand determined by laboratory and *in situ* tracer tests   35

3. Groundwater pollution by inorganic contaminants   **49**

S. Lee & D. A. Spears: Potential contamination of groundwater by pulverized fuel ash   51

R. J. Andrews, J. W. Lloyd & D. N. Lerner: Sewage sludge disposal to agriculture and other options in the UK   63

I. Davey, I. Moxon & D. Hybert: Investigation of contamination at a public supply borehole in Hertfordshire, UK   75

G. G. Bowen, C. Dussek & R. M. Hamilton: Pollution resulting from the abandonment and subsequent flooding of Wheal Jane Mine in Cornwall, UK   93

P. J. K. Sadler: Minewater remediation at a French zinc mine: sources of acid mine drainage and contaminant flushing characteristics   101

4. Groundwater pollution by hydrocarbons   **121**

D. Banks: Migration of dissolved petroleum hydrocarbons, MTBE and chlorinated solvents in a karstified limestone aquifer, Stamford, UK   123

L. Clark & P. A. Sims: Investigation and clean-up of jet fuel contaminated groundwater at Heathrow International Airport, UK   147

R. G. Clark: Remediation of a hydrocarbon leakage from a service station at Wansford, Cambridgeshire, UK   159

J. M. W. Holden & N. Tunstall-Pedoe: Remediation of a petroleum spill to groundwater at a fuel distribution terminal (Long Island, USA) using pump-and-treat and complementary technologies   165

5. Groundwater pollution by chlorinated solvents   **181**

D. G. Muldoon, P. J. Connolly, A. W. Makovitch, J. M. W. Holden & N. Tunstall-Pedoe: Groundwater remediation of chlorinated hydrocarbons at an electronics manufacturing facility in northeastern USA   183

B. D. R. Misstear, R. P. Ashley & A. R. Lawrence: Groundwater pollution by chlorinated solvents: the landmark Cambridge Water Company case, UK   201

B. D. R. Misstear, P. W. Rippon & R. P. Ashley: Detection of point-sources of contamination by chlorinated solvents: a case study from the Chalk Aquifer of eastern England   217

P. K. Bishop, D. N. Lerner & M. Stuart: Investigation of point source pollution by chlorinated solvents in two different geologies: a multilayered Carboniferous sandstone–mudstone sequence and the Chalk   229

CONTENTS

6. Groundwater pollution by radionuclides   **253**

V. LGOTIN & Y. MAKUSHIN: Groundwater monitoring to assess the influence of injection of   255
liquid radioactive waste on the Tomsk public groundwater supply, Western Siberia, Russia

I. N. SOLODOV: The retardation and attenuation of liquid radioactive wastes due to the   265
geochemical properties of the zone of injection

7. Groundwater pollution by exotic organics: acid tars, pesticides and phenols   **281**

D. BANKS, N. L. NESBIT, T. FIRTH & S. POWER: Contaminant migration from disposal of acid   283
tar wastes in fractured Coal Measures strata, southern Derbyshire

B. C. GORE & I. M. CAMPBELL: Great Bridge Marl Pit: a case study in the prevention of   313
contaminant migration

P. J. CHILTON, A. R. LAWRENCE & M. E. STUART: Pesticides in groundwater: some preliminary   333
results from recent research in temperate and tropical environments

J. SWEENEY, P. A. HART & P. J. McCONVEY: Investigation and management of pesticide   347
pollution in the Lincolnshire Limestone aquifer in eastern England

Index   361

Preface

This book is the offspring of an unholy alliance between three different projects. Many of the papers are selected from Professor John Mather's highly successful yearly 'Groundwater Pollution' conferences organized in London by IBC Technical Services Ltd in association with Royal Holloway, University of London. A second group of papers was presented in July 1995 at the University of Sheffield during a meeting convened by Mark Fermor on behalf of the Geological Society's Hydrogeology Group. The topic of the meeting was 'Groundwater Remediation'. The remainder of the papers are case studies, coaxed out of the private and regulatory sectors by Dave Banks and Steve Dumpleton, from a desire to bring some of the often confidential data from contamination incidents into the spotlight of the wider professional arena.

We hope that you like the resulting Special Publication and that we have managed to impose some coherence on the broad variety of papers from a huge range of sources: private consultancy, academia and public authorities. It is perhaps surprising that only now, well into the second 'century' of the series, has a Special Publication solely devoted to pollution of the geological environment finally been produced. This reflects the very tough 'market' in environmental hydrogeology in the UK today, where time to prepare scientific papers is a luxury that few can afford. Bearing this in mind, we extend great thanks to our authors and referee panel for bringing the publication to fruition.

The book presents a typically UK perspective on groundwater contamination, 'warts' and all. It gives the international reader an opportunity to assess the standing of British contaminant hydrogeology in the world today. We are, however, glad that the publication is not exclusively British. In particular, we thank authors who present us with an international perspective: Scott Wilson CDM for two papers on organic contamination at industrial sites in the USA (**Muldoon et al.**; **Holden & Tunstall-Pedoe**); Piers **Sadler** for his description of minewater pollution from France and Drs Igor N. **Solodov**, Viktor Aleksandrovich **Lgotin** & Yuri Vasilevich **Makushin** for their presentation of data, previously unavailable in English, from the major liquid radioactive waste disposal operation in Tomsk, Central Siberia.

We hope that this is a worthwhile compilation which will enable both the experienced hydrogeologist and the interested scientist to discover how we have polluted, investigated and attempted to clean up humankind's most precious geological resource: groundwater.

John Mather, Egham, UK
Dave Banks, Trondheim, Norway
Steve Dumpleton, Keyworth, UK
Mark Fermor, Shrewsbury, UK

The Scientific Review Panel

Section 1: Groundwater pollution: policy and legislation in the UK

The UK has often been described as the 'dirty old man' of Europe: its environmental policies have been seen as too pliable and polluter-friendly by many European neighbours. Defenders of the UK policy cite a pragmatic, rather than obstructively ideological, stance and a willingness on the part of regulators to co-operate with industry to solve groundwater pollution problems. The UK is also playing a leading role in developing a European framework for the use of risk assessment techniques which are acceptable to both industry and regulators (see e.g. Quint *et al.* 1996). It is also the case that the UK Environment Agency represents one of very few European regulatory agencies which have realised the EU's objective of integrated river-basin management of both the quality and quantity of surface water and groundwater resources. Some of the UK's fiercest critics, such as the ideologically rigid Norway, are still years from achieving such a goal. Readers can judge for themselves, as Bob **Harris** explains how the UK implements European groundwater protection policy, Kathy **Mylrea** guides us through the Byzantine labyrinths of the UK legal system and highlights the apparent perversity of some judgements (e.g. that storage of chemicals can constitute a 'natural' use of land, as in the case *of 'Cambridge Water Company versus Eastern Counties Leather'*. See **Misstear, Ashley & Lawrence**, and **Bishop, Lerner & Stuart** for further details in section 5). Finally, Paul **Ashley** examines one of the implications of the 'Eastern Counties Leather' case, namely that liability for a historic groundwater pollution incident rests upon forseeability of damage.

QUINT, M. D., TAYLOR, D. & PURCHASE, R. (eds) 1996. Environmental impact of chemicals: assessment and control. *Royal Society of Chemistry Special Publication*, **176**.

Protection of groundwater quality in the UK: present controls and future issues

R. C. HARRIS

National Groundwater and Contaminated Land Centre, Environment Agency, Olton Court, 10 Warwick Road, Solihull, West Midlands B92 7HX, UK

Abstract: Groundwater pollution and its prevention are discussed in the context of the currently perceived issues in the UK and Europe and the future challenges, particularly in addressing historical pollution within the existing regulatory framework. Contamination from industrial processes and facilities is considered to be the most serious point source but its significance has not been appreciated because of the preoccupation with landfills. The influences of legislation, education and policy are reviewed and the growing impact of liability issues discussed. Progress in the development and implementation of the Environment Agency's Groundwater Protection Policy is described. Developments in the use of groundwater modelling and risk assessment techniques are considered in the light of the growing need to examine the cost-effectiveness of remedial treatment of historical pollution.

Over the past five years groundwater issues have been higher on the UK environmental agenda than at any time previously. However, the awareness of the public in general, and industry in particular, about the importance of groundwater as a water resource and the need for its protection remains at a relatively low level in comparison to other developed countries. The publication and dissemination of a national policy for the protection of groundwater in 1992 has helped to raise the profile (NRA 1992). European initiatives have placed groundwater on the political agenda and, through the production and implementation of directives, affected particular businesses, most notably those in agriculture, waste disposal and property development. Nevertheless it remains a truism that 'out of sight is out of mind' and there is still a need for better education about the potential for many different land users to affect the underlying groundwater environment.

The growing realization of the extent of groundwater pollution under industrial sites coupled with a better technical understanding of the hydrogeological and biogeochemical processes that govern pollutant transport in underground strata will lead to interesting challenges for regulators and industry alike over the next few years if we are to ensure that our water resources are secure for use by future generations.

Groundwater quality issues

There is no catalogue or database of groundwater pollution for the UK. The lack of collated data and a national view has handicapped regulators from focusing on the main activities that have caused groundwater pollution. A preoccupation with landfill has also diverted attention and resources away from other important sources of pollution.

One problem has been the poor standard of our groundwater quality monitoring network. For historical reasons this has been based largely on public supply sources and has never been designed according to specifically identified objectives. Thus what data have been collated give either an optimistic impression, based on public supply abstractions which have received protection over the years from potential sources of pollution, or the opposite where it is based on monitoring around pollution sources such as landfills for specific legislative reasons. It is therefore difficult for the Environment Agency, apart from any other interested party, to gain a good understanding of the state of the nation's groundwater. The lack of resources invested in basic groundwater monitoring over the years will need to be redressed, perhaps at the expense of our surface water surveillance network. The National Rivers Authority (NRA) reviewed the situation in its latter years and the Environment Agency will draw up an overall monitoring strategy.

Diffuse Pollution

The majority of groundwaters that are considered to be contaminated have been affected by diffuse pollutants from agricultural land use practices. Nitrate concentrations in abstracted

HARRIS, R. C. 1998. Protection of groundwater quality in the UK: present controls and future issues. *In*: MATHER, J., BANKS, D., DUMPLETON, S. & FERMOR, M. (eds) *Groundwater Contaminants and their Migration*. Geological Society, London, Special Publications, **128**, 3–13.

groundwater for many outcrop areas of our Major Aquifers will exceed the maximum allowable concentration for drinking water supply ($50 \, \text{mg} \, \text{l}^{-1}$ as NO_3) early in the 21st century.

Nitrate

Concentrations in pore waters leaving the soil zone from under intensively managed agricultural land can exceed $50 \, \text{mg} \, \text{l}^{-1}$ by several times. It is only the dilution afforded by mixing with low nitrate water from non-agricultural areas in the same catchment and older water at depth which allows the water companies to continue to utilize groundwaters in the traditional way. Even so it is necessary to treat groundwater supplies in some areas in order to maintain blended water in the distribution system below the legal limit. The first full-scale nitrate removal plant became operational in 1990 in the Lichfield area (Woodward 1994). Others have followed and if the current rise rates are maintained it will become an increasingly common feature in those central and eastern parts of the country where a heavy reliance on groundwater coincides with intensive agriculture and low rainfall. Figure 1 shows a typical trend for a groundwater source abstracting from the Triassic sandstones in Shropshire.

Solvents

Similarly, diffuse pollution in those urban areas which coincide with Major Aquifers has had a significant impact on groundwater. Studies of Birmingham and Coventry in particular have shown the widespread existence of chlorinated solvents in around 80% of sampled boreholes (Lerner & Tellam 1992). In these areas the ubiquitous use of solvents in the motor and associated industries has resulted in large numbers of discrete sources coalescing to give rise to a diffuse pollution problem. Chlorinated solvents are also a significant source of pollution in rural areas from point sources. At least 14 public supply boreholes need treatment in order to keep supplies to domestic users below the very low legal limits ($30 \, \mu\text{g} \, \text{l}^{-1}$ for trichloroethlyene and $10 \, \mu\text{g} \, \text{l}^{-1}$ for tetrachloroethylene) (Harris 1993).

Pesticides

Similar very low acceptable pollutant levels have made the case against the third category of pollutants, pesticides, difficult to establish. Maximum allowable concentrations for individual pesticides are as low as $0.1 \, \mu\text{g} \, \text{l}^{-1}$ (i.e. five orders of magnitude difference from the concentration for nitrate). For many pesticides analytical techniques have not been devised and for many others detection levels and maximum acceptable levels are similar. Sampling protocols have to be exceedingly stringent at such low concentrations and it is doubtful whether much confidence can be placed in positive data unless there is evidence of repeated exceedance and concentrations significantly in excess of the detection limit.

However, it *is* certain that the non-agricultural use of the herbicides atrazine and simazine

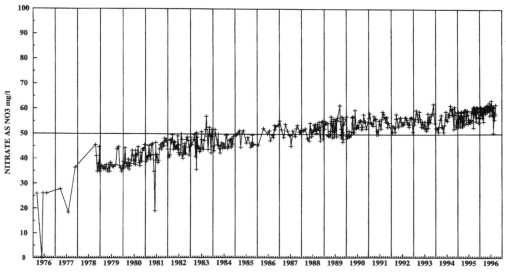

Fig. 1. Extract from the National Groundwater Nitrate Database showing increasing nitrate concentrations in groundwater from a source abstracting water from the Triassic sandstones in Shropshire.

presents a very real threat to groundwater. Several borehole supply sources have become significantly contaminated, some severely enough to be taken out of supply or to necessitate treatment. Several examples can be found of the affected source being situated close to a railway line where periodic spraying has taken place onto a largely soil-free surface, sometimes within a cutting and therefore closer to the water-table.

There are fewer examples of significant contamination from the purely agricultural use of pesticides. The use of soakaways for the discharge of washings or wastes may lead to an exceedance of the attenuating capacity of the soil or underlying strata and will be a significant threat, particularly on fissured aquifers, but much research remains to be carried out before we can quantify the risks. Concerns have recently been raised about the increasing planting of maize since the early 1990s and the consequent use of atrazine as a pre-emergent weedkiller on a more widespread basis.

Point sources

Herbicide spraying on railway track can be considered a linear source of pollution while soakaways are classical point sources. Such discrete sources of pollution can give rise to very significant effects which by their very nature are geographically confined. Because the processes which attenuate pollutants once they enter the ground can be readily overloaded by high concentrations of pollutants, groundwater contamination may spread for some distance as a well-defined plume.

The Department of the Environment carried out a very broad overview of groundwater pollution in England and Wales in 1988 (DoE 1988) but a more comprehensive study was undertaken for the NRA in 1995 (De Hénaut et al. 1996). Although the data for the study were derived from limited sources, the results give the best available indication of the nature and extent of point sources of groundwater pollution in England and Wales.

Landfill sites are numerically the most significant category of land use identified in the study, since the data collected are biased towards those land use categories which are more highly regulated. However, in terms of their actual impact on groundwater they are considered to be somewhat less of a problem than other sources and types of pollutants.

Industrial activities have undoubtedly given rise to the most significant examples in the UK.

Incidents are increasingly coming to light as companies carry out evaluations of their property, in connection with land sales, new development or environmental audits, which involve an investigation of groundwater quality. It is now apparent that beneath the majority of industrial premises that handle, manufacture or store organic chemicals in liquid or soluble form, the groundwater will be found to be polluted to some degree. The extent of localized contamination at some locations can be significant with percentage levels of some soluble compounds encountered at some distance below the water-table.

The main point source groundwater contaminants are organic chemicals. These have many uses such as fuels (hydrocarbons) and degreasants (chlorinated solvents), or in various manufacturing processes (e.g. cutting oils, a variety of raw chemicals). Other sources relate to the deposition of waste chemicals, e.g. the products of coal-gas and coke manufacture (creosote and acid tars). Significant cases from controlled landfill are difficult to find, although where the attenuation capacity both within and outside the landfill has been exceeded, substantial groundwater pollution has been recorded. Examples include landfills at Helpston, Cambridgeshire and Pakefield, Suffolk where pesticide disposal has impacted on groundwater resources, with the former affecting a public supply (Anon 1993a) .

Significant pollution from inorganic chemicals (apart from nitrate) is generally much rarer because of the differing scales of concentration considered to be a problem and the specialized conditions in which they will be mobile. For instance, heavy metals are rarely a problem in the groundwater environment since they can readily adsorb onto the rock matrix unless there are particularly acidic conditions, or other conditions that give rise to increased solubility.

Incidents of groundwater pollution from point sources have been difficult to identify since reliance must be placed on contamination reaching a monitoring location, e.g. a borehole or spring. Because groundwater moves so slowly within aquifers, it can often take many decades for the effects of polluting activities to be detected. By then there is a substantial volume of water, and rock, that is affected and clean-up is very difficult and expensive. Where incidents are identified which present a long-term threat to public water supplies derived from groundwater, the approach of the Environment Agency is usually to require the occupant to carry out a site investigation and identify the extent of the pollution. Once sufficient data are available, a modelling study can be undertaken to determine the likely fate of the pollution plume, the timescales involved and the risks to public supplies or other potential discharge points such as watercourses. A scheme of remediation is then developed and tested against the model. Experience to date shows that larger companies tend to be willing to co-operate in such approaches, although even with these the large costs of a groundwater site investigation and potential clean-up can be difficult to finance, particularly in times of recession. There is more reluctance with smaller companies or landowners who lack the necessary funding, and also where the pollution results from a previous use of the site such that the liability is inherited.

However, there are relatively few instances of highly significant impacts on groundwater users. The UK still awaits an example on the scale of Love Canal, which did so much to raise public awareness in North America and galvanize legislation, both in terms of prevention and clean-up. The example that has done most to raise issues of historical liabilities is the legal case between Eastern Counties Leather and Cambridge Water Company, which was eventually determined by the House of Lords (Anon 1993b). This involved a claim for compensation in respect of a public supply source that was contaminated, and subsequently had to be abandoned, due to contamination by trichloroethlyene which had entered the groundwater as a result of practices at the leather company. The Law Lords determined that, although there was no doubt of the source of pollution, the polluter could not be held responsible for his historical actions when it could not have been foreseen that they would give rise to the problems that occurred. These new concerns over liability for contaminated land and groundwater, fuelled by the debate over new and proposed contaminated land legislation and the Eastern Counties Leather case, have made industry consider groundwater pollution much more seriously than in the past. The insurance profession may therefore impart controls that regulators cannot.

The regulatory framework

The Environment Agency is the primary agency responsible for regulation of the water environment in England and Wales. In Scotland and Northern Ireland the role is undertaken by the Scottish Environmental Protection Agency (SEPA) and the Department of the Environment respectively. Other bodies have an indirect regulatory role, most notably local authorities. Control is exercised through a combination of legal requirements, statutory and non-statutory codes of practice, published policy and guidance documents, together with general advice and education.

Legal controls

Groundwater protection was not specifically addressed in UK legislation until the Water Resources Act of 1963, and only then for the very narrow activity of discharging effluent direct to groundwater by means of wells, pipelines or boreholes. Although various other activities were brought under control in the Control of Pollution Act 1974, these all related to point sources of potential pollution and it was not until the Water Act 1989 that legislation designed specifically to control diffuse pollution, from agricultural land use, was introduced.

The single most important influence on groundwater quality legislation and pollution prevention practice has been the European Directive on the Protection of Groundwater from Dangerous Substances, implemented in 1981. It also only relates to point sources of pollution but requires the protection of *all* groundwater from specific substances or groups of substances (List I and II compounds) regardless of present or future use and the extent of the aquifer within which it is contained.

The relevant legislation under the jurisdiction of the Environment Agency is mostly contained within the Water Resources Act 1991 and effectively takes forward those provisions within Part II of the Control of Pollution Act 1974 and the Water Act 1989 which relate to the water environment. Further powers are contained in the Environment Act 1995, most notably those relating to the clean-up of contaminated land and the enhanced powers to require remediation of polluted groundwater. It is an offence to pollute controlled waters, including groundwater. Powers have been available to prosecute the owners of sites for allowing or knowingly permitting pollution to take place from their land (Section 85 of the Water Resources Act 1991). However, it is difficult and expensive to gather sufficient evidence to prosecute a polluter since the drilling of boreholes is both costly and uncertain. This is reflected in the fact that only three prosecutions regarding groundwater pollution events were taken by the NRA in its lifetime. Powers have also been available under Section 161 of the Water Resources Act for the NRA/Environment Agency, to forestall or remedy pollution where necessary by carrying out relevant works itself and reclaiming the costs of so doing from the site owners. Although S.161 powers have often been invoked in the case of surface water pollution incidents, they have rarely, if ever, been used in promoting the clean-up of historically contaminated groundwaters, or in cases where the groundwater is interacting with surface waters and causing a consequential deterioration in quality. This was primarily because the high expense of undertaking remediation works was not funded within the regulatory authority, particularly when there was in most cases little likelihood of successful cost recovery.

The introduction of Section 57 and Schedule 22 (para. 162) of the Environment Act is therefore a welcome legislative addition to the regulator in this field and should provide an impetus to the further improvement of controlled waters. However, few powers are available to the Environment Agency to *prevent* pollution. Most of its powers are retrospective once pollution has occurred. This presents a problem for groundwater since clean-up is a very difficult and long-term process, highly expensive and rarely completely effective.

Section 92 of the Water Resources Act 1991 enables the Secretary of State to make regulations to control any activity and thereby prevent pollution. The only regulations introduced so far relate to the storage of farm slurries and agricultural fuel oil. The vast bulk of potentially polluting activities remain uncontrolled for the purposes of avoiding pollution of controlled waters.

Section 93 approaches pollution prevention from a different perspective by allowing for the definition of statutory zones within which various prescribed activities can be prohibited or only permitted under formal consent of the Environment Agency. Statutory zones have been a common feature of European pollution control practice for many decades and there are isolated examples of their past use in local situations in the UK. Although they have a place in surface water pollution control, they are not considered appropriate in the UK today, for general groundwater pollution prevention purposes, because of the inherent uncertainty in zone definition. However, statutory groundwater protection zones have effectively been introduced by the back door in relation both to Nitrate Sensitive Areas (NSAs) and the designation of land as Nitrate Vulnerable Zones (NVZs), set up for the control of diffuse pollution of nitrate from agricultural land use practices.

Processes that are prescribed under the Environmental Protection Act 1990 are controlled by authorizations granted by The Environment Agency. The authorization process considers possible discharges to all media, including groundwater and can be a powerful preventative tool in requiring such measures as bunding around storage tanks, above-ground distribution pipework, etc. However, there are many processes which are not covered under the Integrated Pollution Control (IPC) procedures and often only specific activities on a large site are controlled by this means.

Indirect controls

Apart from direct legislation, other legal controls and requirements can be highly effective in preventing pollution or minimizing its impact. The most important, and often undervalued, of these are the controls within the Town and Country Planning Act 1990 and related legislation. The policies in statutory development plans are particularly important in that they set out the framework for land use change and provide the key reference in determining development applications. Many developments, ranging from large industrial estates to graveyards, which pose direct or indirect threats to groundwater require planning permission. The Environment Agency is a consultee of local planning authorities (LPA) over new development, and measures to limit the effect of the particular activity can be requested in any planning permission granted where no other legislation exists to control it.

Another example of an indirect control concerns the requirement for underground petroleum storage tanks to be licensed under the Petroleum (Regulation) Acts 1928 and 1936 by the relevant local authority (fire authorities in Metropolitan areas). Although the Environment Agency plays no formal part in this, its aims and those of the licensing authority are similar: there should be no leakage.

Codes of practice have been developed for many facets of industry and agriculture. These are produced by government departments or industry and are a powerful way of promoting ideals and achieving consistency. Relevant examples are the Code of Good Agricultural Practice for Water, the series of Waste Management Papers, published by MAFF and DoE respectively, and the Institution of Civil Engineers Guide to Contaminated Land: Investigation, Assessment and Remediation. However, unless there is a statutory requirement to follow the guidance it is unlikely to be universally applied. Education and promotion of the guidance has a vital role to play in achieving its aims.

Sometimes it is sufficient to draw the attention of the user or manufacturer of the particular chemical to the problems that result, since they are often unaware of the impact of their actions/ product. The use of atrazine as a herbicide on the permanent way is a good example. This was taken out of use as a result of informal discussions between the water and railway industries and has since been prohibited for non-agricultural use in general.

Education

Public understanding of groundwater and the need for its protection is low within the UK. To some extent this reflects our relatively low usage of groundwater for drinking water (around 30–35% of the total) compared to some other developed countries (Denmark 98%, Germany 89%, Netherlands 67%, USA 50%) and there have not been any significant pollution events to catch the public imagination. In Germany and the Netherlands, for example, it is a common sight to see road signs delineating the boundary of a water protection zone. Besides having the effect of making the area off-limits for the transport of potential contaminants, this is a simple device for increasing public awareness.

Occasionally groundwater issues are used by local objectors in an attempt to help them resist particular developments which they oppose for different reasons more immediate to themselves. Invariably on these occasions a purist viewpoint is taken and strict interpretations of the EC Groundwater Directive are promoted without wishing to understand the complexities of the many factors that the professional hydrogeologist has to take into account. The concept of risk is particularly poorly understood. The professional is as much to blame for this situation as anyone, since there has been little previous attempt to promote understanding beyond a narrow group of specialists.

The United States in particular has a high level of public involvement and consultation in local decision-making. Some States have promoted groundwater awareness campaigns within local communities, aiming much information at schoolchildren. The annual Childrens' Groundwater Festival sponsored by the Nebraska Groundwater Foundation is a good example. (The Nebraska Groundwater Foundation is a non-profit-making educational foundation dedicated to educating the public about the conservation and management of groundwater.) They are attempting to promote better understanding in a popular way including demonstrations, visual aids and even folk songs. 'Hey Mister, that's my aquifer' may not get to number one in the charts but it is a novel way of getting a

message across to those who may be resistant to other means. Educating children is surely the best way of helping ensure that those in future charge of substances that can cause harm to the groundwater environment will at least be aware of their responsibilities. It is an area that has received limited recognition in the UK and one that the Environment Agency would do well to address in future. The publication and promotion of the Groundwater Protection Policy has started this process in a general way but much remains to be done.

Groundwater protection policy

Policies to protect groundwater sources are not a new idea. Many examples can be found in the UK water supply industry of protected areas, relating to the perceived source catchment, within which activities were restricted or banned. This was in some ways easier than today because the management of local water supply was often in the hands of the same body that controlled development, the local authority. Sometimes bye-laws were used. In the Margate Act 1902, the water authority was given the power to control drains, closets, cesspools, etc., over an area of 1500 yards from any well or adit. Brighton Corporation also obtained similar powers in 1924 over an area with a radius of two miles around individual sources abstracting from the Chalk (Thresh & Beale 1925). In more recent times the formation of the Regional Water Authorities in 1974 gave an opportunity to develop more widespread policies across water supply boundaries and catchment divides. The first of these was published by Severn-Trent Water Authority in 1976 (Selby & Skinner 1981).

The formation of the NRA provided a further opportunity for national policy development and a document, published in 1992, has been adopted by the Environment Agency. (The Rivers Purification Boards in Scotland produced a draft Scottish policy based on the NRA document and the DoE Northern Ireland are considering something similar.) It sets out a framework for groundwater protection decision making, particularly in land use planning. One objective was to make other regulatory bodies aware of the NRA's concerns and approach. It also attempted to raise awareness of groundwater matters and enable a greater internal consistency of approach by the Authority. The national policy is based upon the concept of groundwater vulnerability in order that the greatest protection is given to those groundwater resources most at risk (NRA 1992).

Vulnerability mapping

The Environment Agency has published a series of groundwater vulnerability maps which show in general terms for groundwater pollution where the safest and most risky areas are for the development of potentially polluting activities. The maps take into account the large part that soils can play in attenuating the effects of surface loadings of pollutants and also the generalized geology divided into the three broad categories of Major, Minor and Non-Aquifers. These represent the importance of particular rock types for water resources and the intrinsic permeability of the strata. Many other factors also affect groundwater vulnerability in any particular location, such as the depth of unsaturated zone, the presence and nature of overlying Drift deposits and the nature of the contaminants. Since these are so site-specific, it is important to recognize the maps as planning tools to be used primarily as a filtering mechanism for new development. Site-specific studies will always be required when considering detailed proposals (NRA 1995a) .

A total of 53 maps has been published in both paper and CD Rom formats at a scale of 1:100 000, giving complete coverage of England and Wales. The maps have already found acceptance by planning authorities as an aid in the drawing up of structure plans and have also been used by industry in prioritizing actions for site investigation and clean-up on sites spread over wide geographical areas. Other uses include the planning of sewage sludge application to agricultural land and in routeing new transport infrastructure. They should ultimately gain a permanent place in land use planning.

Groundwater protection zones

Protection for individual abstractions is aided by the definition of three annular zones around each borehole and spring source, which are based on 50-day, 400-day travel times and the whole catchment area, in order of decreasing risk to the abstraction. The zones have been produced by the Environment Agency using proprietary steady-state, two-dimensional model codes (FLOWPATH in most cases) with currently available data. In some cases data availability is limited or the hydrogeological situation too complex for the model to produce zones in which a high degree of confidence can be placed. For these situations zones have had to be produced manually according to defined protocols because of the difficulty of modelling. Examples can be found in karstic aquifers and for spring sources. This is one reason why the Environment Agency has not sought to prescribe the zones in statute since with the provision of additional or better data and subsequent remodelling more accurate shapes may be produced. Borehole pumping rates may also vary and, particularly in heavily exploited aquifers, the resulting changes in catchment shapes can have knock-on effects throughout a series of abstraction sources over a wide area (NRA 1995b).

The primary use of groundwater protection zones is therefore, like groundwater vulnerability maps, as a screening tool, giving broad indications about the potential risks to groundwater. Decision-making about specific sites will always require more detailed appraisals of the risks to groundwater, which must also take into account the risk limitation that can be introduced by engineering or management techniques. There is also considerable benefit in water companies and regulators knowing roughly the area from where abstractions draw their water, in order that existing pollution risks can be identified and action taken where appropriate.

The programme of groundwater protection zone definition embarked on in 1992 by the NRA, and completed in 1998, is the most ambitious of any European country. Around 2000 individual sources have had their zones defined and published. Manuals have been published describing the methodologies used, thus allowing those who wish to refine any zone to do so using a similar approach but with additional data.

A number of zones have been defined specifically for the purpose of delineating areas where controls on agricultural land use will have most benefit for the reduction of nitrate leaching. These have been set up under the Nitrate Sensitive Area (NSA) voluntary schemes and also for the purposes of defining Nitrate Vulnerable Zones (NVZs) under the EC Nitrate Directive. For the most part, NVZs are rather smaller areas than the equivalent groundwater protection zones. This is because they have been defined with an inner area of confidence, where they relate to individual groundwater sources, within which there is a greater certainty that the changes in land use will impact on the abstracted water in the longer term (Fermor et al. 1996).

EC Groundwater action programme

The EC Groundwater Directive has been in force since 1981. It has been recognized that there are many deficiencies in it; not the least that by excluding groundwater resource issues and diffuse pollution, it is very narrow in its application. An EC Council Resolution in 1992 followed a Ministerial seminar in the Hague the previous year and has led to the establishment of a Groundwater Action Programme. A plan was published in July 1996 (European Union 1996). This has the objective of establishing 'a programme of actions to be implemented by the year 2000 at national and Community level, aiming at sustainable management and protection of fresh water resources'. It will also provide some key elements for a future framework directive on water.

Specific objectives are as follows:

(i) to maintain the quality of polluted groundwater,
(ii) to prevent further pollution,
(iii) to restore where appropriate, polluted groundwater.

The recognition of the need for an integrated approach such that groundwater and surface water should be managed as a whole, paying equal attention to both quality and quantity aspects, will improve our legislative base. The final plan and a revised directive will be a welcome emphasis of the importance of groundwater and its protection, and should help to promote consistency within the European Union.

Future issues and challenges

As we learn more about the state of our groundwater resources the biggest challenge will not be so much in protecting it from further deterioration as a result of point source pollution but rather how to address the historical legacy from industry and deal with the diffuse sources from agricultural land use and atmospheric deposition.

Historical legacy

A strategy is required for tackling the legacy of groundwater pollution which emanates from a time when the significance of groundwater pollution was largely unrecognized and the concentrations of pollutants which now give rise to concerns about human health were not able to be detected by analytical techniques. In the USA, naivety about the complexity of groundwater flow mechanisms and the efficiency of groundwater clean-up led to widespread remediation targets for drinking water quality which have been difficult if not impossible to achieve, particularly for non-aqueous phase liquids (NAPLs). This philosophy is now being challenged.

Active remediation of groundwater pollution is clearly at a low level in England and Wales (De Hénaut *et al.* 1996). Of the point sources of pollution identified in the NRA study only 44% were having some form of remedial action applied, and of these only 25% (11% of the total) seemed to be positive schemes involving techniques other than surface capping or excavation of overlying soils. Pump and treat operations were being carried out in only 8% of the occurrences identified.

Table 1 is reproduced from a US National Research Council study on groundwater cleanup, and illustrates, from an evaluation of some 80 pump and treat operations, the dependence on geological conditions for a range of contaminants (National Research Council 1994). The Chalk and the Triassic sandstones have been placed into their most likely categories. It is clear that for some aquifers polluted with certain chemical species, remediation will be almost impossible and we will have to look to different strategies.

The UK attitude has always been a pragmatic one based on what is practically feasible and economically necessary. This is reinforced by the 'suitable for use' approach of government to remediation of contaminated land (and groundwater) which requires that remedial action should take place only where:

(i) the contamination poses unacceptable risks to health or the environment; and
(ii) there are appropriate *and cost-effective* means available to do so, taking into account the intended or actual use.

This approach cannot be adopted in the prevention of pollution since the current EC Directive requires the protection of *all* groundwaters but where there has been significant historical impact the appropriateness of clean-up or the extent to which it should be employed is a major factor. The Groundwater Protection Policy of the Environment Agency recognizes this in respect of urban contamination, such as in Birmingham where the aquifer has effectively been abandoned as a potable water resource. Policy statement D6 states 'In areas where historical industrial development is known to have caused widespread groundwater contamination, the Agency will review the merits and feasibility of groundwater clean-up depending upon local circumstances and available funding.' Such decisions cannot be taken lightly so there is a need to gather considerable amounts of information in advance. Apart from a relatively few examples the standard of site investigation in the UK is currently poor.

Table 1. *Relative ease of clean-up for a range of contaminants in different geological environments (after National Research Council 1994)*

Hydrogeology	Mobile, dissolved (degrades/ volatilizes)	Mobile, dissolved	Strongly sorbed, dissolved (degrades/ volatilizes)	Strongly sorbed, dissolved	Separate phase LNAPL	Separate phase DNAPL
Homogeneous, single layer	1*	1–2	2	2–3	2–3	3
Homogeneous, multiple layers (*Triassic sst*)	1	1–2	2	2–3	2–3	3
Heterogenous, single layer	2	2	3	3	3	4
Heterogenous, multiple layers	2	2	3	3	3	4
Fractured (*Chalk*)	3	3	3	3	4	4

*Relative ease of clean-up, where 1 is easiest and 4 most difficult

The setting of groundwater clean-up values

The US experience also cautions us against the use of generic clean-up standards. Drinking water standards are clearly unachievable in many situations and other goals may have to be set depending on circumstance. In almost all situations involving organic compounds there will be a residue left within the pore spaces of the rock, absorbed onto the rock matrix or simply dissolved at low concentrations in the relatively immobile porewater. Natural processes of biodegradation have an important role to play and may, in the right conditions, reduce residual pollutants to background concentrations given enough time. This may be sufficient to prevent any impacts on the biosphere. However, our knowledge of such processes is still extremely limited and it will be difficult for regulators to accept such mechanisms as the reason for inactivity over remediation, without sound research evidence.

There has been considerable work carried out within the DoE R&D programme to consider clean-up values for soils to protect various end users. The former NRA also undertook work to consider appropriate remediation values to protect the water environment, and the Environment Agency is currently working on a methodology for setting remediation targets in respect of soils that are continuing to affect controlled waters and also groundwaters that are historically contaminated. There are some difficult legal issues which need to be resolved in respect of the latter.

Developers, site owners and industry are increasingly aware of their liabilities in having

given rise to, or owning land relating to, groundwater pollution. Whether clean-up is driven by Section 57 or Section 161 legislation, by redevelopment, or by concerns over civil liabilities, the Agency is being asked for advice about the level of remediation that should take place. This prompted a project which was undertaken by the Water Research Centre on behalf of the NRA.

The Agency is keen not to be over-prescriptive regarding the setting of targets, and the methodology allows for a site-specific approach to be adopted. This involves selecting a target/receptor of concern (borehole abstraction, spring or watercourse) and considering the desired water quality that it is required to be maintained; for example, drinking water standards at a public water supply abstraction, or Environmental Quality Standards/Water Quality Objectives for a watercourse. The groundwater quality to be achieved at the place where the groundwater is known to be polluted (i.e. within the plume of groundwater contamination) can then be back-calculated, given some basic information about the characteristics of the aquifer in that particular location.

Groundwater clean-up is very expensive and will be long-term in its application if highly exacting standards are to be achieved. In the case of smaller firms, funding may not be available. The method is intended to make use of the physical effects of dilution and dispersion and the natural biochemical attenuation processes that can occur as groundwater flows through underground strata. It therefore allows for the balancing of costs and benefits and the adoption of a pragmatic approach.

A potential problem arises since the approach is based on the protection of an ultimate receptor which has a known use (borehole for drinking; river for drinking/fishing/recreation, etc.). In order to gain the maximum benefit from the natural clean-up processes, and balance the costs and benefits, the plume of contamination may be allowed to continue to migrate down the groundwater gradient and pollute, currently unpolluted, groundwater. The situation is dealt with in the methodology by inserting a virtual target/surrogate receptor downgradient. This was done to allow for a degree of attenuation without writing off too large tracts of aquifer.

New development and risk assessment

While the Groundwater Protection Policy provides a framework for decision-making over new development, it is most effective at the primary planning stage. Once individual proposals are put forward for detailed consideration, site-specific issues will always need to be evaluated. Hydrogeologists have usually done this according to their own perceptions and knowledge and often not in entirely consistent ways. Their judgements are not always understood by others and as the groundwater knowledge base is increased amongst the community at large there is a need for standard methodologies that can be applied in a uniform manner. This codification of contaminant hydrogeologists' thinking can be classified under the broad heading of risk assessment. Risk assessment techniques are finding increasing favour in helping the professional in the decision-making process and also making the decision more understandable for the lay person. However, tried and tested methodologies have still to be developed for many areas.

One system (LandSim) recently published by the Environment Agency, following work carried out for the DoE and NRA, relates to proposed landfill sites (Gronow & Harris 1996). As landfill design has become increasingly complex, with civil engineering measures reducing leakage rates, the ability of regulators to assess proposals for their acceptability in both short and long term scenarios has decreased. The new methodology will allow regulators and operators alike to test designs against set quality criteria for the target water body/user most at risk. Similar systems will be applicable for assessing the degree to which contaminated land should be cleaned up to avoid continuing water pollution.

Risk assessment techniques can also be applied to diffuse pollutants. GIS-based methodologies are being developed to help assess the risks to catchments, both surface and groundwater, from pesticide usage on farmland and for non-agricultural purposes. Such techniques will be extremely useful in the assessment of the impact of new chemicals on water quality, and help in focusing analytical suites on those compounds likely to be present in the receiving water. However, it must be recognized that for groundwater matters risk assessment techniques can never be a substitute for decision making. They should always be regarded as purely tools to assist the expert in coming to the best technical decision given the information available.

Groundwater modelling

The role of groundwater models has increased over the years. Formally used mainly in aiding our management of resources as purpose-built, unwieldy one-off projects, models are increasingly finding routine application in groundwater pollution problems as new user-friendly software becomes more widely available. They are particularly helpful in the assessment of risk, both for new proposals and where pollution has already occurred. The groundwater protection zoning exercise, which uses proprietary software, has exposed more hydrogeologists to the experience and opportunities afforded by relatively simple modelling. It has also made them aware of the problems of uncertainty and sensitivity inherent in any modelling exercise, which will hopefully encourage a healthy scepticism in the results. Modelling is a highly effective way of encouraging thinking but total reliance on the output is dangerous.

One particular advantage of modelling is that it allows the conceptualization of ideas in a way that is not possible by other means. This is particularly useful in explaining situations to non-specialists. However, there is still room for improvement in the visualization of modelled results. Computer graphics techniques have advanced dramatically in recent years and their adoption in the field of groundwater modelling is long overdue. The combination of the two will allow us to literally 'see' underground and do much to dispel the 'out of sight, out of mind' attitude that has handicapped hydrogeologists in making fellow professionals, industry and the public in general understand the complexities of the problems and the solutions.

Conclusions

As we move into the 21st century the need to maintain and preserve the quality of our water resources will increase. The challenges for regulatory bodies will be in ensuring that controls are adequate to minimize further deterioration whilst not requiring preventative measures that are uneconomic to put in place.

The extent to which we have already polluted our groundwater will become much clearer over the next few years and decisions will need to be taken over how we deal with historical contamination. The new contaminated land legislative regime will undoubtedly be the regulatory driving force but will require a period for all stakeholders to understand and work with it. Remedial options range between active and passive, with an alternative of letting the abstractor undertake clean-up at the point of use.

To some extent the water industry has accepted the latter of the options above, particularly with respect to the problem of nitrate. The water companies are the biggest operators of 'pump and treat' in the country. It is accepted that the agricultural industry in general, exhorted by government to increase productivity, has been the critical factor. No individual polluters can be identified and so the burden falls on society, through the water rate. Can the same argument be used with respect to industrial pollution, especially for chlorinated solvents? It is difficult to be clear on the legal situation since much of the British legal system rests on case law. There have been few examples where the legislation has been tested with respect to groundwater. Arguments in the recent *Eastern Counties Leather* v. *Cambridge Water Co.* case had to refer to 19th century examples for precedent (*Rylands* v. *Fletcher*) and very few prosecutions have been taken on groundwater pollution matters. Much of this needs to be clearer. The opportunities afforded by the creation of the Environment Agency and the prominence given to historical pollution in the contaminated land provisions of the legislation which sets up this body should provide the impetus to build a clear strategy.

References

ANON, 1993*a*. Water consumer or polluter – who pays for aquifer contamination? *ENDS Report*, **224**, 19–21.

——, 1993*b*. Key ruling on civil liability by House of Lords. *ENDS Report*, **227**, 43–44.

DE HÉNAUT, P., HARRIS, R.C., VERNON, C. & HAINES, T. 1996. Evaluation of the extent and character of groundwater pollution from point sources in England and Wales: In: *Proceedings of the 32nd Annual Conference of the Engineering Group of the Geological Society*, Conference on Contaminated Land and Groundwater-Future Directions.

DoE 1988. *Assessment of groundwater quality in England and Wales*. HMSO, London. Prepared for Department of the Environment by Sir William Halcrow and Partners Ltd in association with Laurence Gould Consultants Ltd.

EUROPEAN UNION 1996. Proposal for a European Parliament and Council Decision on an Action Programme for Integrated Groundwater Protection and Management, OJ No C 355, 39. 1–18.

FERMOR, M. M., MORRIS, B. L. & FLETCHER, S. W. 1996. Modelling source catchments in the UK: implementation and review of a qualitative method for representing uncertainty. In: *Subsurface fluid flow (groundwater and vadose zone) modelling*. ASTM STP 1288.

GRONOW, J. & HARRIS, R. C. 1996. LandSim: a regulatory tool for the assessment of landfill site design. *Wastes Management*, February 1996, 30–32.

HARRIS, R. C. 1993. Groundwater pollution risks from underground storage tanks. *Land Contamination and Reclamation*, **1**, 197–200.

LERNER, D. N. & TELLAM, J. H. 1992. The protection of urban groundwater from pollution, *Journal Institution of Water and Environmental Management*, **6**, 28–37.

NATIONAL RESEARCH COUNCIL 1994. *Alternatives for Groundwater Cleanup*. National Academy Press, Washington DC.

NRA 1992. *Policy and Practice for the Protection of Groundwater*. HMSO, London.

—— 1995*a*. *Guide to Groundwater Vulnerability Mapping in England and Wales*. HMSO, London.

—— 1995*b*. *Guide to Groundwater Protection Zones in England and Wales*. HMSO, London.

SELBY, K. H. & SKINNER, A. C. 1981. Management and protection of the quality of groundwater resources in the English Midlands: In: van Duijvenbooden, W., Glasborgen, P. and van Lelyveld, H. (eds) *Quality of Groundwater*, Proceedings, International Symposium, Netherlands, 1981, Studies in Environmental Science, **17**, Elsevier.

THRESH, J. C. & BEALE, J. F. 1925. *The Examination of Waters and Water Supplies*, 3rd edition, J. and A. Churchill, London.

WOODWARD, A. J. 1994. Removing nitrates from potable water. *Institute Water and Environmental Management Yearbook 1994*, 39–45.

Recent UK legal developments relating to pollution of water resources

KATHY MYLREA

*Environmental Law Department, Simmons & Simmons, 21 Wilson Street,
London EC2M 2TX, UK*

Abstract: The new UK statutory contaminated land provisions are expected to come into
effect in 1997. Land will be contaminated for the purposes of the legislation if the presence
of substances in, on or under it mean that significant harm is being caused or there is a
possibility of such harm being caused or pollution of controlled waters is being caused or
is likely to be caused. In a separate legal development, powers contained in the Water
Resources Act 1991 for dealing with water pollution are also being strengthened. There is
some concern that there is potential for overlap or conflict between these two regimes.

Developments at EU level include a communication of European Community water
policy. It is expected to lead to a framework directive on water resources. The European
Commission is shortly expected to take some action on the issue of liability for Environmental Damage.

The period 1995–1996 saw a number of developments, at both UK and EU level, in environmental law and policy specifically relevant to
groundwater protection. There have also been
a number of more general developments in
environmental law which are relevant to an
understanding of the legal framework applicable
to groundwater.

Foremost among developments in the UK are
three provisions contained in the Environment
Act 1995. The first is the creation of a new unified
Environment Agency (the Agency) which, on 1
April 1996, took over, *inter alia*, the pollution
control and water resource management functions of the National Rivers Authority (NRA).
The other two provisions (discussed in this
paper) are the introduction of a specific statutory
regime for dealing with contaminated land, and
the insertion into the Water Resources Act
1991 of a new Section 161A which provides for
service by the Agency of water pollution 'works
notices'. Although neither of these latter provisions is yet in effect, it is expected that they will
be implemented in 1997.

At EU level, in June 1995, the Council and the
Environment Committee of the European Parliament called for a fundamental review of Community water policy. This led, on 21 February 1996,
to the European Commission presenting a communication on European Community water
policy. It is expected that the communication
will lead to the tabling of a framework directive
on water resources in due course. In addition to
the proposal for a water resources framework
directive, the Commission is committed to presenting a Groundwater Action Programme.

1995 also saw a number of developments in
both the civil and criminal law. In June 1995,

the decision of the Queen's Bench Division in
Graham v. *Re-Chem International Limited* was
released. That case involved consideration of
some important issues in the law of nuisance
and negligence, over the course of an extremely
lengthy and complex trial. After extensive
expert evidence, the firm conclusion reached by
the Judge was that the plaintiffs had not established the necessary connection between the
operation of the incinerator and the ill-health
of their cattle. The plaintiff's claim failed.

On the criminal front, there have been fewer
water cases of particular legal significance
during 1995–1996. The 1995 decision in *Attorney
General's Reference No. 1 of 1994*, and the 1994
House of Lords' decision in *NRA* v. *Yorkshire
Water* articulate clearly the points of principle
on the issue of 'causing' water pollution. The
case law on what constitutes 'knowingly permitting' remains less well developed. This has particular ramifications for the new contaminated
land regime; these are discussed below.

UK developments

Environment Act 1995 Contaminated Land Provisions

On a date yet to be decided (probably in early
1997) the new statutory regime dealing with contaminated land will come into force in England,
Scotland and Wales. These provisions are contained in Section 57 of the Environment Act
1995, and will be inserted into the Environmental
Protection Act 1990 as Part IIA. The new part
will comprise more than twenty sections of complex and densely drafted provisions, the purpose

MYLREA, K. 1998. Recent UK legal developments relating to pollution of water resources. *In*: MATHER, J., BANKS,
D., DUMPLETON, S. & FERMOR, M. (eds) *Groundwater Contaminants and their Migration*. Geological Society,
London, Special Publications, **128**, 15–21.

of which is to ensure the identification and also the remediation of contaminated land. Much further material, in the form of regulations and statutory guidance, will need to be issued before the provisions can come into force.

The contaminated land regime will usually be administered by local authorities. For certain properties, referred to as special sites, jurisdiction will pass to the Agency. It is likely that the Agency will only deal with land which poses particularly difficult problems or problems where the Agency may be expected to have particular expertise. (In light of the transfer of the NRA's functions to the Agency, it is reasonable to consider that in cases where water pollution is the concern, the Agency can be considered to have particular expertise.)

A statutory definition of 'contaminated land' is contained in Section 78A. Local authorities will be under a duty to inspect their area from time to time for the purpose of identifying land which meets that definition. Contaminated land for the purposes of the legislation means land which appears to the authority to be in such a condition, by reason of substances in, on or under it, that either:

(a) significant harm is being caused or there is a significant possibility of such harm being caused; or

(b) pollution of controlled waters is being caused or is likely to be caused.

In making the determination as to whether land appears to be contaminated, the local authority must act in accordance with statutory guidance issued by the Secretary of State. That guidance will also include guidance on what harm is to be regarded as 'significant', what degree of possibility of harm is to be regarded as 'significant' and whether pollution of controlled waters is being or is likely to be caused. The formal consultation period on the guidance was between September and December 1996. The draft guidance only contained a brief section on pollution of controlled waters. The draft guidance introduced the concept of risk assessment into the definition of contaminated land. Before a local authority can identify a site as contaminated, it must have identified a contaminant (a source), a receptor being or likely to be polluted by that contaminant (a target), and a pathway by or through which the target could be affected by the contaminant.

In the case of pollution of controlled waters, the relationship between the provisions on contaminated land and other relevant environmental law needs to be clear. The contaminated land provisions do not apply in relation to land where there is a current waste management licence (unless the contamination has migrated onto the site from somewhere else or existed prior to the licensed activities). Nor can the contaminated land provisions be used in such a way as to prevent a person making a discharge to controlled waters pursuant to a discharge consent given under the Water Resources Act 1991 or, in Scotland, under the Control of Pollution Act 1974. Where the contaminated land provisions apply, the statutory nuisance provisions in Part III of the Environmental Protection Act 1990 do not apply.

An issue of concern relates to the quite separate introduction of water pollution 'works notices' into the Water Resources Act 1991 (discussed below). That amendment, tucked away in Paragraph 162 of Schedule 22 to the Environment Act 1995, allows the Agency to serve a 'works notice' in cases where it appears to them that polluting matter has entered, or is likely to enter, controlled waters. The notice may require works to prevent the polluting matter reaching controlled waters or may require the cleanup of water which has already been affected. There is clearly a possibility of overlap between the powers. It is to be hoped that the Government will clarify the circumstances when they would choose to proceed by the works notice route.

The person responsible for cleanup under the contaminated land provisions is referred to as an 'appropriate person' or persons. The question of who is the appropriate person is governed by Section 78F. Primary responsibility will always rest with the person or any of the persons who caused or knowingly permitted the contaminating substances to be in, on or under the land. Case law on 'causing' water pollution has established that a person can be held to have caused such pollution without any fault or knowledge of the event caused being imputed to them. Courts can be anticipated to take a similar approach in the contaminated land context.

Knowingly permitting is a more difficult concept and, although the term has long been part of water pollution law, there is very little case law on its meaning in that context. The concern in this context is that if someone becomes the owner or occupier of land which already has contaminating substances in, on or under it and that the person knowingly allows them to remain there, they can be liable for having 'knowingly permitted' their presence.

However, where, after reasonably enquiry, no person who caused or knowingly permitted (the 'original polluter') has been found then the current owner or occupier of the contaminated land will be the appropriate person.

One limitation on the extent of liability is that the original polluter can only be made responsible for remedial action which is referable to substances that they have caused or knowingly permitted to be there. If they are responsible in a capacity as owner or occupier of land then the concept of referability is not relevant.

From an owner's or occupier's point of view there is at least some good news where the contamination relates to pollution of controlled waters. From the point of view of potentially creating a large number of 'orphan groundwaters', the situation is less good. Section 78J essentially provides that an owner or occupier (who is only responsible where the original polluter cannot be found) cannot be required to do anything which he could not have been required to do if the water pollution limb of the definition of contaminated land did not exist. The object of the restriction was to avoid any additional liability accruing to an owner or occupier of land beyond that which could already attach in relation to the cleanup of water pollution under Section 161 of the Water Resources Act 1991. The extent to which the limitation provides protection to an owner or occupier will depend upon the interpretation placed on the words 'knowingly permitting'.

Where there are two or more persons who could be appropriate persons, the authority is required to determine the matter in accordance with the Secretary of State's guidance. The guidance will be the means by which the potential harshness of what is essentially a regime of joint and several liability is reduced as it is expected to set out the circumstances where someone who meets the definition of original polluter is excluded from liability. It contains a number of 'Exclusion Tests' which will enable some of the appropriate persons to be excluded from liability in certain circumstances. Application of the Exclusion Tests can never result in a situation where there is no-one left who is, at least theoretically, liable.

Remediation notices can require the taking of steps for the purpose of assessing the condition of the contaminated land and also any controlled waters affected by that land. The notice can require the doing of works to prevent or minimize or remedy or mitigate the effects of any significant harm or pollution of controlled water or for restoring land or waters to their former state. Remediation notices can also require the making of subsequent inspections from time to time for the purpose of keeping under review the condition of the land or waters. When considering the contents of the remediation notice, the local authority may only require things to be done by way of remediation which it considers reasonable having regard to the likely cost involved, the seriousness and the harm or pollution in question and having regard to guidance issued by the Secretary of State. In theory, however, a major groundwater cleanup could be required by service of a remediation notice.

Before serving a remediation notice the local authority is required to make reasonable endeavours to consult the person on whom the notice is to be served, the owner and any apparent occupier, concerning what is to be done by way of remediation. A three-month period must be allowed for such consultation before any remediation notice is served. Remediation notices are placed on a public register.

The local authority is precluded from serving a remediation notice in certain prescribed circumstances. One of these is if it is satisfied that appropriate things are being or will be done. In some cases, the appropriate person or persons may agree or choose to voluntarily carry out works of remediation. In such a case the person responsible for the remediation must prepare and publish a remediation statement recording the things to be done, who is to do them and within what period. That statement goes on the public register so the fact that the site has been identified as contaminated will enter the public domain.

There is provision for appeal against a remediation notice and regulations will be made setting out the grounds of appeal. Failure to comply with a remediation notice is a criminal offence punishable on summary conviction by a fine of up to £20 000 plus a daily fine of £2000 (where the contaminated land to which the notice relates is industrial, trade or business premises).

Water pollution works notices

The NRA had powers to exercise pollution control measures and recover the costs thereof under Section 161 of the Water Resources Act 1991. These powers were generally regarded as inadequate to deal with complex and contentious problems such as those presented by major groundwater cleanup or by contaminated land.

Although Section 161 has not been replaced, it has been supplemented by a much more powerful means of securing cleanup of controlled waters. The Agency will, when the provisions are brought into effect, be able to serve a 'works notice' requiring works to be carried out remediating or preventing pollution of controlled waters. The notice can be served on any person who has caused or knowingly permitted poisonous, noxious or polluting solid waste matter to be present at the place from where it is likely, in the opinion of the Agency, to enter any controlled waters or caused or knowingly permitted the matter in question to be present in any controlled waters.

Works notices can require works or operations for the purpose of removing or disposing of the poisonous, noxious or polluting matter, remedying or mitigating any pollution caused by its presence in the waters, and, so far as it is reasonably practicable to do so, of restoring the waters, including any flora and fauna dependent on the aquatic environment of the waters, to their state immediately before the matter became present in the waters. There are not the same requirements to have regard to cost, seriousness and guidance as there are under the remediation notice provisions.

Failure to comply with any requirements of the works notice will constitute a criminal offence. Furthermore, failure to comply entitles the Agency to carry out the works and recover from that person any reasonable costs or expenses and, where criminal proceedings are considered to be an ineffectual remedy, take proceedings in the High Court for the purpose of securing compliance with the notice.

The detailed procedural requirements in relation to the form and content of works notices, requirements for consultation, grounds of appeal and other procedural matters remain to be specified in regulations. However, it is clear that, subject to identifying the person who has caused or knowingly permitted, the difficulty with Section 161, i.e. the Agency having to expend resources and then seek to recover them, has been removed. It remains to be seen if the Agency will consider it appropriate to serve works notices in relation to groundwater contamination cases.

If one considers the water pollution limb of the definition of contaminated land and the factors which entitle the Agency to serve water pollution works notices then it is clear that there is a possibility of someone being served with both a remediation notice and a works notice. It is also possible that the notices could contain different requirements. Co-operation between the relevant local authority and the Agency will go a long way to reducing this possibility. One would suspect that in many cases of water pollution the local authority would actually prefer the Agency to serve a works notice on the basis that they could then say they are satisfied that appropriate things are being done by way of remediation and

not be under a duty to serve a remediation notice. However, it should be noted that the Agency is not under a duty to serve works notices: they merely have the power to do so. In the case of special sites the Agency is already the enforcing authority so the chances of lack of Cupertino should be virtually eliminated.

European developments

European Commission Communication on European Community Water Policy

The Communication (COM (96)59) aims to bring about the restructuring of EC legislation on water and an increased integration of the various existing water policies. It is hoped that the Communication will lead to the tabling of a framework directive on water resources in due course.

The Communication contains an outline of the proposed directive, which suggests that the following directives would remain largely unaffected by it (although the Commission would consider transferring to the framework directive some of the definitions, monitoring requirements and other relevant elements of these directives):

The Bathing Water Directive;
The Dangerous Substances Directive;
The Drinking Water Directive;
The Information Exchange Decision;
The Urban Waste Water Treatment Directive;
The Nitrates Directive;
The Reporting Directive; and
The Integrated Pollution and Prevention Control Directive.

In addition, the following directives would be repealed and replaced by the framework directive:

The Surface Water Directive;
The Fish Water Directive;
The Shellfish Water Directive;
The Groundwater Directive; and
The proposed Ecological Quality of Water Directive.

The framework directive would prescribe a high level of Community water protection and it would establish common definitions for use in all EC water policy. It would require the integration of water resource management with the protection of the natural, ecological state and functioning of the aquatic environment; the integration of water quality and water quantity management (including provisions for the establishment, where necessary, of a water abstraction licensing scheme); the integration of surface water management (including coastal waters) with groundwater management; and the integration of control of pollution measures, such as emission controls, with environmental quality objectives.

The framework directive would require integrated water management planning on a river basin basis, which would involve the following: monitoring water quantity and quality; an assessment of the water needs of society and of the impact of human activities on the water bodies concerned; the setting of objectives, the establishment and implementation of a programme of measures designed to achieve these objectives; public consultation in the decision-making process; and, monitoring and reporting on the implementation of the Directive.

The Groundwater Action Programme, adopted on 10 July 1996 by the European Commission, is intended to present a framework for action within which Member States will have responsibility for defining and implementing measures to pursue the objective of sustainable use of groundwater. In particular, the Action Programme aims to maintain the quality of unpolluted groundwater, prevent further pollution and restore polluted groundwater where this is appropriate.

The proposed Action Programme recommends four particular lines of action:

(1) establishment of elements for integrated planning and management of water resources;
(2) establishment of authorization systems and general rules for abstraction of freshwater resources where appropriate; artificial recharge to be subject to a suitable authorization system;
(3) establishment of authorization systems and general rules for point sources of pollution from discharges and where activities or installations affect or potentially affect groundwater quality;
(4) establishment of measures to be taken to protect groundwater against pollution from diffuse sources, including codes of good practice.

The proposed Action Programme envisages Member States drawing up national action programmes over the coming years. Immediate actions such as mapping, monitoring, review of existing structures and performance of the water sector are suggested as a first step. Other actions may require more appropriation or depend on an assessment of the groundwater situation based upon the results of mapping and monitoring. It is anticipated that the necessary actions and implementation measures will differ between Member States depending on the groundwater situation, existing legal and administrative structures, and on measures already taken or decided.

Commission proposal on environmental liability?

In May 1993, the European Commission presented a Green Paper on Remedying Environmental Damage. Whilst the Green Paper was very much a preliminary discussion document, it is clear from reading it that the Commission favoured a strict civil liability regime for dealing with environmental damage. Issues such as retrospectivity, industries to which the regime would apply, and the treatment of damage to the unowned environment were all raised and discussed in the paper.

The Commission's objective for stimulating a debate on the issue was successful: over 100 written comments were sent to the Commission from several Member States (including the UK), private associations and environmental non-governmental organizations. A public hearing was organized by the Parliament. The European Parliament adopted a resolution in April 1994 calling on the Commission to submit a proposal for a directive on civil liability in respect of future environmental damage.

In addition to informal meetings with experts from the Member States, two formal studies were undertaken to investigate how liability systems dealing with environmental damage operate in the Member States and on the economic implications of different liability systems.

The results of the legal study raise three particular points:

(1) a general lack of mechanisms within the Community for compensation and restoration of natural resource damage;
(2) lack of any obligation under civil law for those recovering compensation to expend it for actual restoration purposes;
(3) many differences in environmental liability regimes within the Community.

The results of the economic study indicate that liability has some comparative advantages in dealing with both accidents or gradual pollution where causation can be shown at a reasonable cost. Key economic issues are the potential benefits of civil liability regarding transboundary pollution; possible effects on the Single Market of differing liability regimes in Member States;

and the potential of civil liability for fostering better compliance with existing regulations.

The Commission has to date emphasized that they are looking at a prospective system and that they are not seeking to establish retrospective liability for damage which has already been caused. There is a concern amongst legal practitioners about how practicable it is to establish a simply prospective system and a concern that any system which only addresses prospective ecological damage will leave enormous areas of environmental damage unrectified.

The European Environment Commissioner, Mrs Bjeregaard, is personally very committed to this topic. She has indicated that she believes action at European level is necessary, but the form of such action is still being considered. There appear to be three main options: signature of an existing Council of Europe Convention on Civil Liability for Damage to the Environment; proposing a new Council directive on the topic; and carrying out further research. There is, however, pressure to end the uncertainty over this proposal one way or another within the reasonably foreseeable future. This would suggest that the idea will either be taken up following more detailed study or be dropped entirely.

Court activity – civil cases

There have been a number of important civil law cases in the last two years which particularly demonstrate the importance of expert evidence in civil cases.

In *Graham* v. *Re-Chem International Limited*, Queen's Bench Division, 16 June 1995, some important issues in the law of nuisance and negligence were considered. The plaintiffs alleged contamination of their farmland by aerial transmission and deposition of toxic chemicals from the incinerator owned and operated by the defendant. It was the plaintiffs' case that their dairy cattle had been poisoned as a result of ingesting the chemicals while grazing on the contaminated land and, as a consequence, had died or had to be disposed of. The case gave rise to a very large number of factual and technical issues of great scientific complexity.

In order to have established their case in nuisance, the plaintiffs would have to have demonstrated that the relevant damage was foreseeable at the time. The Court found that at the time the incinerator was in operation it had been known for some time that destruction of waste by incineration resulted in the emission of organic pollutants such as dioxins and furans in fly ash carried in the flue gases. It was

also well known that some of these compounds were very toxic. Significantly, the Judge held that this knowledge rendered foreseeable any harm or damage caused by the emission of those toxic substances from the incinerator, even though the precise chemical processes which led to the formation of many of those substances were not known at the time.

The central issue in nuisance was therefore simply whether the relevant substances were actually emitted from the incinerator and whether those chemicals actually caused the alleged damage to the plaintiffs' land and cattle. To succeed in the negligence action, the plaintiffs would have had to go on to demonstrate the breach by the defendant of a duty of care owed to them.

The Judge was satisfied that the symptoms of ill-health in the herd were not indicative of dioxin toxicity and that the plaintiffs had not established the necessary connection between the operator of the incinerator and the ill-health of the cattle. He found that there was no single explanation for the symptoms of ill-health experienced by the herd, but was satisfied that the main reason for the health problems was an unrelated condition known as 'fat cow syndrome'. The plaintiffs failed in both their nuisance and negligence claims.

The issue of foreseeability also arose in the Court of Appeal decision *Margerson and Hancock* v. *J W Roberts Ltd* (2 April 1996). It upheld the November 1995 High Court ruling on the liability of operators of an asbestos factory located at Armley, Leeds, to local residents affected by asbestos dust.

The first plaintiff claimed damages for personal injury in respect of mesothelioma caused by environmental exposure to asbestos dust emitted from the factory at Armley, Leeds. He had grown up close to the factory from 1925 to 1943 and had lived there from 1948 to 1957. During the course of the proceedings, he died and the action was maintained by his widow. The second plaintiff also claimed to have contracted mesothelioma as a result of childhood exposure to asbestos in the vicinity of the factory between 1938 and 1951.

The case did not turn on causation but rather on whether at the material time there was, or could be, foreseeability – what the Judge described as 'the foresight of injury that is fundamental to tortious liability'. Following *Cambridge Water Company* v. *Eastern Counties Leather plc* it was conceded by the plaintiffs that proof of foreseeability of damage of a relevant type was a prerequisite for liability in nuisance and under *Rylands* v. *Fletcher*.

The Court of Appeal emphasized that the defendant's actual or constructive knowledge at the time of exposure that preliminary damage could result from environmental exposure rendered it liable even for the unanticipated or serious condition, mesothelioma, suffered by the plaintiffs. In this case the Judge relied principally on the extensive lay evidence as to the dust conditions at the relevant time rather than on expert reports.

Criminal law

There have been very few recent prosecutions for groundwater pollution. However, although it only resulted in a £2500 fine, the NRA was successful in prosecuting Shell UK Limited for causing groundwater pollution from a petrol leak at a service station. In that case, Shell pleaded guilty to causing polluting matter to enter groundwater.

The Agency published an Enforcement Policy Statement in May 1996 which sets out the general principles and approach that the Agency will adopt towards enforcement. The need for a 'quick and effective' response to serious breaches and a 'discriminating and efficient' approach to other breaches is noted. The Agency's four enforcement principles are proportionality, consistency, targeting and transparency.

It remains to be seen if the Agency will adopt a more rigorous enforcement policy towards groundwater polluter offences than its predecessor. An Environment Agency prosecution policy is expected to emerge in due course. However, the Agency state that they will be seeking to raise Court's awareness of the gravity of pollution offences and encourage them to make full use of their powers.

Conclusions

The law and policy regarding protection of groundwater continues to develop at both EU and UK level. Both new and existing laws are likely to be tested in the Courts in future. The next few years will be an active period and good technical advice will be essential to enable those writing and applying the laws to ensure that the laws deliver the desired effects.

Foreseeability of environmental hazards: the implications of the Cambridge Water Company case

R. PAUL ASHLEY

*Mott MacDonald, Demeter House, Station Road,
Cambridge CB1 2RS, UK*

Abstract: The House of Lords decision, in the case of *Cambridge Water Company* v. *Eastern Counties Leather and Others*, introduced the concept of foreseeability as one test of civil liability for environmental pollution. In this paper the main criteria for foreseeability of pollution hazards are considered, including:

- recognition of the acute and chronic toxicity of a potential pollutant;
- availability of suitable detection and analysis methods;
- understanding of migration and transport mechanisms in the environment.

Examples of the application of foreseeability criteria are discussed, such as chlorinated hydrocarbon solvents (as in the Cambridge Water Company case), methyl tertiary butyl ether and mine drainage waters.

Prior to the *Cambridge Water Co.* v. *Eastern Counties Leather* case (Misstear *et al.* 1998), it was commonly held that the absence of fault could not be used as a defence against a claim under the rule in *Rylands* v. *Fletcher*; in other words, that strict liability would apply. In a civil claim for damages caused by pollution, for example, this would mean that if the pollution was caused by inadvertent spills or leaks, then the polluter would still be held liable.

The peculiar characteristics of hydrogeology have stretched the application of this principle to an extreme case. For surface waters, a pollutant's lifespan may be measured in terms of hours or days, although the damage it causes may last longer. For groundwaters, however, where bacteria are less common, the oxygen supply is sparse, volatilization is restricted and photo-oxidation is non-existent, a pollutant's lifespan may be measured in years or decades, possibly remaining undetected for much of that time. During that interval the world outside may change, and, in particular, laws regulating the presence of such pollutants in the environment may be introduced.

The case of *Cambridge Water Co.* v. *Eastern Counties Leather* allowed this situation to be definitively examined. The House of Lords, while upholding the principle of strict liability, in that the absence of fault still does not constitute a defence against a claim of this type, has now set out a fall-back defence of 'reasonable foreseeability': if the polluter could not have foreseen the harm that could be caused by his activities, then he is not liable in civil law.

It is the purpose of this paper to examine what foreseeability may mean in practice, and provide some guidelines to those who need to identify foreseeable pollution risks. Three main criteria are proposed as determining whether a risk is foreseeable:

- a substance must be recognized as being harmful to humans or the wider environment;
- it must be possible to analyse for the substance at the concentrations of concern;
- the mechanisms of migration and transport of the substance in the environment must be understood.

These criteria shall be considered in their application to actual pollutants: chlorinated solvents, methyl tertiary butyl ether and, briefly, acidic mine drainage waters.

Chlorinated hydrocarbon solvents

Trichloroethene (TCE) was first manufactured early in this century, and introduced on a large scale as a commercial solvent in 1929. Tetrachloroethene (perchloroethylene, or PCE) was introduced as a less volatile alternative to TCE in 1950 (Whim 1984; Herbert *et al.* 1986). In comparison with older petroleum-based solvents, TCE and PCE had the great advantage of being non-inflammable and non-explosive. At an early stage in their use, it was recognized that TCE and PCE at high exposure levels were acutely toxic to humans in vapour or liquid form (Stuber 1932; Rowe *et al.* 1952), and from the early 1960s TCE and PCE were progressively replaced by the less toxic 1,1,1-trichloroethane and 1,1,2-trichlorotrifluoroethane. The acute toxicity of TCE and PCE is recognized

ASHLEY, R. P. 1998. Foreseeability of environmental hazards: the implications of the Cambridge Water Company case. *In*: MATHER, J., BANKS, D., DUMPLETON, S. & FERMOR, M. (eds) *Groundwater Contaminants and their Migration*. Geological Society, London, Special Publications, **128**, 23–26.

in the UK Occupational Exposure Limits (Anon 1992).

From the mid 1970s onwards, there was concern that TCE and PCE were potential carcinogens at trace concentrations in drinking water, expressed in a report of the US Environmental Protection Agency (Anon 1975), arising from tests on rats and mice. Although a clear mechanism relating TCE and PCE to carcinogenicity in humans was not identified, the concern was sufficient in 1984 for the World Health Organisation (Anon 1984) to recommend Tentative Guideline concentrations of $30 \mu g \, l^{-1}$ and $10 \mu g \, l^{-1}$ for TCE and PCE in drinking water. Before that, the European Community (Anon 1980) had set a Guide Level of $1 \mu g \, l^{-1}$ for all organochlorine compounds in drinking water, a level implemented in the UK with other EC drinking water standards in 1985. In 1989, the revised UK drinking water standards (Anon 1989) included the WHO standards for TCE and PCE as Maximum Concentrations.

These developments in water quality standards parallel the development of chemical analysis techniques. Gas–liquid chromatography (GLC, or just GC) was developed in the 1950s (James & Martin 1951) as a means of separating and measuring organic components in a mixed sample. Wide-bore packed chromatography columns were used initially to separate the samples, but had insufficient resolution for the detection of low concentrations, particularly in small samples, and high-resolution open-tubular capillary columns were introduced in the late 1950s (Golay 1958). Chlorinated organic compounds remained especially difficult to detect at low concentrations until the invention of the electron capture detector (ECD) by Lovelock (1961). By the late 1950s (Holmes & Morrell 1957), mass spectrometry (MS) could be combined with GC to identify more reliably the separated components of an organic mixture, but it was the mid to late 1980s before GC/MS became a routine analysis technique.

Apart from the introduction of the ECD, the improved instrumentation resulted not so much in lower detection limits, but rather in the greater ease with which GC and GC/MS could be applied to routine analysis of water samples. Chromatographs in the 1960s required extreme care in setting up, calibration and operation, and the poor resolution of wide-bore columns meant that complex environmental samples could not be handled. Since the early 1970s, however, miniaturized electronics, combined with commercially manufactured capillary columns, have reduced the cost of chromatographs, increased their resolving power, and made them easier to operate and more likely to generate reproducible data. Even the notoriously unstable ECD can now be used for on-site operations with an acceptable degree of confidence. The GC/MS system, close-coupled with a computer database of mass spectra, can now be used to trawl water samples for all detectable constituents, whether or not they are the specific subject of the sampling programme.

As analyses became easier and cheaper, a clearer picture of the distribution of trace organic pollutants in the environment was produced. The US EPA (Anon 1975) listed TCE and PCE in a survey of organic contaminants in drinking water, as did Fielding *et al.* (1981) for the Water Research Centre in the UK. A survey of organic contaminants in UK groundwater was completed only in 1985 (Kenrick *et al.* 1985), but there was sufficient concern at the environmental hazards of TCE, PCE and related compounds that research into their behaviour in groundwater was commenced in the early 1980s. The principal mechanisms controlling organic pollutant behaviour (solution, degradation, sorption/retardation) were recognized at an early stage. The investigations conducted by Anglian Water Authority and the British Geological Survey at the ECL site, however, were one of the first detailed field studies of how these mechanisms actually operated in the Chalk. The conclusions of the work of BGS (Lawrence *et al.* 1992), building on the research into nitrate in Chalk groundwater in the 1970s, provided a framework of understanding for the behaviour of dense chlorinated solvents in this, the most vulnerable of the major UK aquifers.

A complex relationship can thus be seen, connecting the improvement in analytical techniques, the recognition of the potentially harmful effects of chlorinated solvents in the environment, and the growth of hydrogeological knowledge. Lower detection limits allowed the implications of toxicological tests to be examined in the environment; the consequent discovery of widespread water contamination by trace organic chemicals supported the demand for legislation to control them which, in turn, generated pressure for cheaper and more reliable chemical testing to support the new compliance testing industry.

The question arises: at what moment would a user of chlorinated solvents become reasonably able to foresee the potential harm they could cause in the aquatic environment? In the *Cambridge Water Co.* v. *Eastern Counties Leather* case, the House of Lords decided that that moment was some time after 1976, the end of the most likely period when PCE entered the ground at the ECL site. From a scientific and technical perspective, it is suggested that that moment was when concern about the presence in groundwater at low concentrations of PCE and TCE as potential carcinogens began to receive publicity in professional and trade literature (as opposed to scientific journals), which corresponds with the mid to late 1970s. From an alternative, more legalistic, perspective, since the House of Lords decided that Cambridge Water Co. suffered loss only when regulations were implemented which made the water from its well unfit for drinking (18 July 1985), foreseeability may possibly be said have commenced when the regulations were first proposed (1980).

Methyl tertiary butyl ether

This paper has concentrated on chlorinated solvents because of the importance of the *Cambridge Water Co.* v. *Eastern Counties Leather* decision. However, it is of interest to consider how the same criteria may apply to other potential pollutants. One such is methyl tertiary butyl ether (MTBE), an anti-knock additive in unleaded petrol. A full review of the use and properties of MTBE is given in a paper by Symington *et al.* (1994) from which much of the following information has been obtained.

MTBE was introduced in the US in 1979, and in the UK from the mid 1980s. In the UK, petrol is permitted to contain up to 10% of MTBE and, although current practice is to keep the proportion at 1% or less, it is expected to rise. The increasing use of unleaded petrol will also increase the total consumption of MTBE.

The toxicity of MTBE is believed to be relatively low, although, at high vapour concentrations, it can cause headaches and nausea. MTBE is detectable in drinking water by taste and odour down to as low as $6 \mu g \, l^{-1}$, and possibly down to $2 \mu g \, l^{-1}$. There are no specific drinking water quality standards for MTBE imposed by UK, EC or US legislation, although its effects are covered by taste and odour limits.

MTBE is analysed in water down to a detection limit of $0.1 \mu g \, l^{-1}$ by high-performance liquid chromatography (HPLC), a technique that has been available at least since the date of introduction of MTBE to the UK.

The behaviour of MTBE in groundwater systems is still the subject of research, but, because of the frequency of spills and leaks of petrol, there is concern that it may become a ubiquitous contaminant of groundwater. It is believed to be resistant to degradation in groundwater, is highly soluble (in excess of $50\,000 \, mg \, l^{-1}$), and is not significantly sorbed onto solids in the aquifer. Thus it is likely to travel at approximately the same rate as the flow of groundwater itself. In addition to its detrimental effects on drinking water, there is some concern that MTBE can enhance the solubility of BTEX compounds (benzene, toluene, ethyl benzene and xylene, the main hydrocarbons in light fuels such as petrol). The distribution of MTBE in groundwater has been investigated in the US at least since 1986, and in recent years the National Rivers Authority has commissioned research into its distribution in UK groundwater.

MTBE is sufficiently volatile to be removed from raw water supplies by aeration in a stripping tower and by activated carbon filtration, although the latter is only of limited effectiveness because of the low sorption properties of MTBE.

It is clear that the presence of MTBE, even if considered non-toxic, is undesirable in drinking water, and thus also undesirable in the groundwater which may be used for drinking water supplies. From a scientific and technical perspective, the criteria of foreseeability would appear to have been satisfied almost from the time of its introduction to the UK. It was known to have adverse effects in drinking water, it was possible to analyse for it at low concentrations, and it was known to be persistent in groundwater systems.

From a legal perspective, MTBE remains unrecognized. However, Anglian Water has recently initiated a civil claim against the US Air Force, claiming damages in compensation for the cost of treating borehole water to remove MTBE. This action may in turn lead to a change in the regulatory position.

Mine drainage waters

The case of mine drainage waters is mentioned briefly here as a contrast to the case of MTBE, which remains unregulated by specific water quality standards. The case of the highly polluting waters which drained into the Carnon River from a failed containment system at the abandoned Wheal Jane tin mine in 1991 received considerable public attention. The waters contained high concentrations of heavy metals and other contaminants, at low pH, and caused a highly visible plume of discoloured water in the river.

A potentially greater set of problems will occur when dewatering ceases at former coal mines. However, as has been pointed out by the counsel for ECL (Vallance 1994), the drafters of Clause 89(3) of the Water Resources Act 1991 (repeating similar provisions in the Control of Pollution Act

1974, Clause 31) did actually foresee this problem, and relieved those responsible for old mine workings of statutory liability for pollution of controlled waters. It will be interesting to see whether this position can be maintained in the light of steadily increasing concern from the public and environmental regulators.

Conclusions

Criteria for judging the foreseeability of environmental risks have been presented, considering:

- the recognition of the toxic potential of the substance in question;
- the ability to detect the concentration of the substance at levels of concern;
- the understanding of the substance's behaviour in the environment.

It is important to recognize that these criteria are independent of the involvement of the person or company using the substance: that is, a risk may be foreseeable even if an individual polluter is not aware of the fact. It is thus in the interest of users to keep abreast of current knowledge of environmental hazards arising from the substances for which they are responsible.

This paper is based on a presentation given to the IBC conference 'Controlling Environmental Pollution' in London on 22 June 1994.

References

ANON, 1975. *Preliminary assessment of suspected carcinogens in drinking water.* US EPA report to Congress.

——, 1980. Directive relating to the quality of water intended for human consumption (80/778/EEC). *Official Journal of the European Communities,* L229.

——, 1984. *Guidelines for drinking water. Volume 1: Recommendations.* World Health Organisation.

——, 1989. *Water supply (water quality) regulations 1989.* Statutory Instrument SI 1147, HMSO.

——, 1992. *Occupational exposure limits 1992.* Health & Safety Executive. Publication EH40/92.

FIELDING, M., GIBSON, T. M., JAMES, H. A., McLOUGH-LIN, K. & STEEL, C. P. 1981. *Organic micropollutants in drinking water.* Water Research Centre, Technical Report TR159.

GOLAY, M. J. E. 1958. *In:* DESTY, D. H. (ed.) *Gas Chromatography 1958 (Amsterdam Symposium).* Butterworths, London.

HERBERT, *et al.* 1986. The occurrence of chlorinated solvents in the environment. *Chemical Industry,* **24,** 861.

HOLMES, J. C. & MORRELL, F. A. 1957. *Applied Spectrometry,* **11,** 86.

JAMES, A. T. & MARTIN, A. J. P. 1951. Liquid–gas partition chromatography. *Biochemical Journal,* **48.**

KENRICK, M. A. P., CLARK, L., BAXTER, K. M., FLEET, M., GIBSON, T. M & TURRELL, M. B. 1985. *Trace organics in British aquifers.* Water Research Centre, Technical Report TR 159.

LAWRENCE, A. R., STUART, M. E., BARKER, J. A., CHILTON, P. J., GOODDY, D. C. & BIRD, M. J. 1992. *Review of groundwater pollution of the Chalk aquifer by halogenated solvents.* Report prepared by the Hydrogeology Research Group of the British Geological Survey for the National Rivers Authority.

LOVELOCK, J. E. 1961. *Analytical Chemistry,* **33,** 162.

MISSTEAR, B. D. R., ASHLEY, R. P. & LAWRENCE, A. R., 1998. Groundwater pollution by chlorinated solvents: the landmark Cambridge Water Company case. *This volume.*

ROWE, U. K. *et al.* 1952. *Archives of Industrial Hygiene and Occupational Medicine,* **5,** 566.

STUBER, K., 1932. Injuries to health in the industrial use of trichloroethylene and the possibility of their prevention. *Arch. Gewerbepathol. Gewerbehyg.,* **2,** 398.

SYMINGTON, R., BURGESS, W. G. & DOTTRIDGE, J. 1994. Methyl tertiary butyl ether (MTBE) a groundwater contaminant of growing concern. *In: Proceedings of Conference on Groundwater Pollution.* 16–17 March, London, IBC.

VALLANCE, P. 1994. Address to a seminar on *Cambridge Water Company* v. *Eastern Counties Leather,* 13 January, Lincoln's Inn, Berrymans Solicitors, London.

WHIM, B. P. 1984. The changing role of halogenated solvents in industry. *In:* KAKABADSE, G. (ed.) *Solvent problems in industry.* Elsevier, Amsterdam, 121.

Section 2: New perspectives on groundwater contaminants and their migration

Each decade presents its own 'fashionable' groundwater contaminant, in the 1970s it was nitrate, in the 1980s chlorinated solvents, in the 1990s possibly pesticides and methyl tertiary butyl ether (MTBE). MTBE is an additive to petroleum, particularly the unleaded fuels which are meant to be so environmentally friendly. But as **Burgess** *et al.* explain, MTBE, unlike tetraethyl lead, is very soluble and is rapidly transported in groundwater. It also has low volatility, poor biodegradability and imparts a taste to water at the ppb level – a recipe for a particularly tricky contaminant (see papers by **Ashley** in section 1 and **Banks** in section 4 for further details).

Modelling tools are often an essential part of a hydrogeologist's armoury when assessing contaminant transport and fate. They are not always easy to apply, however. Limitations in the equivalent porous medium approach, the available data and our understanding of the physico-chemical processes are particularly acute when addressing fissure flows systems such as those found in the Chalk or in granite, but surely a nice granular glacial aquifer should be straightforward? Not according to **Williams** *et al.* Even in such systems, the vagaries of hydrochemical kinetics and the existence of preferential groundwater flow pathways confound most simplistic modelling approaches and can lead to significant overestimates of breakthrough times.

Methyl tertiary butyl ether (MTBE): a groundwater contaminant of growing concern

W. G. BURGESS, J. DOTTRIDGE & R. M. SYMINGTON

Hydrogeology Group, Department of Geological Sciences, University College London, London WC1E 6BT, UK

Abstract: Methyl tertiary butyl ether (MTBE) is a fuel oxygenate rapidly becoming the preferred alternative to leaded and other, unleaded, additives in petrol. MTBE was introduced in the US as a fuel oxygenate in 1979, and has been used in the UK since the mid 1980s. Environmental legislation is expected to lead to increased use of MTBE in the future. Current UK legislation allows MTBE to constitute up to 10% by volume of unleaded petrol. MTBE has a low toxicity, yet its low odour and taste threshold makes it a problem at concentration levels above $2-3\,\mu g\,l^{-1}$ in water. It is more than ten times more soluble than the BTEX components of fuel, is apparently not retarded by sorption, and is reported to be non-biodegradable. Suggestions that MTBE exhibits co-solubility with BTEX components, thereby increasing their mobility, add to concerns about its behaviour in aquifers. This paper reviews the properties of MTBE and the limited published work on its behaviour as a groundwater contaminant.

MTBE (methyl tertiary butyl ether, $CH_3-O-C(CH_3)_2-CH_3$) was introduced as a fuel oxygenate additive in the US in 1979, and is now in widespread use throughout the world (Owen 1990; Unzelman 1991). Alternative names are tertiary butyl methyl ether (TBME) and 2-methoxy-2-methylpropane. Atmospheric legislation targeting exhaust emissions (e.g. EEC 83/351) has encouraged the use of fuel oxygenates. MTBE is used in preference to the alternative alcohol-based oxygenates because of its lower blended vapour pressure.

MTBE now has the most rapidly increasing production figures of all petrochemicals worldwide (Petrochemical Report 1993). Production in US refineries was 118 000 barrels/day in 1991, and projections are for a capacity of 200 000 barrels/day by 1995 (Unzelman 1991). UK production is about 2320 barrels/day, from one refinery (Institute of Petroleum MTBE file). Legislation in the UK allows up to 10% of MTBE by volume in fuel (EEC 85/536), although current use is estimated by the petroleum industry to be less than 1%. The long-term trend is towards increased use of MTBE in unleaded petrol in the UK.

MTBE is believed to have a relatively low toxicity (Duffy *et al.* 1992). At high levels, inhalation of MTBE vapours causes headaches and nausea (Crow 1993). In the groundwater environment, it is recalcitrant and some of its properties, which differ significantly from the BTEX components of fuel (benzene, toluene, ethylbenzene and xylene), have led to concerns about its behaviour in aquifers. In particular, MTBE is very much more soluble than benzene, the most soluble of the BTEX components, and is apparently not significantly retarded by adsorption. Furthermore, the weight of published evidence suggests that MTBE is not biodegradable (Barker *et al.* 1990; Jenson & Arvin 1990; Sulfita & Mormile 1993). It has been recorded as migrating at the same rate as chloride in groundwater (Barker *et al.* 1990). There have also been concerns that MTBE may be co-soluble with BTEX components and thus may enhance their mobility (Garrett *et al.* 1986).

MTBE has a very low taste and odour threshold (Angle 1991; IWD 1991; Anderson 1993). By taste, MTBE is detectable in water at about $6\,\mu g\,l^{-1}$, and by one report as low as $2\,\mu g\,l^{-1}$. There are no official drinking water standards for MTBE in UK, EC or US legislation. Recommended maximum levels have been quoted as $200\,\mu g\,l^{-1}$ by the Tulsa University Medical Centre (Hartley & Englande 1992) and $50\,\mu g\,l^{-1}$ by the State Toxicologist for Maine USA (Garrett *et al.* 1986). In view of the problem posed by taste, it seems likely that a limit at or below $5\,\mu g\,l^{-1}$ may become an operational requirement. The detection limit of MTBE in groundwater by high-pressure liquid chromatography has been determined as better than $0.1\,\mu g\,l^{-1}$ (Symington 1993). Despite its low toxicity, the presence of MTBE in groundwater at concentrations greater than a few micrograms per litre is a serious situation at public water supply abstractions, and demands remedial action on its own account. Elsewhere, it may be of value to monitor for MTBE as an indicator and warning of the more serious consequences of a fuel spill.

BURGESS, W. G., DOTTRIDGE, J. & SYMINGTON, R. M. 1998. Methyl tertiary butyl ether (MTBE): a groundwater contaminant of growing concern. *In*: MATHER, J., BANKS, D., DUMPLETON, S. & FERMOR, M. (eds) *Groundwater Contaminants and their Migration.* Geological Society, London, Special Publications, **128**, 29–34.

Table 1. *The physical properties of MTBE*

Common names	MTBE (Methyl Tertiary Butyl Ether)	
	TBME (Tertiary Butyl Methyl Ether)	
	2-Methoxy-2-methylpropane	
Physical state	Liquid	(The Merck Index 1989)
Colour	Colourless	(The Merck Index 1989)
Odour threshold	680 ppb	(Angle 1991)
Taste threshold	2–3 ppb	(Tester, pers. comm.)
Water solubility	48 g/100 g at 20 °C	(The Merck Index 1989)
Co-solubility effect	No co-solubility effect	(Stephenson 1992)
Density	0.7404 g ml^{-1} at 20 °C	(The Merck Index 1989)
Vapour pressure	32.66 kPa at 25 °C	(The Merck Index 1989)
Aqueous half-life	540 min	
Absorption	0.004 g per 1.0 g activated carbon	(API 1991)
Henry's Law constant	4.5×10^{-4} (atm m^3 mol^{-1})	(USA EPA 1986)
Molecular mass	88.15	(The Merck Index 1989)
Melting point	−110 °C	(The Merck Index 1989)
Boiling point	55 °C	(The Merck Index 1989)
Log partition coefficient		
Octanol/water (K_{ow})	0.94–1.30	(Veith *et al.* 1983; IWD, 1991;
		Funisaki *et al.* 1985)
Fuel water (K_{fw})	15.5 at 22 °C	(Cline *et al.* 1991)
Drinking water standards:		
EEC, WHO, EPA	None	
Recommendations	200 ppb	(Hartley & Englande 1992)
	50 ppb	(Garrett *et al.* 1986)
Atmospheric half-life	4 days under summer conditions	(Bott *et al.* 1992)
Reactivity (OH)	2.8×10^{-12} cm^3 molecule^{-1} s^{-1}	(Bott *et al.* 1992)

The properties of MTBE

A summary of the principal physical properties of MTBE is given in Table 1 (from Symington 1993).

MTBE is *hydrophilic* in a two-phase (water–MTBE) system. The solubility of MTBE in water at standard temperature and pressure (20 °C and 1 atm) is 48 000 mg l^{-1} (Merck Index 1989). In comparison, benzene, the most soluble aromatic hydrocarbon, has a solubility of 1780 mg l^{-1}; the solubility of toluene and *m*-xylene, two other principal BTEX components, is 515 mg l^{-1} and 170 mg l^{-1} respectively. The solubility of MTBE increases with decreasing temperature (Stephenson 1992; also Fig. 1), so that in shallow groundwaters in the UK, MTBE has a solubility in excess of 50 000 mg l^{-1}. The high solubility of MTBE in water is due to the dipolar nature of its molecule (Brown 1972). This polarity also affects co-solubility properties and volatility.

Due to the high solubility of MTBE in water, it has been postulated that MTBE may effectively increase the solubility of other hydrocarbon components by co-solvency (Garrett *et al.* 1986). Fuel is a complex mixture of over 200 organic compounds, and the concern is that co-solvency could result in an unexpected increase in the solubility of some other components.

Research has, however, demonstrated a slight decrease in solubility for all BTEX components with increasing MTBE content, in a standard reference unleaded fuel with average BTEX content of 15% by volume (API 1991). Other published data confirm this conclusion (Groves 1988; Cline *et al.* 1991). A summary of the API results is illustrated in Fig. 2.

MTBE (Henry's constant 4.5×10^{-4} atm m^3 mol^{-1}) is an order of magnitude less volatile

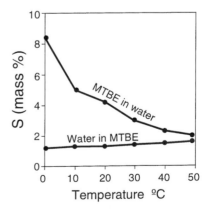

Fig. 1. Mutual solubility of water and MTBE as a function of temperature.

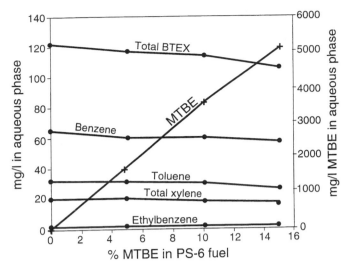

Fig. 2. Co-solubility of BTEX and MTBE, water/fuel ratio 10:1, at 10 °C (after API 1991).

than the BTEX components, making it less susceptible to remediation by volatilization.

In a ternary (fuel–water–MTBE) system, MTBE is *hydrophobic*. At a shallow groundwater temperature of 10 °C, it preferentially concentrates in the fuel phase at about 80% by volume. The fuel–water partition coefficient, K_{fw}, is temperature-dependent (Fig. 3). The octanol–water partition coefficient, K_{ow}, has been determined in a series of laboratory experiments to be in the range 0.94–1.3 (Veith *et al.* 1983; Funisaki *et al.* 1985). This implies that the tendency for MTBE to be retained by organic carbon in the aquifer matrix is two orders of magnitude *less than* for benzene. Ilett *et al.* (1990) and API (1991) have reported on the adsorption of MTBE by activated charcoal. Adsorption occurs at maximum efficiency, 96%, with a charcoal:MTBE ratio of between

6:1 and 8:1. The maximum adsorption capacity of MTBE on charcoal is 0.004 g/g charcoal (API 1991), which is small in comparison to the adsorption of BTEX (0.03 g/g charcoal for BTEX). At the levels of carbon content of the main UK aquifers, the Chalk and the Triassic sandstones, the low adsorption capacity of MTBE might therefore be expected to provide negligible retardation.

MTBE as a contaminant in groundwater

Field observations

The physical properties summarized above suggest that once in the ground as a component of split fuel, MTBE will be a mobile and persistent groundwater contaminant. The mobility of MTBE in groundwater has been demonstrated in field studies from North America (Garrett *et al.* 1986; Barker *et al.* 1990), and has been investigated in an unpublished confidential study in the UK (Symington 1993).

At one site in Maine, MTBE was the first fuel component to be detected in wells up to 300 m from the source of contamination in an unconfined bedrock aquifer of fractured schists (Fig. 4). Controlled field experiments were carried out in an aerobic unconfined sandy aquifer at Camp Borden, Ontario, by Barker *et al.* (1990). The progress of fuel-contacted groundwater, one fuel source spiked with MTBE at 15% by volume, was compared with the movement of chloride as a conservative tracer using a dense network of multi-level piezometers. The results

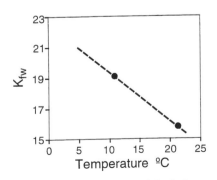

Fig. 3. Temperature dependency of the fuel–water partition coefficient of MTBE (after Cline *et al.* 1991; API 1991).

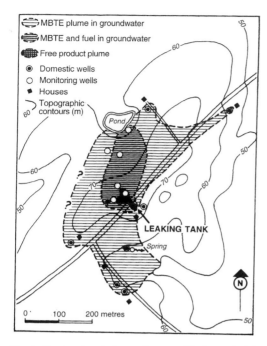

Fig. 4. Contaminant plumes from a spill of petrol with MTBE, Maine, USA (after Garrett *et al.* 1986).

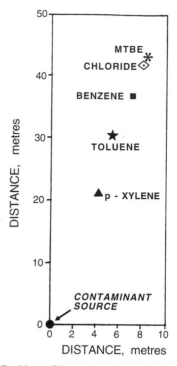

Fig. 5. Positions of the centres of mass for the chloride, MTBE, benzene, toluene and p-xylene contaminant plumes for the Borden experiment, at 476 days after injection (after Barker *et al.* 1990).

show the mobility of MTBE in groundwater to be only slightly less than that of chloride, over a 476-day period and along a flow path of less than 50 m, as illustrated in Fig. 5. The BTEX components of the fuel show the expected chromatographic separation and retardation due to sorption. MTBE formed the outer region of contamination beyond the source zone, being the sole identified hydrocarbon contaminant in over half of the total contaminated area. This observation has been repeated in a recent case study in the UK, demonstrating the mobility of MTBE as a contaminant in groundwater (Symington 1993). Significantly, the Borden study did not show the presence of MTBE to have any observable effect on the mobility of the BTEX components of the fuel at the site. Concentrations of MTBE much higher than 15% by volume would be needed to reduce the sorption of BTEX in a rock matrix with low solid organic carbon content. The co-solvency effect may become significant when hydrocarbon fuel is present as a discrete phase.

Laboratory studies

The results of laboratory studies on the bio-degradability (Jenson & Arvin 1990; Sulfita &

Mormile 1993), fuel–water partitioning (API 1991; Cline *et al.* 1991), octanol–water partitioning (Veith *et al.* 1983; Funisaki *et al.* 1985), and sorption on charcoal of MTBE (Ilett *et al.* 1990; API 1991) are summarized in Table 1.

Discussion and conclusions

MTBE can be the largest single component of unleaded petrol. It is soluble in water, mobile and persistent in the subsurface, and poses a problem of water supply at very low concentrations. MTBE was recently introduced into the UK and has already been detected at public water supply boreholes. It is likely to become a groundwater contaminant of major concern in the future.

The physical properties of MTBE have implications for the likely effectiveness of aquifer and groundwater remediation strategies. MTBE concentrates 80% in the fuel phase. However, it is more soluble, more mobile and less volatile than the BTEX components of fuel. The free product fuel phase (if present), the water–fuel–MTBE phase, and the water–MTBE phase may all require separate remediation approaches. Any individual remediation will have its limitations (e.g. Mackay & Cherry 1989). Skimmer pumps, volatilization methods and pump-and-treat may all be useful according to the pattern of the contamination and the particular hydrogeological conditions. No assessments of the effectiveness of specific remedial measures for MTBE contamination in groundwater have been published.

Water treatment systems for MTBE-contaminated water rely on volatilization (Lowry 1988; API 1990). MTBE removal rates between 55% and 100% have been achieved using air stripping systems at 15 sites reported by API (1990). Activated carbon filtering is considered an expensive option compared with volatilization, because of the low sorption of MTBE on charcoal and the early breakthrough of MTBE compared with other soluble components of fuel.

The physical properties of MTBE have been documented, but the behaviour of MTBE in geological materials, specifically in different types of aquifers, requires much more study. It is uncertain to what extent MTBE may be retained by matrix diffusion in UK aquifers. The nature of MTBE interactions with clays and with oxyhydroxides is unknown. Co-solvency effects on other organic compounds such as organochlorides have not been reported. Toxicity studies are still continuing. The future significance of MTBE as a contaminant in groundwater will be better quantified once these issues have been addressed.

This paper is based on a review prepared as part of an investigation for the National Rivers Authority (prior to the establishment of the Environment Agency). The views expressed are those of the authors and not necessarily those of the Environment Agency. NRA (now EA) officers accept no liability for any loss or damage arising from the interpretation or use of the information or reliance upon the views contained within the paper.

References

ANDERSON, E. V. 1993. Health studies indicate MTBE is safe gasoline additive. *Chemical and Engineering News*, **71**(38), 9–18.

ANGLE, C. D. 1991. Letter to the Editor. *JAMA*, **266**(21), 2986.

API 1990. *A compilation of field-collected cost and treatment effectiveness for the removal of dissolved gasoline components from groundwater.* Document 4525, November 1990. Health and Environmental Sciences, American Petroleum Institute, 1220L Street, Northwest Washington, DC 2005, USA.

—— 1991a. *Cost-effective, alternative technologies for reducing the concentrations of methyl tertiary butyl ether and methanol in groundwater.* Document 4497, May 1991. Health and Environmental Sciences, American Petroleum Institute, 1220L Street, Northwest Washington, DC 2005, USA.

—— 1991b. *Chemical fate and impact of oxygenates in groundwater: Solubility of BTEX from gasoline-oxygenated mixtures.* Document 4531, November 1991. Health and Environmental Sciences, American Petroleum Institute, 1220L Street, Northwest Washington, DC 2005, USA.

BARKER, J. F., HUBBARD, C. E. & LEMON, L. A. 1990. *Proceedings of the NWWA/API Conference of Petroleum Hydrocarbons and Organic Chemicals in Groundwater*, 113–127.

BOTT, D. J., DAWSON, M. W., PIEL, W. J. & KARAS, L. J. 1992. *MTBE environmental fate.* Arco Chemical Company, for Life Cycle Analysis and Eco-Assessment in the Oil Industry, The Institute of Petroleum, London.

BROWN, W. H. 1972. *Introduction to organic biochemistry.* Willard Grant Press, Boston, MA.

CLINE, P. V., DELFINO, J. J. & RAO, P. S. C. 1991. Partitioning of aromatic constituents into water from gasoline and other complex solvent mixtures. *Environmental Science and Technology*, **25**(5), 914–920.

CROW, P. 1993. MTBE question in Alaska. *Oil and Gas Journal*, 15 March 1993, 36.

DUFFY, J. S., DEL PUP, J. A. & KNIESS, J. J. 1992. Toxological evaluation of methyl tertiary ether (MTBE): testing performed under the TSCA consent agreement. *Journal of Soil Contamination*, **1**(1), 29–37.

EEC Environmental Directives. 78/665 Atmospheric Emissions, 83/851Atmospheric Emissions, 85/536 Oxygenates as Fuel Additives.

FUNISAKI, N., HADA, S. & NEYA, S. 1985. Partition coefficients of aliphatic ethers – molecular surface area approach. *Journal of Physical Chemistry*, **89**(14), 3046–3049.

GARRETT, P., MOREAU, M. & LOWRY, J. D. 1986. *Proceedings of the NWWA/API conference of petroleum hydrocarbons and organic chemicals in groundwater*. Dublin, Ohio, 227–238.

GROVES, F. R. JR., 1988. Effect of cosolvents of the solubility of hydrocarbons in water. *Environmental Science and Technology*, **22**(3), 282–286.

HARTLEY, W. R. & ENGLANDE, A. J. JR. 1992. Health assessment of the migration of unleaded gasoline – a model for petroleum products. *Water Science Technology*, **25**(3), 65–72.

ILETT, K. F., LAURENCE, B. H. & HACKETT, L. P. 1990. Alimentary tracts and pancreas. *Journal of Gastroenterology and Hepatology*, **5**, 499–502.

Institute of Petroleum. MTBE File. IP, 61 New Cavendish Street, London W1M 8AR.

IWD 1991. *Canadian water quality guidelines for methyl tertiary-butyl-ether.* Draft Copy, IWD Scientific Series Report, Environmental Quality Guidelines, Water Quality Branch, Inland Water Directorate, Environment Canada, Ottawa, Ontario.

JENSON, H. M. & ARVIN, E. 1990. Solubility and degradability of the gasoline additive MTBE, methyl-tert.-butyl-ether, and gasoline compounds in water. *Contaminated Soils*, **90**, 445–448.

LOWRY, J. D. 1988. Removal of petroleum hydro-carbons and MTBE from water by multistaged POE aeration system. *Proceedings of the FOCUS conference on eastern groundwater issues*, 27–29 Sept., National Water Well Association, Dublin, Ohio, 339–352.

MACKAY, D. M. & CHERRY, J. A. 1989. Groundwater contamination: pump-and-treat remediation. *Environmental Science and Technology*, **23**(6), 630–636.

Merck Index 1989. *The Merck Index*, **11**, 5908–5909.

OWEN, 1990. *Automotive Fuels Handbook*. Published for the Society of the Chemical Industry by John Wiley, Institute of Petroleum Library.

Petrochemical Report 1993. Methanol, MTBE suppliers will likely keep up with rising demand. *OGJ Special, Oil and Gas Journal*, March 1993, 48–52.

STEPHENSON, R. M. 1992. Mutual solubilities: water-ketones, water-ethers, and water-gasoline, alcohols. *Journal of Chemical Engineering Data*, **37**(1), 80–95.

SUFLITA, J. M. 1993. Anaerobic biodegradation of chemicals of environmental concern in the terrestrial subsurface. *International Symposium on Subsurface Microbiology*, Sept. 1993, ISSM, Bath, UK.

—— & MORMILE, M. R. 1993. Anaerobic biodegradation of known and potential gasoline oxygenates in the terrestrial subsurface. *Environmental Science and Technology*, **27**(5), 976–978.

SYMINGTON, R. M. 1993. *MTBE as a Contaminant; With Field Data on a Release of Unleaded Fuel in an Unconfined Fractured Aquifer*. MSc thesis, University of London (UCL).

UNZELMAN, G. H. 1991. US Clean Air Act expands role for oxygenates. Technology, *Oil and Gas Journal*, 15 April, 44–48.

VEITH, G. D., CALL, D. J. & BROOKE, L. T. 1983. Structure–toxicity relationships for the fathead minnow, *Pimephales promelas*: Narcotic Industrial Chemicals. *Canadian Journal of Fish and Aquatic Science*, **40**, 743–748.

Mechanisms for ^{85}Sr migration through glacial sand determined by laboratory and *in situ* tracer tests

G. M. WILLIAMS, P. J. HOOKER, D. J. NOY & C. A. M. ROSS

Fluid Processes and Waste Management Group, British Geological Survey, Keyworth, Nottingham NG12 5GG, UK

Abstract Prediction of solute transport based on the laboratory measurement of radionuclide/ sediment sorption parameters has been compared with field tracer tests involving the release of ^{85}Sr and a conservative tracer ^{131}I into a radially divergent flow field around a recharge well in a confined glacial sand aquifer. Tracer breakthrough was monitored in multi-level sampling installations. Two tests at different imposed rates of groundwater flow were conducted. At average flows up to 5 m day^{-1}, the retardation of ^{85}Sr at a given sampling point was lower than that observed with an average flow rate of 0.5 m day^{-1}, suggesting a kinetic control on sorption. Sequential selective leaching was also undertaken to study the geochemical phases responsible for sorption. In a core consisting predominantly of medium sand, approximately 80% of the ^{85}Sr was present in ion exchange sites, with 12% in the carbonate phase. The remaining 8% was recovered easily by extraction with distilled water. Kinetic data for Sr sorption on sand and silt core material suggested a two-stage process which could be fitted by two empirical sorption/desorption rate constants. These were used to simulate the effect of kinetics of sorption on the breakthrough curves for strontium. The retardation of ^{85}Sr at the lower flow rate was in relatively good agreement with laboratory values derived from batch sorption experiments lasting several days. At the high flow rate, longitudinal dispersivity α_L for ^{85}Sr was larger than that for ^{131}I, but this discrepancy reduced at the lower flow rate where some convergence to the ^{85}Sr values was observed. A possible explanation is that at the high flow the opportunity for ^{131}I diffusion into relatively static porewater was limited; whereas with solutes moving at a lower rate there was a greater opportunity for diffusion into 'dead end' pores.

The need to predict contaminant migration in groundwater has prompted researchers to carry out *in situ* tracer experiments to confirm the validity of model predictions based on *a priori* characterization of the aquifer. Aquifers are heterogeneous and many uncertainties appear in model predictions because of the complexity of the physical and chemical processes that take place during flow. Conservative tracers which do not undergo any chemical interaction with the aquifer minerals reflect the effects of physical heterogeneity in the flow regime, while comparing conservative tracers with tracers which chemically react with the aquifer provides an indication of the chemical heterogeneity.

One of the simplest interactions modelled is the reversible sorption of a solute on the solid phase. This leads to a retardation of the solute relative to the average velocity of the advecting groundwater which can be predicted from the distribution coefficient (K_d). The K_d is the equilibrium distribution of the solute between aqueous and solid phases, and providing that the sorption reaction is instantaneous and reversible and the K_d is constant over the range of aqueous solute concentrations, it can be used to calculate the retardation factor from the following:

$$R = 1 + \left(\frac{1-\theta}{\theta}\right)\rho K_d \qquad (1)$$

where R is the retardation factor, θ is the porosity, ρ is the grain density, and K_d is the distribution coefficient. It should be recognized that K_d (ml g^{-1}) is the distribution coefficient at true thermodynamic equilibrium and is thus rarely obtained. Where equilibrium is not established but where measurements approach the K_d, the distribution ratio R_d is used.

Values of R_d can be determined from simple batch sorption tests. These involve shaking an excess of groundwater spiked with radionuclide, with a small, and possibly non-representative, sample of disturbed rock, which may be considerably different from conditions prevailing within the aquifer. In general, measurements of R_d have revealed considerable discrepancies between different laboratories (Relyea & Serne 1979) which may be due to a number of factors, e.g. a kinetic control on sorption.

In order to assess the value of batch sorption experiments, a field tracer experiment was undertaken to study the migration of a reactive species, ^{85}Sr, with a conservative tracer ^{131}I, introduced together into a radially divergent flow field around a recharge well in a confined unconsolidated glacial sand aquifer. These breakthrough curves were used to determine retardation factors, for comparison with laboratory-derived values based on the distribution ratio.

WILLIAMS, G. M., HOOKER, P. J., NOY, D. J. & ROSS, C. A. M. 1998. Mechanisms for ^{85}Sr migration through glacial sand determined by laboratory and *in situ* tracer tests. *In*: MATHER, J., BANKS, D., DUMPLETON, S. & FERMOR, M. (eds) *Groundwater Contaminants and their Migration*. Geological Society, London, Special Publications, **128**, 35–48.

To take into account possible kinetic effects on sorption/desorption of the reactive tracer, tests were conducted at two imposed flow rates. In addition, the mineralogical phases responsible for Sr sorption were investigated by sequential leaching experiments on core material contacted with Sr in the laboratory.

Experimental details

Description of the test site

The tests were conducted in a shallow glacial sand horizon interbedded between clays. The borehole array (Fig. 1) consists of a recharge well, a tracer release well and three multi-level sampling installations with sampling ports spaced at 15-cm intervals. Conventional water injection and water level recovery tests were undertaken in the wells with screens over the full thickness of the aquifer which provided

values for hydraulic conductivity in the range 2.9–4.7 $(\times 10^{-6})\,\mathrm{m\,s^{-1}}$. Porosity measurements on core material ranged from 23 to 40%, and grain density was $2.65\,\mathrm{g\,cm^{-3}}$.

Aquifer mineralogy and groundwater chemistry

The glacial sand aquifer consists of fine to medium ferruginous quartz with minor amounts (10%) of feldspar and calcite. The clay fraction consists of mica, kaolinite and chlorite, with trace amounts of smectite. Organic matter is present mainly in the form of allochthonous coal fragments. Surface area measurements averaged $22\,\mathrm{m^2\,g^{-1}}$ by the BET method. Groundwater was analysed for major ion and trace components, and was found to be slightly acidic with a relatively high organic content composed principally of fulvic material (Table 1).

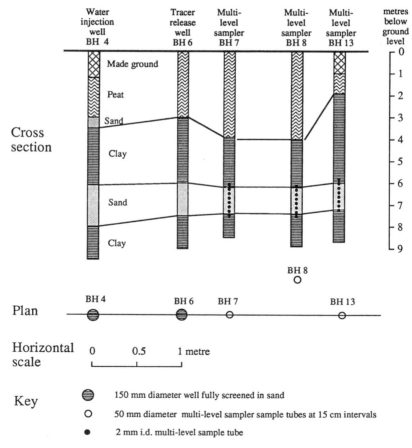

Fig. 1. Details of the borehole array.

Table 1. *Composition of groundwater from the test site*

pH	6.25
Eh	0–150 mv
EC	550 μS
Na	30 mg l^{-1}
K	3.6 mg l^{-1}
Ca	68 mg l^{-1}
Mg	16 mg l^{-1}
Fe (total)	0.41 mg l^{-1}
Mn	1.1 mg l^{-1}
NH$_4$	<1 mg l^{-1}
Cl	31 mg l^{-1}
SO$_4$	28 mg l^{-1}
NO$_3$	<2 mg l^{-1}
HCO$_3$	134.2 mg l^{-1}
SiO$_2$	14 mg l^{-1}
Al (total)	<0.1 mg l^{-1}
TOC	13 mg l^{-1}

Details of the tracer tests

Once steady-state water levels had been established by injecting groundwater into the recharge well, phials containing the tracers (74 MBq ^{131}I as NaI, and 185 MBq ^{85}Sr as SrCl$_2$) were broken simultaneously in the tracer release well and mixed in the water column using a small submersible pump. Water samples were taken simultaneously from the tracer release well and multi-level samplers (Hitchman 1988), and the activity in 5-ml samples was determined using a 2 inch (5 cm) NaI well-type detector with a Davidson multichannel analyser (Model 2056-4K).

Longitudinal dispersivity was determined from the breakthrough curves using a curve-fitting method based on the 1-D analytical solution to the advection–dispersion equation with an exponentially decaying source (van Genuchten & Alves 1982). The solution is as follows:

$$\frac{C}{C_0} = \frac{1}{2} \exp(-t_R \cdot R^*)$$

$$\times \sum_{i=1}^{i=2} \left[\exp\left\{ \frac{Pe}{2} \left(1 + (-1)^i \sqrt{1 - \frac{4R^*}{Pe}} \right) \right\} \right.$$

$$\left. \times \operatorname{erfc}\left\{ \frac{1 + (-1)^i t_R}{2} \sqrt{\frac{Pe}{t_R}} \right\} \right] \qquad (2)$$

where Pe is the Vx/D (Peclet Number), t_R is the Vt/Rx (Dimensionless time) and R^* is the $\lambda Rx/V$. Also, V $(= V_d/\theta)$ is the pore velocity; V_d is the Darcy velocity; θ is the effective porosity; λ is the source decay constant; t is the time; x is the distance; R is the retardation factor; D $(= \alpha_L \cdot V)$ is the dispersion coefficient; and α_L is the longitudinal dispersivity.

Although this equation is for 1-D uniform flow, Sauty (1980) has shown that a 1-D solution to the advection–dispersion equation is valid for radial flow from wells for Peclet numbers greater than 10. Peclet numbers in the test are an order of magnitude greater than this. As the tracer release is some way from the water injection well, the conditions become even closer to those of a uniform flow velocity, further justifying the use of the van Genuchten and Alves analytical solution.

Laboratory sorption experiments

Batch sorption studies
Batch sorption experiments were carried out on discrete samples from specific sections of a core which fully penetrated the test zone. A homogenized sample of the complete aquifer core was also used for sorption experiments. In each case, approximately 0.5 g of sediment was placed in a 30-ml polypropylene centrifuge tube with 10 ml of 0.45 μm filtered groundwater spiked with radionuclide and stable carrier. The tubes were either gently agitated or left static at two temperatures for 7 days. They were then centrifuged and 5-ml aliquots of the supernatant removed for analysis by gamma spectrometry using a Philips PW 4580 batch counter. The spiked groundwater samples were counted concurrently with distilled water blanks. The same procedure was applied to groundwaters not contacted with sediment, which acted as controls.

Sequential leaching experiments
In order to study the geochemical phases responsible for the uptake of Sr, core material that had been contacted with ^{85}Sr in the laboratory was subjected to a sequential leaching procedure adapted from Ross (1980) (Table 2). A core fully penetrating the aquifer was sectioned into subsamples every 15 cm. After being homogenized, 5 g of each sub-sample were dried to constant weight to determine moisture content. Duplicate samples (2 g) were placed in polypropylene centrifuge tubes to which 10 ml of groundwater (spiked with approximately 1.9×10^{-15} g ml^{-1} of ^{85}Sr) were added; the tubes were agitated at ambient temperature (18 ± 3°C) for 24 h. The tubes were then centrifuged at 1370 g for between 10 and 20 min until the supernatant was visibly clear. Aliquots (5 ml) of the supernatant were counted on a Philips PW 4580 batch gamma counter. The solid samples were then extracted as outlined.

A silt horizon from within the core which had been used initially for a study of sorption kinetics was also subjected to this sequential leaching scheme.

Table 2. *Details of sequential leaching scheme (After Ross, 1980)*

Phase extracted	Extraction procedure
Interstitial fluid	Centrifuge and count supernatant
Soluble fraction	Add 10 ml distilled water; shake for 90 min; centrifuge and count supernatant
Exchangeable cations	Add 10 ml 1M ammonium acetate; shake for 90 min; centrifuge and count supernatant
Carbonate fraction	Add 15 ml 1M acetic acid; shake for 90 min; centrifuge and count supernatant

Table 3. *Results of tracer tests.*

Sample point	Retardation factor		$\alpha_L{}^{131}I$ (m)		$\alpha_L{}^{85}Sr$ (m)	
	Slow	Fast	Slow	Fast	Slow	Fast
7-5	–	–	0.0083	0.0020	0.0029	0.0083
7-6	26.1	9.2	0.0041	0.0020	0.0177	0.0041
7-7	29.9	10.4	0.0177	0.0029	0.0188	0.0110
7-8	34.9	10.0	0.0133	0.0041	0.0177	0.0170
7-9	16.0	19.0	0.0023	0.0041	0.0013	0.0290

'Slow' and 'Fast' refer to the tracer tests at the two imposed flow rates.

Sorption kinetics

Sr sorption kinetics were determined on two sand samples (T7 and one from core M5) and a silt sample (from core M5) using a water:rock ratio of 10:1. The reaction vessels were maintained at ambient conditions, and only occasionally agitated. Strontium uptake on the sediment was measured by abstracting 3-ml or 5-ml aliquots at known times, centrifuging and gamma counting the ^{85}Sr activities. Contact times were 58 days for the silt and 24 h for the sand samples. Some of the constituents of the groundwaters were measured by ICP-OES analysis for both silt and sand experiments carried out for the M5 core.

Kinetic tests carried out on sand T7 with 0.45 μm filtered groundwater and with unfiltered water showed no significant differences. The aqueous phase was analysed by ICP-OES to ensure that the ^{85}Sr activity was directly proportional to the total strontium concentration, and to confirm that no other reactions such as precipitation or mineral dissolution were occurring.

Results and discussion

Tracer tests

A summary of the results and selected breakthrough curves are given in Table 3 and Figs. 2–5. In the fast flow test, ^{85}Sr was retarded

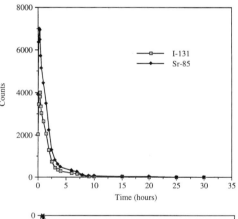

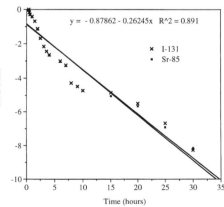

Fig. 2. Changes in tracer concentration with time in the release well for the fast flow test.

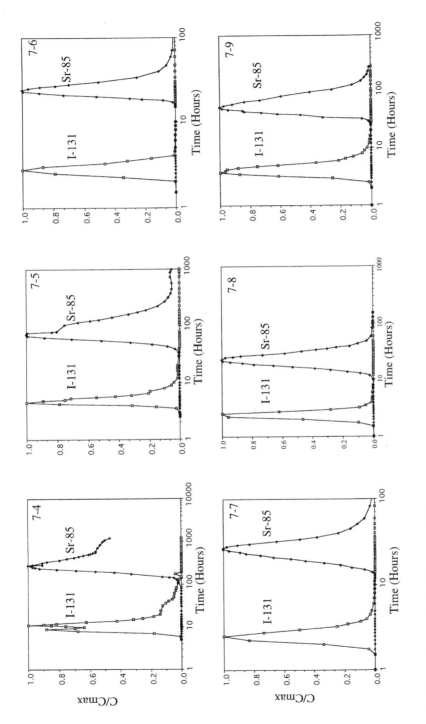

Fig. 3. Breakthrough curves for fast flow tracer test.

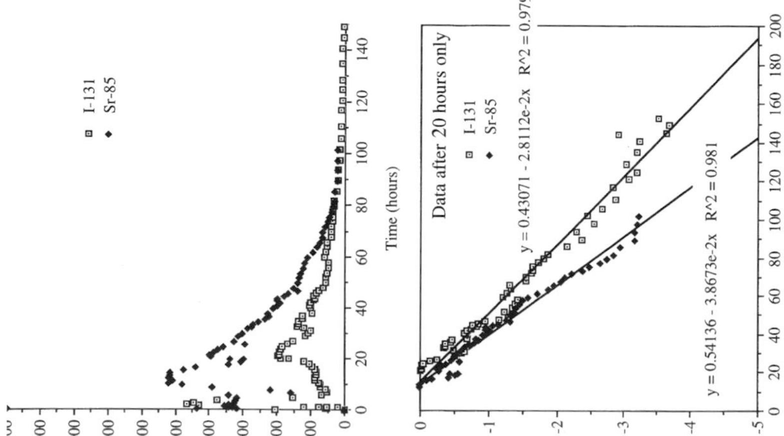

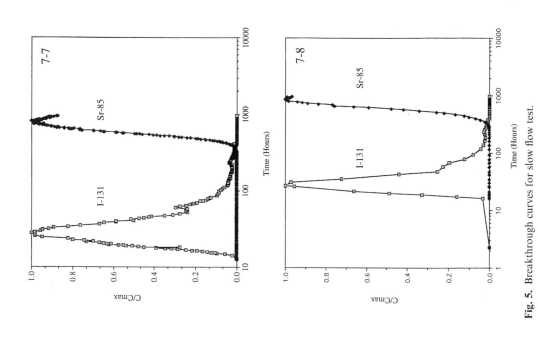

Fig. 4. Changes in tracer concentration with time in the release well for the slow flow test.

Fig. 5. Breakthrough curves for slow flow test.

with respect to ^{131}I by factors of between 8.7 and 25.6 (for all the sampling points), which give R_d values in the range 0.86–2.77 ml^3 g^{-1}. For a given sample point, dispersivities based on [85]Sr breakthrough were larger than those calculated from ^{131}I breakthrough, by factors of between 3 and 5. This was also the finding of Pickens *et al.* (1981), who reported a mean dispersivity α_L of 1.9 cm for [85]Sr and a mean of 0.8 cm for ^{131}I, calculated from breakthrough curves obtained under a fast flow condition, and attributed the effect to nonequilibrium adsorption effects for [85]Sr.

Repeating the tracer test at a flow rate approximately ten times lower resulted in an increase in the observed retardation factors and a better agreement between the dispersivities obtained for ^{131}I and [85]Sr at the same sample point. These dispersivities were, however, similar to those found for [85]Sr in the previous 'fast flow' experiment, suggesting that under relatively high imposed flows the apparent dispersivity for the conservative iodide tracer is lower. In the 'slow flow' experiment, monitoring of the injection well composition was poor over the 2-h period after breaking the individual glass phials containing ^{131}I and [85]Sr in the borehole, and the injection characteristics are less well constrained (Fig. 4).

Averaging the breakthrough curves for the multi-level installation (BH 7) gives the overall curve in Fig. 6, which displays the relative contributions from different samplers, and simulates the expected breakthrough curve that would have been obtained from a fully penetrating screened well. In this case, the breakthrough curve has multiple peaks and it is not possible to infer a single dispersivity value. However, if the averaged curve is split into two peaks the dispersivity calculated correlates approximately with that of the individual sets of dispersion values. This illustrates the effect that the monitoring installation can have on the estimation of aquifer dispersivity and suggests that the heterogeneity of the formation must be taken into account in the design of the experimental conditions, e.g. setting the vertical scale over which the solute concentrations should be measured.

Being a radially divergent test, the tracers were not recovered in the field so it was not easy to estimate how much activity was sorbed onto the sediment. Attempts to assess [85]Sr recovery by comparison with ^{131}I suggested that the tracers were not well mixed in the injection borehole and this further complicated recovery estimates.

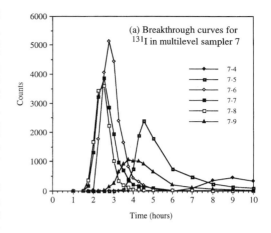

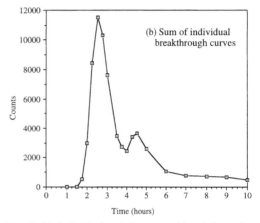

Fig. 6. (a) Individual and (b) summed breakthrough curves for Borehole 7.

Laboratory sorption results

Distribution ratios
The distribution ratios determined from batch sorption experiments under a range of conditions and the calculated retardation factors are summarized in Table 4. For the core material higher values of distribution ratio were obtained when the reaction tubes were shaken in comparison to when they were static. Retardation factors range from 22 to 40 for core material while a value of 68 was obtained for the unrepresentative homogenized core sample.

Sorption isotherms
The homogenized core (Fig. 7) shows approximately linear [85]Sr sorption over the concentration range studied, with an R_d of approximately 7.6 ml^3 g^{-1}. This is not very representative of the field situation since homogenizing the aquifer mixes silt/clay lenses with sand

Table 4. *Summary of distribution ratios (R_d) determined from batch sorption experiments.*

Sample	Conditions	R_d range (average) ml g^{-1}	Retardation factor (calculated)
Homogenized sample from full thickness of aquifer	shaken 18 °C	7.6	68
Discrete cores M3	static 9 °C	1.98– 3.57 (2.37)	22
Discrete cores M3	shaken 9 °C	3.66– 4.83 (4.41)	40
Discrete cores M3	static 18 °C	2.37– 3.92 (2.92)	26.9
Discrete cores M4	static 9 °C	1.54– 5.74 (2.51)	23.3
Discrete cores M4	static 18 °C	2.03– 4.92 (2.88)	26.5

and increases the overall sorption capacity of the material. In choosing representative core material, emphasis should be placed on defining the most permeable and/or least sorbing flow path for predicting the fastest breakthrough times.

Mineralogical distribution of ^{85}Sr

A summary of the association of ^{85}Sr with mineralogical phases in samples of sand and silt is given in Table 5. The soluble fraction accounted for between 7.4 and 11.5% of the total ^{85}Sr in the sediment, with an average of 8.5%. The R_d for desorption involving distilled water (containing no solutes) was always higher than that for the case with unspiked groundwater, since the removal of ^{85}Sr from carbonate and sulphate phases, although limited by their solubilities, will increase with the volume of distilled water added.

The ^{85}Sr released by extraction with ammonium acetate accounted for between 74.2 and 82.0% (average 79.2%) of the ^{85}Sr in the sediment. This is equivalent to a cation exchange capacity of 0.0028 meq per 100 g of sediment. Since Sr accounts for 0.24% of the total cation concentration in the aqueous phase (i.e. $\sum$(Na + K + Ca + Mg + Sr) = 5.95 meq l^{-1}) then the total cation exchange capacity of the sediment is likely to be about 1 meq per 100 g, assuming that the ratio of cations in the aqueous phase is similar to their ratio in the solid phase. Since the measured cation exchange capacity of the sand gave a similar range of between 0.3 and 0.75 meq per 100 g, then it is likely that selectivity in the uptake of Sr over the competing cations is negligible.

The carbonate fraction, extractable by acetic acid, accounted for between 9.8 and 12.9% (average 11.6%) of the total, which represents most of the remaining Sr in the sediment. The residual

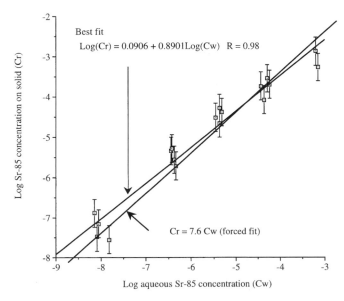

Fig. 7. ^{85}Sr isotherm with homogenized core material.

Table 5. *Summary of Sr distribution in core material (%)*

Lithology	Soluble	Exchangeable	Carbonate	Residual
Sand (20 samples)	8.5 ± 1.25	79.12 ± 2.62	11.62 ± 1.62	0.55 ± 2.51
Silt (4 samples)	3.84 ± 0.41	58.25 ± 1.26	6.68 ± 0.16	31.23 ± 1.59

fraction (calculated by difference) ranges from -2.2 to $+3.3\%$ (average of 0.5%), and probably occurs in silicate or oxyhydroxide phases (Jackson & Inch 1980, 1983).

In contrast, the distribution of Sr in the silt was significantly different from the sand, with less than 70% recovered by the extractants used. The remaining 30% is likely to be irreversibly sorbed onto the sediment. A direct correlation between R_d and particle size distribution, as observed by others (e.g. Patterson & Spoel 1981; Jackson & Inch 1989), is therefore to be expected.

Sorption kinetics

Changes in ^{85}Sr activity in the aqueous phase as a function of time are shown for both sand and silt samples (Fig. 8). For the sand sample (T7), the R_d increased to about $3 \, \text{ml}^3 \, \text{g}^{-1}$ during the first 16 h after which sorption continued at a much reduced rate. The sand (core M5) with a slightly higher silt content, had a higher sorption capacity (38% of ^{85}Sr taken up after 24 h), and consequently a larger final R_d value. For the silt, after 58 days, 62% of the ^{85}Sr was sorbed giving a final R_d value of about $14 \, \text{ml}^3 \, \text{g}^{-1}$.

The relationship between the ^{85}Sr activities and the total strontium concentrations is shown in Fig. 9. Departure from the ideal $1:1$ relationship with longer contact times may be explained as follows:

(1) ^{85}Sr had been isotopically exchanged for stable strontium ions from the sediments giving an apparently higher uptake;

(2) a dissolution reaction had taken place to augment the soluble stable Sr fraction in opposition to the sorption process.

At present, it is difficult to decide the relative significance of these mechanisms, although geochemical modelling of the groundwater suggests that it is undersaturated with respect to sulphate and carbonate minerals of Sr and Ca (including calcite, a phase observed to be present in the sediments). Thus a slight increase in the aqueous Ca and Sr concentrations may be expected. A dissolution process for calcite would explain the slight increase observed for Ca as the kinetic experiments proceeded, and if trace amounts of

Sr were released simultaneously the departure from the ideal $1:1$ relationship seen in Fig. 9 would be established. This demonstrates the general lack of equilibrium between sediment and groundwater samples when they are brought together in laboratory experiments.

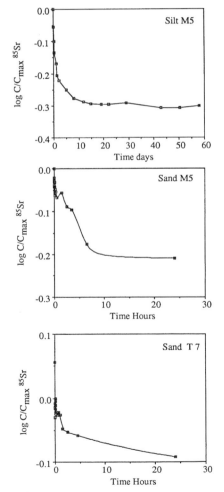

Fig. 8. ^{85}Sr uptake on sand and silt as a function of time.

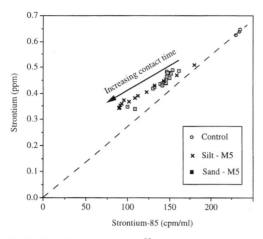

Fig. 9. Relationship between [85]Sr activity change and total strontium concentration change.

Modelling

Comparison of predicted with observed results

Predicted travel times for a conservative tracer between release well and monitoring points estimated from empirically determined water injection rates, from hydraulic conductivity measurements in fully penetrating wells, and from porosity measurements on cores, have over-estimated the observed travel times of the conservative tracer ^{131}I by a factor of five (Table 6). This is because the small-scale heterogeneity in hydraulic conductivity demonstrated by the smaller-scale tracer breakthrough data is ignored. Similarly, the homogenized and unrepresentative core material produced a high R_d, giving a retardation factor about seven times that observed.

Several questions arise from a comparison of the two tests at different flow rates:

(1) Why is the dispersivity of the aquifer calculated from ^{131}I breakthrough at the high flow rate lower than that calculated from the low flow rate?

(2) Why is the aquifer dispersivity calculated from [85]Sr breakthrough curves generally higher than that calculated from ^{131}I curves for the same monitoring point?

(3) Why is the retardation of [85]Sr at the high flow rate less than that at the low flow rate?

Assuming that ^{131}I behaves conservatively, any changes in the breakthrough behaviour at the different flow rates must reflect physical mechanisms during transport. Processes such as

Table 6. *Predicted and observed breakthrough times in the fast flow test.*

Radionuclide	Predicted	Observed
[131]I	12.5 h	2.5 h
[85]Sr*	34 days	1 day
Retardation factor	68	10

* Based on homogenised core material.

anion exclusion could influence the dispersivity of the iodide anions if they are repelled by the negatively charged surfaces of the minerals thus inhibiting dispersion. This phenomenon was invoked by Ogard *et al.* (1988) to explain the behaviour of [36]Cl relative to tritium movement in alluvium. An alternative factor that may affect the shape of the breakthrough curve is the interchange of a conservative solute between 'mobile' and 'immobile' porewater. Goltz & Roberts (1988) assigned sorption sites to the immobile region to simulate qualitatively the spatial solute concentration data for the tracer tests carried out in sand at Borden, Ontario. The effect of this mechanism on the dispersivity estimated from ^{131}I curves is considered in the following section.

In addition to these purely physical mechanisms, the migration of a reactive solute can be influenced by other factors, including reversible or non-reversible sorption, non-linear sorption, precipitation–dissolution, redox, complexation reactions, and isotopic dilution. The laboratory results suggest, however, that the most likely reactions affecting [85]Sr migration will be sorption kinetics, precipitation–dissolution, and isotopic dilution.

Dual porosity model

To study the physical processes giving rise to variations in longitudinal dispersivity experienced by ^{131}I at the two flow rates, the 'dual porosity' model contained within the data fitting programme CXTFIT of Parker & van Genuchten (1984) has been used. This model assumes that the porewater is divided between 'mobile' and 'immobile' regions. Convective and dispersive transport of solutes takes place in the mobile water, while transfer of solutes in and out of the immobile regions is assumed to be by diffusion at a rate proportional to the difference in concentration between the two regions. For the purposes of this interpretation the flow velocity (v_m) for the fast and slow tracer tests was taken as 5 m day^{-1} and 0.5 m day^{-1}, respectively. The programme CXTFIT was used to obtain values

Table 7. *Best-fit parameters using CXTFIT model.*

Test	D (m^2 s^{-1})	β	ω	α_L (mm)	R	$\alpha^* L/\theta$ (m s^{-1})
^{131}I fast	1.14×10^{-7}	0.86	0.27	1.98	1	1.34×10^{-5}
^{131}I slow	3.21×10^{-8}	0.55	4.03	6.05	1	1.18×10^{-5}

for the parameters D, R, β and ω, where

$$\beta = \frac{\theta_m + f \rho K_d}{\theta + \rho K_d}$$

$$\omega = \alpha^* \left(\frac{L}{q} \right)$$

$D =$ dispersion coefficient $(\alpha_L \cdot v_m)$; $R =$ retardation factor; $\theta =$ total porosity; $\theta_m =$ mobile phase porosity; $\rho =$ bulk rock density; $K_d =$ equilibrium sorption coefficient; $f =$ fraction of sorption sites that equilibrate with the mobile phase; q $(= v_M \cdot \theta_m) =$ water flow rate; $\alpha^* =$ rate constant for diffusion into immobile porewater; $L =$ tracer path length.

No sorption is assumed for ^{131}I and the retardation factor (R) is taken to be unity. Under this condition, β is just the fraction of mobile water. The expression for ω can also be rearranged to give a value for $\alpha^*[L/\theta]$, which might be expected to be a constant for the rock.

The differences between β $(= \theta_m/\theta)$ in the ^{131}I tests (Table 7) imply that 86% of the total porosity was mobile in the fast test, while in the slow test this fell to 55%. The result of the dual porosity model is shown in Fig. 10.

Modelling flow with sorption kinetics

Kinetic rate constants

Ignoring the actual mechanism of sorption/exchange (Jackson & Inch 1989), the kinetic data suggest a two-stage sorption process, as

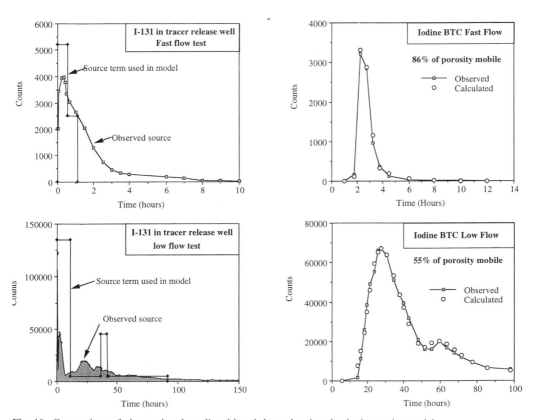

Fig. 10. Comparison of observed and predicted breakthrough using the dual porosity model.

described by Nyffeler *et al.* (1984) and used by Hooker *et al.* (1986).

$$C \underset{k_{r1}}{\overset{k_{s1}}{\rightleftharpoons}} S^* \underset{k_{r2}}{\overset{k_{s2}}{\rightleftharpoons}} S \qquad (3)$$

where C = Sr concentration or ^{85}Sr specific activity in water; S^* = hypothetical intermediate Sr-surface complex; S = concentration of Sr sorbed in the solid phase; k_{s1} = forward reaction rate constant for first stage reaction; k_{r1} = backward reaction rate constant for first stage reaction.

Likewise, k_{s2} and k_{r2} are the forward and backward reaction rate constants, respectively, for the second stage reaction. These first-order rate constants can be calculated by solving the following differential equations:

$$dC/dt = k_{r1}S^* - k_{s1}C$$

$$dS^*/dt = k_{s1}C - (k_{r1} + k_{s2})S* +k_{r2}S$$

$$dS/dt = k_{s2}S* -k_{r2}S$$

Initially, $C = C_0$ and $S^* = S = 0$, and at equilibrium $C = C_e$, $S^* = S_e^*$ and $S = S_e$. Therefore

$$S_e^* = k_{s1}/k_e \cdot C_e = r_1 C_e \qquad (7)$$

$$k_{s1}C_e - (k_{r1} + k_{s2})S_e^* + k_{r2}S_e = 0 \qquad (8)$$

$$S_e = k_{s2}/k_{r2} . S_e^* = r_2 r_1 C_e \qquad (9)$$

Also

$$(S_e + S_e^*) = (C_0 - C_e) \qquad (10)$$

and

$$r_2 r_1 C_e + r_1 C_e = (C_0 - C_e) \qquad (11)$$

That is,

$$r_1(r_2 + 1) = (C_0/C_e - 1) \qquad (12)$$

Thus, knowing C_0 and C_e constrains r_1 and r_2. Also, since r_2 must be positive, there will be a maximum value for r_1. The relative magnitude of r_1 to r_2 determines where the change from 'fast' to 'slow' sorption takes place. Having chosen r_1 and r_2, one is still free to choose the absolute values of, say, k_{s1} and k_{s2}. The two-stage sorption process was used as a model to fit the stable Sr concentration/time data for the silt and sand samples from core M5 and to fit the ^{85}Sr activity/time data for sand T7 (Fig. 8). The rate constants calculated to fit these data are given in Table 8, and it is clear that the rate constants are system-specific.

Two-stage kinetic sorption model
The rate constant parameters for two sand samples were introduced into a 1D finite difference transport model involving time-stepping by Gear's method. A grid size of 2.5 cm was used

Table 8. *Summary of Sr kinetic rate constants* $(\times 10^{-5} s^{-1})$

Rate constant	M5 Silt	M5 sand	T7 sand
k_{s1}	0.5	20.0	9.5
k_{r1}	1.43	200.0	55.0
k_{s2}	0.03	6.0	2.0
k_{s2}	0.077	1.5	1.67

with a longitudinal dispersivity of 2.5 cm. Calculations were initially carried out at two flow rates of $5\,m\,d^{-1}$ ($5.787 \times 10^{-5}\,m\,s^{-1}$) and $0.5\,m\,day^{-1}$ ($5.787 \times 10^{-6}\,m\,s^{-1}$) corresponding to the two field tracer test conditions of 'fast' and 'slow' rates. Further computations were carried out at lower flow rates until there was no visible change in the curve shape, indicating the approach to equilibrium sorption. This occurred when the flow had been decreased to $5 \times 10^{-5}\,m\,day^{-1}$ ($5.8 \times 10^{-10}\,m\,s^{-1}$). The results of the transport calculations (Fig. 11) are plotted

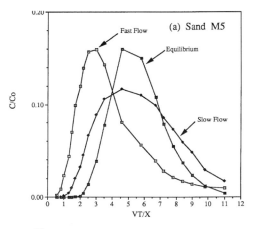

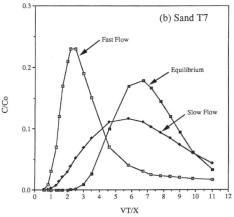

Fig. 11. Predicted breakthrough with Sr kinetics: (a) sand M5 (b) sand T7.

as relative concentrations versus dimensionless time defined as (vt/x), where v is the flow rate, t is the time, and x is the observation distance (53 cm). The concentrations are relative to the peak of the source curve which decays exponentially from an initial value. The decay rate for this curve is $1.777 \times 10^{-4}\,s^{-1}$ as observed in the high flow rate test and $1.777 \times 10^{-5}\,s^{-1}$ inferred for the low flow test.

The shape and height of the breakthrough curve under these conditions turn out to be a complex function of the flow rate. This is due to the interaction between the flow rate and the rates of the two-stage sorption reactions. From the curves it can be deduced that under the *in situ* conditions used in the field experiments the sorption processes are not at equilibrium, and the ^{85}Sr tracer experiences much less retardation than would have been anticipated by an equilibrium sorption model.

The curves shown in Fig. 11 are probably not suitable for direct comparison with the field data if an additional retardation mechanism such as diffusion into dead-end pores is operating, or if there is evidence of stratification in the aquifer when an intra-layer model would be applicable (e.g. Gillham *et al.* 1984; Valocchi 1988; Goltz & Roberts 1988). They do demonstrate that the breakthrough times can be influenced by the kinetic processes and that these could be used to explain the form of the breakthrough curves.

Conclusions

Simple estimates of tracer transit times from injection to monitoring wells based on an average value of porosity and the imposed injection rate, do not agree with observed results if the physical heterogeneity of the aquifer is ignored.

However, there is some agreement between the retardation factors (or distribution ratios) of ^{85}Sr determined by laboratory and field tracer tests. At average flows up to $5\,m\,day^{-1}$ (fast flow test), the retardation of ^{85}Sr at a given sampling point was, in general, lower than that observed with an order of magnitude lower average flow rate. Empirical kinetic rate constants for sorption derived from laboratory experiments, while not necessarily implying any specific sorption mechanism, have predicted the difference in retardation when used in a two-stage kinetic transport model. The retardations at lower flows, where sorption kinetics become less important, were in relatively good agreement with laboratory values derived from batch sorption experiments under static conditions lasting several days.

Longitudinal dispersivities estimated from breakthrough curves at each monitoring point, were in the range 6–29 mm, and almost two orders of magnitude less than the horizontal distance used in the measurement. The dispersivity derived from ^{131}I breakthrough curves at a given monitoring point at the high flow rate was lower by a factor of three than at the lower flow rate. A dual porosity model containing mobile and immobile porewater can explain this difference, and suggests that at the high rate of flow 86% of the porewater is mobile, whilst at the low flow only 55% is mobile. Averaging the breakthrough curves at a multilevel sampling installation to simulate the breakthrough expected from a fully penetrating screened well at the same location, yielded a complex double-peaked function which could only be analysed by assuming a two-layered aquifer with differing hydraulic properties. However, the dispersivities so derived were in agreement with the two sets of data obtained from multi-level samplers.

The results show that in small-scale field tracer tests, kinetic effects and disequilibrium between water and rock can be significant, and will probably lead to a pessimistic value for radionuclide sorption and retardation. Averaged hydraulic measurements do not allow prediction of groundwater flow rates to be determined in heterogeneous aquifers, and in such systems the scale of the hydraulic measurement should be matched to the scale of the heterogeneity.

The authors are grateful to the UK Department of the Environment and the Commission of the European Communities who funded this work, and to British Nuclear Fuels plc who allowed the tracer experiments to be conducted at the Drigg site. Numerous BGS staff contributed to this work and thanks are extended to A. Cook, S. P. Hitchman, G. P. Wealthall, A. Forster, P. Jackson, L. S. Martin, A. Bloodworth and B. Vickers; and to Jenny Higgo for helpful discussions. This paper is published by permission of the Director of the British Geological Survey (Natural Environment Research Council).

References

GILLHAM, R. W., SUDICKY, E. A., CHERRY, J. A. & FRIND, E. O. 1984. An advection–diffusion concept for solute transport in heterogeneous unconsolidated geological deposits. *Water Resources Research*, **20**(3), 369–378.

GOLTZ, M. N. & ROBERTS, P. V., 1988. Simulations of physical nonequilibrium solute transport models: application to a large-scale field experiment. *Journal of Contaminant Hydrology*, **3**, 37–63.

HITCHMAN, S. P. 1988. A collection manifold for multilevel groundwater sampling devices. *Ground Water*, **26**(3), 348–349.

HOOKER, P. J., WEST, J. M. & NOY, D. J. 1986. *Mechanisms of sorption of Np and Tc on argillaceous materials*. CEC Report EUR 10785 EN.

JACKSON, R. E. & INCH, K. J. 1980. *Hydrogeochemical processes affecting the migration of radionuclides in a fluvial sand aquifer at the Chalk River Nuclear Laboratories*. Environment Canada, Inland Waters Directorate Scientific Series No. 104, National Hydrology Research Institute.

—— 1983. Partitioning of strontium-90 among aqueous and mineral species in a contaminated aquifer. *Environmental Science and Technology*, **17**(4), 231–237.

—— 1989. The in-situ adsorption of ^{90}Sr in a sand aquifer at the Chalk River Nuclear !Laboratories. *Journal of Contaminant Hydrology*, **4**, 27–50.

NYFFELER, U. P., LI, Y.-H. & SANTSCHI, P. H. 1984. A kinetic approach to describe trace-element distribution between particles and solution in natural aquatic systems. *Geochimica et Cosmochimica Acta*, **48**, 1513–1522.

OGARD, A. E., THOMPSON, J. L., RUNDBERG, R. S., WOLFSBERG, K., KUBIK, P. W., ELMORE D. & BENTLEY H. W. 1988. Migration of chlorine-36 and tritium from an underground nuclear test. *Radiochimica Acta*, 44/45, 213–217.

PARKER, J. C. & VAN GENUCHTEN, M. Th, 1984. *Determining transport parameters from laboratory and field tracer experiments*. Virginia Polytechnic Institute and State University Bulletin 84-3.

PATERSON, R. J. & SPOEL, T. 1981. Laboratory measurements of the strontium distribution coefficient K_{dSr} for sediments from a shallow sand aquifer. *Water Resources Research*, **17**(3), 513–520.

PICKENS, J. F., JACKSON, R. E., INCH, K. J. & MERRITT W. F. 1981. Measurement of distribution coefficients using a radial injection dual-tracer test. *Water Resources Research*, **17**(3), 529–544.

RELYEA, J. F., & SERNE, R. J. 1979. *Controlled sample programme publication No. 2, Interlaboratory comparison of batch Kd's*. PNL 2872.

ROSS, C. A. M., 1980. Experimental assessment of pollutant migration in the unsaturated zone of the Lower Greensand. *Quarterly Journal of Engineering Geology*, **13**, 177–187.

SAUTY, J. P. 1980. An analysis of hydrodispersive transfer in aquifers. *Water Resources Research*, **16**(1), 145–158.

VALOCCHI, A. J. 1988. Theoretical analysis of deviations from local equilibrium during sorbing solute transport through idealised stratified aquifers. *Journal of Contaminant Hydrology*, **2**, 191–207.

VAN GENUCHTEN, M Th. & ALVES, W. J., 1982. *Analytical solutions of the one dimensional convective dispersive solute transport equation*. US Department of Agriculture, Technical Bulletin No. 1661.

Section 3: Groundwater pollution by inorganic contaminants

The papers in this section deal mainly with nitrogen compounds and heavy metals, derived from a variety of sources.

The papers by **Sadler** and **Bowen** *et al.* describe the contamination of groundwaters flowing into, and subsequently emerging from, mines by metal-rich acidic mine drainage. In both these papers, the mines in question are metals mines, but it is worth noting that ferruginous waters derived from coal-mine abandonment are currently also giving great cause for concern in the UK (Younger 1995), especially in the light of the regulatory vacuum that has existed around this issue. **Lee & Spears** examine the opposite end of the mining spectrum: PFA, the highly alkaline residue, rich in semi-volatiles, remaining in flues after coal has been combusted and the acidic gases driven off. For further reading on this topic, readers may consult Lee & Spears (1995).

Andrews *et al.* and **Davey** *et al.* describe contaminant sources which, at first sight, seem to be anything but inorganic; namely sewage sludge and pig slurry! But in addition to being potential sources of organic carbon and microbiological contamination, they may, under some circumstances, give rise to significant nitrate or ammoniacal nitrogen contamination. They also contain heavy metals (Spears 1987) derived either from industrial effluent or from metals supplements fed to domestic animals. The potential consequences of injudicious use or disposal of sewage or agricultural slurries are now becoming apparent in lands such as Holland, Denmark and Lithuania (Klimas & Paukstys 1993), where significant ammonium contamination of unconfined groundwaters is developing in some areas.

The paper by **Davey** *et al.* illustrates a common problem encountered by hydrogeologists working on contamination issues in the UK. The intense land-use means that, for a given contaminant find, there may a number of plausible sources. In a rather heterogeneous aquifer such as the Chalk, it can be surprisingly difficult to identify the real source, even after several rounds of hydraulic testing and chemical sampling.

KLIMAS, A. & PAUKSTYS, B. 1993. Nitrate contamination of groundwater in the Republic of Lithuania. *Norges Geologiske Undersøkelse Bulletin*, **424**, 75–85.

LEE, S. & SPEARS, D. A. 1995. The long-term weathering of PFA and implications for groundwater pollution. *Quarterly Journal of Engineering Geology*, **28**, S1–S15.

SPEARS, D. A. 1987. An investigation of metal enrichment in Triassic Sandstones and porewaters below an effluent spreading site, West Midlands, England. *Quarterly Journal of Engineering Geology*, **20**, 117–129.

YOUNGER, P. L. 1995. Hydrogeochemistry of minewaters flowing from abandoned coal workings in County Durham. *Quarterly Journal of Engineering Geology*, **28**, S101–S113.

Potential contamination of groundwater by pulverized fuel ash

S. LEE[1] & D. A. SPEARS[2]

[1]*School of Life Sciences, The Catholic University of Korea, San 43-1, Yokkok 2-Dong, Wonmi-Gu, Puchon 422-743, Korea*
[2]*Department of Earth Sciences, Dainton Building, University of Sheffield, Brookhill, Sheffield S3 7HF, UK*

Abstract: Disposal sites containing pulverized fuel ash (PFA) generate leaches which could contaminate surface and groundwaters. PFA has elevated concentrations of soluble salts and trace elements, including As, B, Cd, Cr, Mo, Ni, Pb, Se, V and Zn, which are associated with the surfaces of PFA particles. Batch and column leaching tests in the laboratory have been used to predict the environmental impact of PFA. However, laboratory testing may not reproduce natural conditions in the field. The majority of the field-based studies have been concerned with the relatively rapid release of trace elements from the surfaces of the PFA particles. Less information is available on the longer-term leaching behaviour of field-disposed PFA, where elements may be originating from the silicate glass and minerals. Two recently completed field-based studies suggest that leachate from older PFA could potentially contaminate groundwater over a period of many years.

Pulverized fuel ash (PFA), a by-product from coal-fired power stations, consists of particles sufficiently small to be extracted from flue gases by electrostatic precipitators. PFA particles are mainly derived from the melting of mineral matter and the residue of organic combustion. Approximately 80% of the ash produced is PFA (Roy *et al.* 1981). The remainder of the ash is furnace bottom ash (FBA), which consists of larger particles, including molten material which fell from the walls of the furnace.

Due to the high temperature of combustion (1000–1500 °C), PFA is predominantly composed of spherical aluminosilicate glass particles, which are of silt to sand-size. Minerals are also present, including mullite as acicular crystals within the glass, iron oxides as discrete particles derived from diagenetic iron minerals, and partially fused original quartz grains. According to Simons & Jeffery (1960), British PFA contains between 50 and 90% glass, 9–35% mullite, 1–6.5% quartz and up to 5% magnetite/hematite. The chemical composition of PFA depends upon several factors such as the rank and chemical composition of feed coal, and the type and design of boiler (Adriano *et al.* 1980). Table 1 is a compilation of selected PFA chemistries from the USA, which demonstrates a much greater variation than in the UK, probably because the latter are all of Carboniferous age.

The mineral matter present in coal is a major contributor to the composition of PFA. In general, elements are more concentrated in PFA than in coal because of the loss of carbon on ignition. For those elements mainly associated with detrital minerals, concentrations in the PFA are comparable with other geological materials such as shale or mudstone. However, other elements that are associated with either the organic matter or the minerals formed in the coal, such as pyrite, are concentrated in PFA at much higher levels than in most geological materials (data from: Lindsay 1979; Adriano *et al.* 1980; Ainsworth & Rai 1987). Many of these trace elements (As, B, Cd, Cr, Mo, Ni, Pb, Se, V and Zn) are concentrated on the surfaces of PFA particles as a consequence of volatilization at high combustion temperatures and subsequent condensation onto ash particles (Natusch & Wallace 1974; Smith 1980).

Due to the enrichment of the trace elements and their surface association, PFA has been recognized as a potential contaminant in a number of studies, which aimed to understand the possible chemical toxicity of PFA and the environmental impact of the leachate produced from the PFA on surface and groundwater. Some of these studies were conducted in the laboratory using batch and column leaching tests. Although these experimental studies provide information on the maximum extractability of elements from the PFA samples, they do not necessarily accurately reflect the behaviour of PFA in the field. Field-based studies are fewer in number and these have been mainly concerned with the PFA lagoons in which leachates are characterized by rapid reactions of the soluble fractions of the surface associated elements in the PFA. Little information is available from field sites on the long-term weathering of PFA, which may be

LEE, S. & SPEARS, D. A. 1998. Potential contamination of groundwater by pulverized fuel ash. *In*: MATHER, J., BANKS, D., DUMPLETON, S. & FERMOR, M. (eds) *Groundwater Contaminants and their Migration*. Geological Society, London, Special Publications, **128**, 51–61.

Table 1. *Compositional variations of PFA, USA*

Al	Ca	Mg	Na	K	Fe	Si
0.10–20.85%	0.11–22.30%	0.04–7.72%	0.01–7.10%	0.17–6.7%2	1.00–27.56%	1.02–31.78%
S	As*	Ba	B*	Cd*	Cr*	Cu*
0.04–6.44%	2–440 ppm	1–138 00%	10–5000 ppm	0.1–130 ppm	4–900 ppm	33–2200 ppm
Pb*	Mn*	Zn*	Mo*	Ni*	Se*	V*
3–1200 ppm	25–3000 ppm	14–3500 ppm	1–140 ppm	2–4300 ppm	0.2–130 ppm	12–1180 ppm

Sources: Adriano *et al.* (1980); Hansen & Fisher (1981); Roy *et al.* (1982); Ainsworth & Rai (1987).

due in part to the difficulties of monitoring ash mounds (Carlson & Adriano 1993).

In 1990, 48.6×10^6 tonnes of coal were consumed in coal-fired power stations in the UK (Central Statistical Office 1992) and about 10×10^6 tonnes of PFA were produced. Up to half of the PFA is commercially utilized as materials for the construction industry, such as in road bases, structural fills, cement/concrete mixtures and soil amendments. However, the excess, non-marketed production of PFA in the UK amounts to approximately 6.5×10^6 tonnes per year, which is currently being disposed of in lagoons, land-fill sites and mounds. In the USA about 69×10^6 tonnes of PFA and FBA were produced in 1984 and this amount is expected to rise to approximately 120×10^6 tonnes by the year 2000 (USEPA 1988). In the future there may be less PFA in the UK to be disposed of because of a decline in coal-fired power stations, and disposal sites will be carefully engineered to reduce the risk of groundwater contamination. In the past, disposal practice was not always so sound and there is the possibility of long-term weathering reactions influencing groundwater composition. Elsewhere in the world the production of coal for power generation is rapidly increasing, particularly in China and India, and there is the threat of both short- and long-term weathering reactions adversely influencing surface and groundwater compositions. In 1991, Powell *et al.* estimated that PFA production in India was approximately 64×10^6 tonnes, and in China 250×10^6 tonnes. They report that some rivers in India are becoming clogged by ash and millions of people drink the water in which the ash is deposited.

This paper briefly reviews laboratory and field studies, including results from recently completed work on two disposal sites in the UK in which PFA was disposed of 16 and 40 years ago. The focus of this work is to establish the potential PFA-leachate contamination of groundwater.

Experimental studies

Batch leaching test

On contact with water, major and trace elements in PFA are released into solution in amounts controlled by the location of elements in the ash and their respective solubilities. Many studies have used batch leaching tests (Mattigod *et al.* 1990) to assess element mobility in ash disposal sites. These studies are influenced by several factors, such as the nature of the ash, the solid-solution ratio, the duration of the experiment, and the intensity and method of agitation during extraction. The results of batch leaching tests consequently do not accurately reflect the actual leaching behaviour of PFA in the field (Mattigod *et al.* 1990).

The results of the leaching experiments show large variations in the water-soluble concentrations of all major and trace elements, depending on the factors mentioned above. Experimental data from batch leaching tests using different conditions are compiled in Table 2. Acid extraction dissolves proportionally more of an element from the PFA, sometimes extracting nearly all the available Al, Ca, Na, Mg and K; whereas in water less than 10% of all major elements are extracted except for Ca, Na and S. Concentrations of Ca, Na and S in leachates range from tens to thousands of milligrams per litre. Magnesium and K concentrations also reach up to several hundred milligrams per litre. In contrast to these concentrations, Si and Fe in PFA water extracts show maximum values of 46 and $3 \, \text{mg} \, l^{-1}$ respectively. This indicates that significant fractions of Ca, Mg, Na and K in PFA are in a highly soluble form, such as oxides and sulphates. The pH of the water extract from PFA worldwide is extremely variable, ranging from 3.3 to 12.3, whereas the pH obtained from UK PFA is much narrower, ranging between 6 and 9. The trace elements that are known to be enriched on the surface of PFA particles, such as As, B, Cd, Cr, Cu, Mo

Table 2. *Concentrations of elements in the extract from batch leaching tests (extracted by water or acids, unit: $mg\,l^{-1}$)*

pH	Al	Ca	Mg	Na	K
3.3–12.34	0.12–62	67–634	<0.05–118	1.87–2008	0.72–191
Fe	Si	S	As	Ba	B
<0.005–3	0.05–46	33–3583	<0.08–14	0.05–2.0	0.1–109
Cd	Cr	Cu	Pb	Mn	Zn
<0.01–1.8	0.02–44	<.01–24	<0.05–38	<0.001–290	<0.01–121
Mo	Ni	Se	V		
0.01–6.8	<0.01–8.5	<0.05–0.4	<0.003–1.1		

Sources: Dreesen *et al.* (1975); Hansen & Fisher (1981); van der Sloot *et al.* (1982); Ainsworth & Rai (1987)

and Zn, show high concentrations in extracts (Theis & Wirth 1977; Hansen & Fisher 1980; Ainsworth & Rai 1987).

Nearly all the results shown in Table 2 for leaching tests were based on fresh ash. In recent work weathered and fresh ashes have been extracted using distilled water (Lee 1994) and the results revealed that the extractability of Fe, Al and Si was little different between fresh and weathered ash, whereas concentrations of soluble elements such as Ca, Na, K, Mg and S were much lower in weathered PFA than in fresh ash. These data give support to the idea that a significant fraction of the Ca, Mg, Na and K is highly soluble and surface-associated.

Batch leaching tests are conducted on a relatively short-term basis and under conditions, such as acid extraction and mechanical agitation, that do not conform to natural conditions in the field. Therefore the data lack quantitative and thermochemical constraints but nevertheless provide characterization of individual PFA samples, including maximum extractability under certain conditions, and also information on an element location within the ash particles.

Column leaching test

Column leaching tests have also commonly been adopted to simulate PFA leaching under field conditions. The water (porewater) to solids proportion is more comparable with the natural situation than batch leaching where there is proportionally a larger volume of water. Some experiments have been over a time period of several years (Brown *et al.* 1976; Dudas 1981; Hjelmar 1990). Mattigod *et al.* (1990) summarized the leaching trend from continuous column leaching tests into two categories. Calcium, Na and S are in the first group, in which elements show higher concentrations in

the initial leachate, and thereafter the concentrations decline rapidly before reaching a steady state. In the second group elements such as Al and Si are released slowly in the initial leaching stages, but concentrations increase as leaching progresses. Magnesium, K and Fe showed the first pattern in the study of van der Sloot *et al.* (1982) but followed the second pattern according to the data of Dudas (1981) and Warren & Dudas (1984).

Volatile trace elements are enriched on the surfaces of the ash particles and are accessible to the porewaters in the columns. Boron concentrations are high initially and decrease rapidly as the porewaters are replaced (Jones & Lewis 1960) Elements that form soluble anionic species in alkaline solutions, including Cr, Mo, As, Se and V (Hjelmar 1990), are also leached in significant concentrations, which decrease as a function of time.

Previous field studies

Lagoons

Field studies have focused on PFA settling ponds and lagoons, due to the high initial water content and the known rapid release of elements into solution. Studies of ash pond effluents show similar results to those of column leaching tests, with Ca and S (present as SO_4^{2-}) as principal cationic and anionic constituents (Dreesen *et al.* 1977; Talbot *et al.* 1978; Theis *et al.* 1978; Simsiman *et al.* 1987). Groundwater samples collected from an area bordering a PFA settling pond showed increased concentrations of Ca, K, Fe and SO_4^{2-} for the major elements, and As, B, Mn, Mo, Ni, Sr and Zn for the trace elements (Hardy 1981). Concentrations of As, B, Mo and Se were also found to be elevated in effluent water from a PFA pond by Dreesen *et al.* (1977).

Disposal mounds

In monitoring a PFA mound for three years, Simsiman *et al.* (1987) observed large B, Na and SO_4^{2-} plumes in the adjacent groundwater system. Sakata (1988) reported Ca and SO_4^{2-} as major ions in the leachate of weathered PFA from ash mounds and also showed that inorganic elements had infiltrated into the underlying soil.

In addition to high concentrations of B, Le Seur & Drake (1987) detected As and Se over analytical detection limits, and Se exceeding $0.1 \, mg \, l^{-1}$ in the groundwater samples from shallow monitoring wells within the PFA landfill. The infiltration of ash leachate into underlying soil and elevated concentrations of B, Mo, Mn and Sr in shallow groundwater were also reported by Rehage & Holcombe (1990).

Current UK field studies

UK-based studies were undertaken to provide information on the long-term leaching of PFA under natural conditions, a previously neglected area of research. Ash mounds at Drax [NGR SE 655 277] and Meaford [NGR SJ 896 373] Power Stations have been investigated and the results presented in Lee (1994) and Lee & Spears (1995).

The two sites investigated differ in age and design. At Drax Power Station an ash disposal mound, known as the Barlow mound, was started in the mid 1970s. The site is well-managed and engineered, permitting ash of a known age to be sampled. The ash sampled and analysed in detail was emplaced in 1975. The ash mound sampled at Meaford is from a decommissioned power station. The ash from this latter site is substantially older and is thought to be about 40 years old.

PFA samples were taken from the ash mounds by means of boreholes. A hand-auger drill was used to obtain samples at 30-cm depth intervals down to 5.0 m. Porewater samples were extracted from the PFA samples by centrifugation, following the method of Edmunds & Bath (1976), and analysed by ICP-AES, AAS and Ion Chromatography. The chemical composition of the PFA was determined by XRF. The mineralogy was determined by XRD and SEM. Full details of the analytical method are described in detail in Lee (1994) and Lee & Spears (1995).

Porewater chemistry—depth trends

The composition of porewaters in PFA mounds provides a sensitive record of reactions between infiltrating porewaters and PFA. The trend of element concentrations as a function of depth is of particular interest. Increasing concentrations with depth are indicative of continued reaction between PFA and the infiltrating porewater as a function of contact time. As pore fluids move downwards, the older porewater is located at a greater depth compared with younger porewaters, and the greater contact time between PFA and porewaters results in higher solution concentrations. The achievement of constant concentrations with depth provides evidence of equilibrium, whereas decreasing concentrations could be due to the introduction of an element from an external source, with concentrations decreasing either by reaction or by dilution. Another possibility, of no change with depth, might indicate that the material or elements are inert.

On Fig. 1 are plotted the depth variations of major elements in the PFA porewaters from the Barlow mound and the Meaford disposal site. In general, the depth variation from the two sites shows similar trends, although concentrations differ, with higher concentrations recorded at Barlow, which is consistent with the younger age of this material and presumably less-weathered condition. Sodium, K and SO_4^{2-} increase with depth in both porewaters. Calcium generally increases with depth in Meaford porewater but this element appears to achieve a near-constant value in the Barlow porewater. Magnesium concentrations are relatively constant in the Meaford porewater but increase with depth in the Barlow porewater. Silicon concentrations are low, and near-constant values for Barlow suggests that an equilibrium is achieved. Porewater concentrations of Cl^- and NO_3^- in the Barlow PFA decrease with depth and this trend seems to be indicative of an external origin, such as fertilizer, which is known to have been applied to the surface. The Cl^- and NO_3^- in the Meaford porewater also generally decrease with depth, but there are spikes at around 1.5 m and concentrations are lower than at Barlow, presumably because the latter has had more fertilizer applied.

Nearly all trace elements show depth-related trends (Fig. 2). The elements which increase with depth in porewaters from both sites are B, Cr, Li, Mo, Pb, Ni and As. Selenium increases with depth in the Barlow porewater but shows approximately constant values in the Meaford porewater after an initial decrease at shallower depths. Strontium increases with depth in the Meaford porewaters but at Barlow more constant values were recorded. Barium concentrations show less variation but do decrease in the

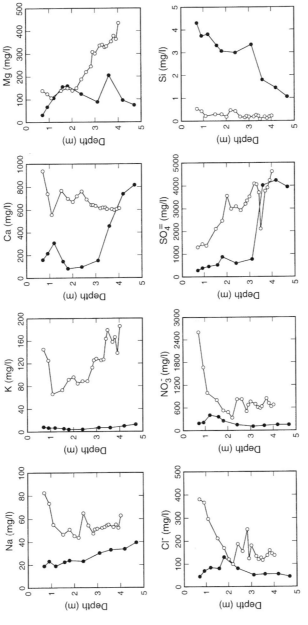

Fig. 1. Variation of major elements in porewaters as a function of depth in the Barlow (*k*) and Meaford (*l*) ash disposal sites.

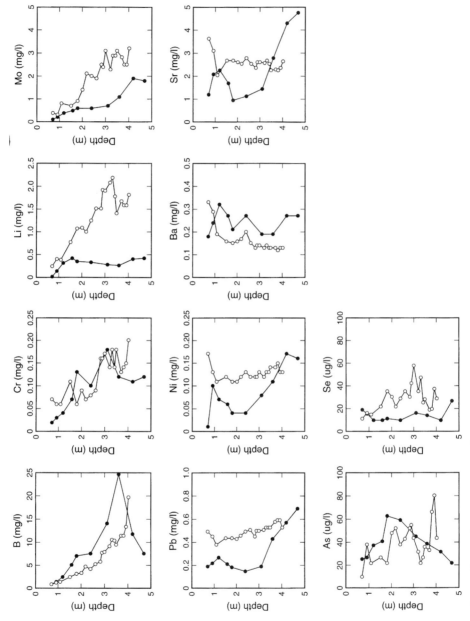

Fig. 2. Variation of trace elements in porewaters as a function of depth in the Barlow (*k*) and Meaford (*l*) ash disposal sites.

Barlow borehole, although the decrease is relatively small for most samples.

The increasing depth trends of porewater are the consequence of the accumulating effect of the elements released from the reaction of PFA during weathering. The depth trends investigated provide evidence of the weathering reaction of PFA involving element release and their downward migration with the infiltrating porewater. Some element concentrations, such as Ca, Si, Ba and Sr, show near-constant values after the initial increase. This trend suggests that these elements have achieved equilibrium concentrations with respect to some specific solid phases. However, the decreasing depth trends of Cl^- and NO_3^- are attributable to an external source.

PFA chemistry

In the previous section, increases were noted in element porewater concentrations as a function of borehole depth. If the changes in porewaters are derived from the weathering of PFA as claimed, then there should be detectable, depth-related changes in the whole-rock PFA composition. However, the total mass of an element removed may be a small fraction of that available in the PFA; therefore the total mass will be difficult to detect, although giving rise to a significant increase in the porewater concentrations. Nevertheless, providing a significant fraction of an element is removed from the PFA, the depth trends in porewater concentrations should be confirmed by the PFA analyses. Element depletion is detected in the PFA near-surface for some elements, including S, Mg, Ca and Na (Table 3). Near-surface depletion is greatest at the Meaford site which is consistent with the older age of the PFA and the longer period available for weathering. However, Meaford PFA has higher average trace-element concentrations compared to Barlow PFA, which is thought to reflect compositional differences in the feed coal.

A comparison of fresh and weathered Barlow PFA (Table 3) does not show substantial differences, except for a few elements, in spite of weathering having been active for 16 years. The PFA chemistry over that time period is believed to have been relatively constant with local mines supplying coal of a uniform composition. The lack of substantial differences in PFA composition with age for most elements is thought to reflect the fact that the amount passing into solution is only a small fraction of that available in the bulk PFA. It is also concluded that the weathered PFA has the potential to release elements into solution over a long time period and generate leachates with significant concentrations as demonstrated on Fig. 1. It is notable that the ratio of FeO to total Fe_2O_3 [FeO/ $(FeO + Fe_2O_3)$] is 0.342 in fresh Barlow PFA and 0.326 in weathered Barlow PFA – a difference of about 5%, indicating that little oxidation of iron has occurred in a 16-year weathering period.

Equilibrium relationships: weathering reactions

Several elements, including Ca, Ba and Sr, show evidence of attaining equilibrium concentrations in the depth profiles. Porewater analytical data from weathered PFA were processed using a computer modelling program, WATEQ4F (Ball et al. 1987), to detect any potential solid phase that might control the solubility of specific elements. Calculated activities are compared with the activities in equilibrium with respect to a specific solid phase, or a single ion activity product is compared with the equilibrium constant of a specific solid phase to determine the saturation state of the porewaters.

Figure 3(a) is a plot of calculated Ca activity against SO_4^{2-} activity, together with a line showing equilibrium with respect to gypsum. Both Barlow and Meaford porewaters achieve saturation with respect to gypsum at depth, but the Meaford porewaters achieve equilibrium at a greater depth than do the Barlow porewaters (Fig. 3(b)). This is consistent with the concept that the older PFA at Meaford is more weathered. Gypsum is predicted as a solubility-controlling solid phase for Ca in PFA porewater.

Plots of the calculated Ba and Sr activities for the calculated SO_4^{2-} activities imply the presence of solubility-controlling solid phases. Felmy et al. (1993) suggested that co-precipitation of Ba and Sr in the form of (Ba, Sr)SO_4 was the possible solid phase which could control the concentrations of the Ba and Sr in PFA porewater. Other predicted solubility-controlling solid phases detected from Barlow and Meaford porewaters are $Fe(OH)_3$ for Fe, $Al(OH)_3$ for Al and CuO for Cu (Lee 1994; Lee & Spears 1995).

Although several potential solubility-controlling solid phases were predicted, it should be noted that these phases have not been detected directly by XRD or SEM. Possibly this is because only minor amounts of the phases could be involved.

Table 3. Chemical analyses of PFA from Barlow and Meaford boreholes (% and ppm of ashed sample, 850°C)

		LOI*	SiO_2	Al_2O_3	$Fe_2O_3^{(+)}$	MgO	CaO	Na_2O	K_2O	SO_3	%
Barlow Fresh PFA	Mean	2.26	53.43	25.85	7.40	1.73	1.90	1.56	3.75	0.35	
	Std. Dev.	1.17	0.64	0.66	0.26	0.09	0.19	0.16	0.17	0.11	
	N = 3										
Barlow Weathered	Mean	4.70	50.74	25.90	9.54	1.48	1.39	1.02	3.71	0.18	
	Std. Dev.	1.12	1.11	0.53	0.43	0.15	0.17	0.13	0.07	0.14	
	N = 8										
Meaford Weathered	Mean	9.89	42.91	24.93	13.14	1.42	3.13	0.67	2.12	0.24	
	Std. Dev.	1.36	1.37	0.55	0.66	0.06	0.14	0.05	0.07	0.10	
	N = 8										

		V	Cr	Mn	Ni	Cu	Zn	Rb	Sr	Ba	Pb	ppm
Barlow Fresh PFA	Mean	255.2	136.6	504.5	113.0	142.2	116.0	165.1	293.4	1063.2	75.1	
	Std. Dev.	14.6	10.2	27.0	11.5	16.9	6.0	13.9	19.6	154.4	9.7	
Barlow Weathered	Mean	275.6	139.3	465.3	113.6	175.5	112.1	165.2	248.0	762.5	76.6	
	Std. Dev.	20.4	7.9	36.7	7.9	22.9	21.0	4.4	15.8	30.5	13.4	
Meaford Weathered	Mean	272	115	969	143	199	959	101	360	1244	239	
	Std. Dev.	8.3	5.6	60.4	12.1	18.8	60.8	6.1	37.9	2.0	36.4	

* LOI, loss of ignition

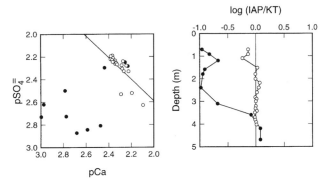

Fig. 3. (a) Measured activities of Ca versus SO_4^{2-} together with the calculated activity line showing equilibrium with respect to gypsum. (k, Barlow; l, Meaford). (b) Variation in saturation indices with respect to gypsum as a function of depth (k, Barlow; l, Meaford).

Long-term porewater evolution in PFA disposal mound and leachate composition

Fresh ash begins to liberate high amounts of soluble fractions of surface-associated elements on contact with water when disposed of in the field or lagoon. The initial release of the soluble salts is followed by a rapid decline as leaching progresses until lower, near-constant values are achieved (Jones & Lewis 1960; Brown et al. 1976; Dudas 1981; van der Sloot et al. 1982; Hjelmar 1990). At the Barlow and Meaford disposal sites the emphasis has been placed on longer-term weathering reactions. The results

Table 4. *Chemical composition (mg l^{-1} of porewaters from Barlow and Meaford PFA mounds compared with World Health Organisation guidelines for drinking water quality (1984)*

Sample	pH	Eh(mv)	Ca	Mg	Na	K
Barlow (4.0 m)	8.26	409.4	608.3	363.2	52.1	137.4
Meaford (4.2 m)	8.94	436.2	738.7	95.9	33.7	10.3
WHO						200

Sample	Fe	Al	B	Ba	Cd	Co
Barlow (4.0 m)	0.09	0.03	13.40	0.10	0.05	0.15
Meaford (4.2 m)	0.29	1.30	11.81	0.27	0.04	0.10
WHO	0.3	0.3			0.005	

Sample	Cr	Cu	Li	Mn	Ni	Pb
Barlow (4.0 m)	0.15	0.01	1.59	0.00	0.13	0.59
Meaford (4.2 m)	0.11	0.02	0.41	0.02	0.17	0.57
WHO	0.05	1.0		0.1		0.05

Sample	Si	Sr	V	Zn	Mo	Cl$^-$
Barlow (4.0 m)	0.17	2.35	0.14	0.02	2.5	144.7
Meaford (4.2 m)	1.5	4.31	0.11	0.04	1.9	55.2
WHO				5		250

Sample	NO$_3^-$	SO$_4^{2-}$	Se*	As*	Hg*
Barlow (4.0 m)	634.4	4023.0	37	80	<5.0
Meaford (4.2 m)	150.7	3930.0	10.0	32.0	<5.0
WHO	45	400	10	50	1

* Units: μg l^{-1}.

demonstrate that porewaters within the PFA mounds contain significant concentrations of major and trace elements. The evolution of these porewaters is consistent with slow infiltration through the mound and continued reaction as a function of contact time.

The chemical composition of PFA porewaters is in general dominated by two major ions, Ca^{2+} and SO_4^{2-} (Theis *et al.* 1978; Kopsick and Angino 1981; Roy & Griffin 1984; Sakata 1987; Lee & Spears 1995). Among the trace elements B is the most abundant in PFA leachate (Cox *et al.* 1978; James *et al.* 1982). Arsenic and Se are also important trace elements in PFA leachate (Le Seur & Drake 1987; Eary *et al.* 1990).

The porewater composition of the weathered Barlow and Meaford PFA collected at depths of 4 m and 4.2 m, respectively, is shown in Table 4. Although the concentration levels are much lower than those of fresh PFA leachate, many elements exceed the level of some drinking water quality standards (Table 4). Furthermore, the leachate concentrations are from relatively shallow depths in the PFA. A longer contact time with PFA would give higher concentrations and this could be achieved by infiltrating through a greater depth of PFA or a slower infiltration rate due to lower permeability.

The porewater chemistry shown in Table 4 could represent the composition of the leachate that would be released from a PFA mound. If an impermeable seal was absent below the PFA mound the leachate would infiltrate into the groundwater. In such a situation the opportunities for dilution are less than if the leachate was discharged into surface waters. In conclusion, it is not only the leachate from fresh PFA which should be of concern but also that produced by older PFA.

The authors wish to thank staff in the former CEGB and latterly National Power plc for allowing access to PFA disposal sites as well as providing valuable information about the sites. A financial contribution towards the research was also provided through contract LC/3/JEP/3001 of the National Power/Power Gen Joint Environmental Programme. The views expressed in this paper, however, are those of authors and not necessarily those of the power generators.

References

ADRIANO, D. C., PAGE, A. L., ELSEEWI, A. A. & CHANG, A. C. 1980. Utilization and disposal of fly ash and other coal residues in a terrestrial ecosystem: A review. *Journal of Environmental Quality*, **9**, 333–344.

AINSWORTH, C. C., & RAI, D. 1987. *Chemical characterization of fossil fuel wastes*. Electric Power Research Institute, Palo Alto, CA, EPRI Report EA-5321.

BALL, J. W., NORDSTROM, D. K. & ZACHMAN, D. K. 1987. *WATEQ4F – A personal computer FORTRAN translation of the geochemical model WATEQ2 with revised database*. US Geological Survey, Open-file Report 87-50.

BROWN, J., RAY, N. J. & BALL, M. 1976. The disposal of pulverized fuel ash in water supply catchment areas. *Water Research*, **10**, 1115–1121

CARLSON, C. L. & ADRIANO, D. C. 1993. Environmental impacts of coal combustion residues. *Journal of Environmental Quality*, **22**, 227–247.

Central Statistical Office 1992. *Annual Abstract of Statistics*. HMSO, London.

COX, J. A., LINDQUIST, G. L., PRZYJAZNY, A. & SCHUMULBACH, C. D. 1978. Leaching of boron from coal ash. *Environmental Science & Technology*, **12**, 722–723.

DREESEN, D. R., GLADNEY, E. S., OWENS, J. W., PERKINS, B. L., WIENKE, C. L. & WANGEN, L. E. 1977. Comparison of levels of trace elements extracted from fly ash and levels found in effluent waters from a coal fired power plant. *Environmental Science & Technology*, **10**, 1017–1019.

DUDAS, M. J. 1981. Long-term leachability of selected extracts from fly ash. *Environmental Science & Technology*, **15**, 840–843.

EARY, L. E., RAI, D., MATTIGOD, S. V. & AINSWORTH, C. C. 1990. Geochemical factors controlling the mobilization of inorganic constituents from the fossil fuel combustion residue: II Review of the minor elements. *Journal of Environmental Quality*, **19**, 202–214.

EDMUNDS, W. M. & BATH, A. H. 1976. Centrifuge extraction and chemical analysis of interstitial waters. *Environmental Science & Technology*, **10**, 467–472.

FELMY, A. R., RAI, D. & MOORE, D. A. 1993. The solubility of (Ba, Sr)SO$_4$ precipitates: thermodynamic equilibrium and reaction path analysis. *Geochimica et Cosmochimica Acta*, **57**, 4345–4363.

HANSEN, L. D. & FISHER, G. L. 1980. Elemental distribution in coal fly ash. *Environmental Science & Technology*, **14**, 1111–1117.

HARDY, M. A. 1981. *Effects of coal-fly ash disposal on water quality in and around the Indiana dunes National lake-shore, Indiana*. USGS, Indianapolis, Water-Resources Investigations 81-16.

HJELMAR, O. 1990. Leachate from land disposal of coal fly ash. *Water Management and Research*, **8**, 429–449.

JAMES, W. D., GRAHAM, C. C., GLASCOCK, M. D. & HANNA, A.-S. G. 1982. Water-leachable boron from coal ashes. *Environmental Science & Technology*, **16**, 195–197.

JONES, L. H. & LEWIS, A. V. 1960. Weathering of fly ash. *Nature*, **185**, 404–405.

KOPSICK, D. A. & ANGINO, E. E. 1981. Effect of leachate solutions from fly and bottom ash on groundwater quality. *Journal of Hydrology*, **54**, 341–356.

Le Seur, S. L. & Drake, E. E. 1987. Hydrogeology of an alkaline fly ash landfill in eastern Iowa. *Ground Water*, **25**, 519–526

Lee, S. 1994. *The Long-Term weathering of PFA and Its Implications for Groundwater Pollution*. PhD thesis, University of Sheffield, UK

—— & Spears, D. A. 1995. The long-term weathering of PFA and implications for groundwater pollution. *Quarterly Journal of Engineering Geology*, **28**, S1–S15.

Lindsay, W. L. 1979. *Chemical Equilibria in Soils*. John Wiley, New York.

Mattigod, S. V., Rai, D., Eary, L. E. & Ainsworth, C. C. 1990. Geochemical factors controlling the mobilization of inorganic constituents from the fossil fuel combustion residue: I Review of the major elements. *Journal of Environmental Quality*, **19**, 188–201.

Natusch, D. F. S., Wallace, J. R. & Evans. C. A., Jr. 1974. Toxic trace elements: preferential concentration in respirable particles. *Science*, **183**, 202–204.

Powell, M. A., Hart, B. R., Fyfe, W. S., Sahu, K. C., Tripathy, S. & Samuel, C. 1991. Geochemistry of Indian coal and fly ash, environmental considerations. In: Sahu, K. C (ed.) *Proceedings, Environmental Impact of Coal Utilization*. IIT, Bombay, 23–38.

Rehage, J. A. & Holcombe, L. J., 1990. Environmental performance assessment of coal ash site: *Little Canada structural ash fill*. Electric Power Research Institute, Palo Alto, CA, EPRI-EN-6532

Roy, W. R. & Griffin, R. A. 1984. Illinois basin coal fly ashes. 2. Equilibria relationships and qualitative modeling of ash–water reactions. *Environmental Science & Technology*, **18**, 739–742.

——, Thiery, R. G., Schuller, R. M. & Suloway, J. J. 1981. *Coal fly ash: a review of the literature and proposed classification system with emphasis on environmental impacts*. Illinois State Geological Survey, Champaign, IL, Environmental geology note No. 96.

Sakata, M. 1987. Movement and neutralization of alkaline leachate at coal ash disposal. *Environmental Science & Technology*, **21**, 771–777.

Simons, H. S. & Jeffery, J. W. 1960. An X-ray study of pulverised fuel ash. *Journal of Applied Chemistry*, **10**, 328–336.

Simsiman, G. V., Chesters, G. & Anderson, A. W. 1987. Effect of ash disposal ponds in groundwater at coal-fired power plant. *Water Research*, **21**, 417–426.

Smith, R. D. 1980. The trace element chemistry of coal during combustion and the emissions from coal-fired plants. *Progress in Energy and Combustion Science*, **6**, 53–119

Talbot, R. W., Anderson, M. A. & Anders, W. A. 1978. Qualitative model of heterogeneous equilibria in a fly ash pond. *Environmental Science & Technology*, **12**, 1056–1062.

Theis, T. L. & Wirth, J. L. 1977. Sorptive behaviour of trace metals on fly ash in aqueous systems. *Environmental Science & Technology*, **11**, 1096–1100.

——, Westrick, J. D., Hsu, C. L. & Marley, J. J. 1978. Field investigation of trace metals in groundwater from fly ash disposal. *Journal of Water Pollution Control Federation*, **50**, 2457–2469.

USEPA 1988. *Waste from the combustion of coal by electric utility power plants*. US Environmental Protection Agency, Washington, DC, USEPA Report 530-SW-88-002.

Van der Sloot, H. A., Wijkstra, J., Van dal, A., Das, H. A., Slanna, J., Dekkers, J. J. & Wals, J. D. 1982. *Leaching of trace elements from coal solid waste*. Netherlands Energy Research Foundation, ECN-120.

Warren, C. J. & Dudas, M. J. 1984. Weathering processes in relation to leachate properties of alkaline fly ash. *Journal of Environmental Quality*, **13**, 530–538.

Sewage sludge disposal to agriculture and other options in the UK

R. J. ANDREWS[1], J. W. LLOYD[2] & D. N. LERNER[3]

[1]*Dames & Moore, Blackfriars House, St Mary's Passage, Manchester M3 2JA, UK*
[2]*Hydrogeology Research Group, School of Earth Sciences, University of Birmingham, Edgbaston, Birmingham B15 2TT, UK*
[3]*Department of Civil Engineering, University of Bradford, Bradford BD7 1DP, UK*

Abstract. Due to the forthcoming European Union ban on sewage sludge disposal at sea and associated environmental legislation which has increased the amount of sewage sludge produced annually in the UK, there is a need for consolidation and expansion of existing sludge disposal outlets and assessment of the suitability of alternative and innovative disposal options. This paper reviews the main elements of sludge disposal in the UK in relation to their environmental sensitivity, sustainability and general security. Much of the additional sludge produced by changes in waste water treatment is likely to be accommodated by an increase in disposal by incineration and application to agricultural land. The effect of sludge application to land has, therefore, come under increasing examination, primarily due to potentially detrimental long-term effects on soil fertility and on nitrate leaching to groundwater. The implications and causes of nitrate leaching are discussed in relationship to sludge types and recommendations made for minimizing potentially adverse impacts to sensitive groundwater.

Concern over the environmental impact of sewage sludge, together with other environmental issues, has given rise to a variety of recent legislation; for example, EC Directives 80/778/EEC (EC 1980) and 86/278/EEC (EC 1986), Sludge (Use in Agriculture) Regulations 1989 (Anon, 1989), the Environmental Protection Act 1990 and The Code of Good Agricultural Practice for the Protection of Water 1991 (MAFF 1991). The introduction of such legislation, together with the likelihood of further restrictions, has meant that approaches to sludge disposal are being reconsidered. The ban on the disposal of sewage sludge at sea (EC 1991) due to come into force after 31st December 1998, and the associated establishment of sewage treatment works for coastal discharges mean that alternative disposal routes are being examined and existing ones safe-guarded (and expanded where practical) to accommodate the additional loads.

Sewage sludge contains a variety of substances that may cause environmental and public health problems. The most important of these are toxic metals (such as Hg, Cu, Ni, Pb, Zn, As, Cd, Se, Mo, Co, Cr and Ag), nitrogen (predominantly in the form of ammonium and organic-N), pathogens, phosphate, greenhouse and acid gases, and organic chemicals (most notably PCBs, organochlorine pesticides, PAHs, phthalate esters, surfactants, furans, chlorobenzenes, solvents, phenols and dioxins) (Rogers 1987; Alcock & Jones 1993). In addition, the biochemical oxygen demand (BOD) of sludge can present a problem if there is a risk of pollution to surface waters. Odour problems associated with sludge may also be of public concern. The extent to which sludge is contaminated with the above materials depends upon its origin. Sludge from a dominantly industrial or urban source is likely to have a higher level of contamination (particularly from metals) than sludge with a purely domestic or rural source.

In the UK, as everywhere else in the world, sewage is processed in a variety of ways; however, the final sludge product can be generally classified as being either liquid undigested sludge (LUS), liquid digested sludge (LDS), undigested sewage cake (USC) or digested sewage cake (DSC). Each of these sludge types has a different physical and chemical make-up and this, together with the degree of contamination, affects the sludge's suitability for particular disposal options. Sewage cake tends to have a higher quantity of metals and organic chemicals per unit volume than liquid sludge, due to a concentration effect from dewatering and adsorption on to the higher percentage of organic matter. The sludge type also has a strong control on the relative proportions of ammonium and organic-N present. Nitrogen in raw sewage is dominantly in the form of organic-N and is largely confined to the solid proportion of the sludge. Ammonium is adsorbed on to organic material in raw sludge and is also found in solution. Once screened and settled, this raw form of sludge is essentially LUS, having a low solids content of 5–10% and a total N content of

ANDREWS, R. J., LLOYD, J. W. & LERNER, D. N. 1998. Sewage sludge disposal to agriculture and other options in the UK. *In*: MATHER, J., BANKS, D., DUMPLETON, S. & FERMOR, M. (eds) *Groundwater Contaminants and their Migration*. Geological Society, London, Special Publications, **128**, 63–74.

about 4% dry solids basis. If LUS is dewatered then a cake is produced – USC. USC has a higher solids content of about 20–25% and, as a result of dewatering, the ammonium content is reduced so that most of the nitrogen occurs as organic-N, the total N content being approximately 3%. Digestion of sludge causes the most easily mineralized organic-N to be converted into ammonium. Liquid digested sludge (LDS) is therefore dominated by nitrogen in the form of ammonium but still retains a similar dry solids and total N content to LUS. The organic-N present represents the least mineralizable fraction. Dewatering LDS produces digested sewage cake (DSC) with a similar dry solids and total N content to USC but which is dominated by more refractory (slow mineralizing) organic-N.

Disposal options other than agricultural

General UK situation

There are numerous potential disposal options for sewage sludge but their practicality, security, environmental impact, energy consumption and cost vary considerably. In most cases more conservative, tried and tested disposal routes are preferred by the UK water supply companies/disposal authorities. Many new innovative techniques either lack any form of testing at a large operational scale or do not present a secure outlet in the short to medium term and so are not seen as currently practical options. Their introduction is further impaired by either uncertain market conditions or the lack of existing developed markets to receive the products. The uncertainty of potential markets is increased by the absence of specific standards for alternative sludge end-uses. Until the issue of quality standards is resolved, market development and expansion will be inhibited.

Current government legislation establishes the principle of an integrated approach to pollution control. As part of this, sludge disposal routes are selected on the basis of a best practicable environmental option (BPEO); that is, the disposal option that for a given set of objectives 'provides the most benefit or least damage to the environment as a whole, at acceptable cost, in the long term as well as the short term' (Royal Commission 1988).

The main practical disposal outlets that are currently available include the following: recycling of sludge to agricultural land; spreading of sludge at dedicated disposal sites; use in reclamation of derelict land, former industrial

sites and coal mining waste tip restoration; forestry; landfill disposal; various forms of incineration; sea dumping; and coastal outflow pipes. The latter two outlets are mostly being phased out by the end of 1998 due to a wish by disposal authorities to be seen to be environmentally friendly. More recent techniques that have been adopted in some programmes or are currently being examined in pilot schemes are pellet/granule production (thermal drying) for use as a combustive fuel source in power generation, a granular fertilizer or simply as a bulk reduction exercise prior to land filling; composting; lime treatment; soil manufacture; oil production; gasification; wet oxidation; cement production; brick production; fat and protein production; underground disposal/land filling and undersea disposal. Of these treatment techniques, thermal drying is currently regarded as having good potential, primarily due to the flexibility of associated outlets.

A survey of England and Wales by the Department of the Environment (Table 1) in 1980 (DoE/NWC 1983) showed that of the main disposal outlets, 43% of sludge went to agriculture (including forestry), 24% was dumped at sea, 17% was landfilled, 5% was used for land restoration, 4% was incinerated, and the remaining 7% was either stockpiled or used for other purposes. In general, the majority of sludge previously dumped at sea will be absorbed by these other main disposal routes. More recent MAFF/DoE assessments have shown that the proportion of sludge going to agriculture was 42% in 1991 (approximately 465 000 tonnes) and that the total was forecast to double by 2006 to 926 000 tonnes (although still remaining at 43% of all sludge disposed). This increase will be driven by the higher standards of treatment required by the EC Urban Waste Water Treatment Directive (EC 1991).

Regional differences in the percentage of sludge disposed of via a particular route (Table 1) have mostly been constrained economically and geographically through the extent of rural and urban land and the amount of coastline available, but also by the degree to which sludge is contaminated by metals. However, current changes in sludge disposal are increasingly affected by strategic concerns. In land-locked and industrialized regions of the UK, such as Severn-Trent and West Yorkshire, incineration already accounts for a significant proportion of sludge disposal. In the Yorkshire region as a whole, it is planned that by the late 1990s incineration of sludge from cities will account for 50% of all sludge disposed; landfill and lagooning will take 23% and the remaining 27% will

Table 1. Disposal routes for sewage sludge (% of total produced) in England and Wales (DoE/NWC 1983; Schroder 1989)

Disposal method	England and Wales (1980)	Northumbria	North West	Yorkshire	Severn Trent	Welsh	Anglian	Thames	Wessex	South West	Southern
Agriculture (including forestry)	43	27*	23	51	47	65	80	52	69	62	64
Sea	24	48	51	5		33	13	40	23	29	15
Landfill	17	25		31	19		7*	8		5	16
Land restoration	5										
Incineration	4			13	17						5
Dedicated land	Significant (7)	(see agriculture)	5		17		(see landfill)				
Others, e.g. stockpiled	Significant (7)		21†		2						

*Includes dedicated land disposal
†Includes land use other than agricultural

go to agricultural land as digested sludge (Walker 1991). In more rural areas the agricultural outlet is large, particularly in the Anglian region. Efforts are currently being made by the UK water supply companies to protect the agricultural outlet from future legislation and to change public opinion due to the obvious importance of this disposal route. The extent of landfill disposal varies regionally but tends to be highest in areas with low sea disposal, Northumbria being a notable exception.

Sea disposal and coastal discharges

The ban on disposal at sea has been driven by concerns over eutrophication, particularly of the southern and eastern North Sea. Direct disposal of sludge to sea has accounted for a far higher percentage of total sludge disposal in the UK than in any other European Community state, due primarily to an extensive coastline. Sea disposal has been particularly prevalent in UK coastal areas with large urban populations and/or with high levels of industrialization where it has offered a simple and relatively cheap solution to the disposal of large quantities of, often contaminated, sludge. Removal of this option has caused problems in northern areas of the UK where acid soils prevent land application of sludge of a suitable quality (predominantly due to potential mobilization of metals) and also countrywide for those sludges in which high metal contents also rule out such an option. Under these circumstances, landfill and incineration currently appear to be the only viable alternatives.

Legislation to prohibit discharge of untreated sewage into coastal waters via outfall pipes has prompted the establishment of new sewage treatment works. The wider environmental impact of such schemes requires consideration of their effect on local ecology, air and water quality, and increase in traffic. The problem of disposing of highly contaminated sludge must also be addressed.

Landfill disposal

The percentage of sludge that is disposed of to landfill in the UK is significantly smaller than in most other European Community states (Powlesland & Frost 1990). Expansion of landfill disposal in the UK is limited due to an increasing shortage of suitably licensed sites and a corresponding rise in costs. In addition, future legislation in the form of EC Directives on landfilling of waste may affect this disposal option. Although landfilling of sludge can be sustained in the short to medium term, these factors may make long-term dependency strategically undesirable. With time, the distance from a sewage treatment works to suitable landfill sites may increase, with associated increases in transport costs.

Due to cost considerations, the landfill option is mostly reserved for sludge that is contaminated by metals to the extent that it is unsuitable for disposal to agricultural land. Sludge disposed to landfill is predominantly in the form of digested cake, as most landfill operators will not accept raw sewage and the reduction in volume from the original liquid sewage reduces landfill and transport charges. Ash from incineration of sludge is also disposed to landfill, again due to the metal content and cost savings from the large volume reduction. An increase in incineration may be accompanied by a reduction in landfilling of digested cake, with cake being substituted by ash. Such a move would reduce both landfill and transport costs, although these costs may be outweighed by incineration pre-treatment costs. Production of sludge granules through direct thermal drying, which produces a volume reduction of approximately 95%, may also reduce landfill and transport costs. Co-composted sludge and municipal waste may also potentially be used in landfills to provide a daily cover.

The overall environmental impact of landfill disposal, including problems associated with the mobilization of the sludge-derived metals in acidic leachate, lowers the value of this disposal option when considered under BPEO criteria, although the enhanced production of energy from landfill gas during sludge breakdown provides cost benefits, and the increased rate of degradation of waste in the landfill has a general and significant impact in reducing risk to the environment. The long-term security of this outlet is low; however, in the short term it remains a relatively practical disposal option.

Incineration

The quantity of sewage sludge disposed of by incineration in the UK is small by comparison to most other European Community states, but is increasing rapidly. Existing incinerators are mainly confined to urban–industrial areas. With the advent of a ban on sea disposal of sludge, the incineration option has come under increasing scrutiny due to the need for a secure, dependable outlet that is capable of dealing with large quantities of sludge of variable quality and degree of contamination. Incineration is seen by the UK water supply companies as a highly practical, medium-secure disposal option with a relatively low overall environmental impact. Additional benefits are derived from bulk reduction of the original sludge; reduced transport costs; removal of various problems associated with raw sewage (e.g. odour, pathogens); a reduction in the amount of greenhouse gases produced as compared to landfill; and the generation of heat from combustion representing 'green energy' which could be used to produce electricity or run the incinerator installation. The attractions of incineration have meant that its use is increasing in the UK, notably in the Thames and Yorkshire regions.

Despite the above-mentioned benefits, there are a significant number of problems associated with the incineration option. Future legislation such as the new EC Directive on Incineration of Hazardous Waste represent a threat to the security of the outlet, particularly with respect to flue gas emissions of heavy metals (such as Pb and Cd), dioxins, hydrogen chloride and NO_x. Compliance with such restrictions should be achievable through the use of improved technology although this will increase operational costs. The security of the outlet is also affected by the incineration method chosen. Methods currently being examined are dedicated incineration of raw or digested sludge, co-incineration with municipal solid waste, co-combustion with coal, co-incineration with chemical waste, and vitrification. The first of these methods is currently most widely practised and also appears to be the most secure option in terms of legislation and ease of achieving planning permission and discharge consents. The other methods suffer from lack of control of the disposal outlet and/ or higher energy demands. In particular, co-incineration of sludge and industrial waste has met with significant planning permission and regulatory/compliance difficulties. Incineration has suffered from a poor public image despite having a generally low overall environmental impact. Problems with odour, traffic, noise and water quality are generally minimized although questions of air quality, visual impact and the dependency of fly ash disposal on landfill remain, as exemplified by recent refusals of planning permission (Anon 1992). Perhaps the most obvious drawback to incineration is that of cost in planning, development and implementation. These factors must be weighed against the perceived benefits when selecting a BPEO for sludge disposal.

Disposal of sludge to land

Over 48% of sludge produced in England and Wales is disposed or 'recycled' to land, which is slightly higher than the average European figure; it is 42% in the UK as a whole. Application of sludge to land is an obvious disposal option due to its nitrogen, phosphate and organic matter content and represents a practical and 'green' alternative to the methods mentioned above. The metal content of many sludges, however, means that careful soil monitoring is required. EC Directive 86/278/EEC (EC 1986) states that sludge disposed to land should be treated to reduce odour nuisance, and the total metals load to soil controlled/limited depending upon soil acidity. Metal levels in sludge have declined over the last decade; however, concerns have recently increased regarding their long-term effect on soil fertility. In particular, Cd and Zn have been implicated as microbial inhibitors in soils with long-term high sludge applications showing reduced soil microbial populations and reduced ability to fix atmospheric nitrogen. They may also play a role in reducing soil fertility. As a consequence, recommendations have been presented to reduce the permitted levels of Cd and Zn applied to agricultural land. In addition to metals, levels of pathogens and residual organic compounds, such as dioxins and PCBs, which may enter the food chain, may also need to be assessed before sludge can be applied. Levels of PCBs in sludge have been shown to be an order of magnitude higher than the average for UK soils (Alcock & Jones 1993), and concerns have arisen regarding potential uptake by dairy cattle and ensuing PCB levels in milk.

Organic matter in sludge acts as a soil conditioner improving the structure of a soil and its water-retaining capacity. Nitrogen and phosphate are both essential plant nutrients. The degree of availability of nitrogen depends upon the sludge type. The availability of phosphate, however, is independent of sludge type, with approximately 50% being available to crops in the first year following application. In general terms, organic-N in sludge is mineralized slowly to ammonium, which can then be nitrified more rapidly to nitrate. Both ammonium and nitrate can be assimilated by plants and nitrate can be leached into groundwater. Sludges with a high proportion of ammonium (such as liquid digested sludge) therefore act as quick-release fertilizers, whilst those dominated by organic-N (such as digested sewage cake) act as slow-release fertilizers. The more resistant (refractory) the organic-N to mineralization, the

Table 2. *N-availability for different sludge types (Hall 1986)*

Sludge type	N-availability in 1st year (% of total N)	Leaching risk
LDS	60	Highest
LUS	35	
USC	20	
DSC	15	Lowest

lower the N-availability of the sludge and the less risk of increasing nitrate leaching. N-availability, and thus the level of risk of increasing nitrate leaching, is a direct consequence of sludge production techniques (Table 2).

In addition to applying sludge independently, a composted mixture of straw and granulated digested sludge may also be used. Composting requires large storage facilities and presents a less versatile alternative than sludge alone. Bulk, and therefore transport costs, are increased and there is a reduction in security of the outlet due to a dependency on a straw supply, whilst the problem of metal contamination remains unaffected. The use of dried sludge as a granular fertilizer is currently being examined by several UK water supply companies as it alleviates storage and transport costs.

Traditionally, most sludge to land disposal has been via dedicated land and agricultural routes, although more recently land restoration and forestry have become significant outlets. With the exception of disposal to dedicated land, the security of the outlets is endangered by the variable quality of sewage sludge and potential public health scares. It is therefore essential that these disposal routes are carefully and sensitively managed.

Dedicated land

Regulations concerning sludge disposal at dedicated sites are less stringent and permit higher soil metal contents than are acceptable on other land types. Dedicated land disposal is a very practical option for water supply companies, since it usually occurs adjacent to sewage works, minimizing transport costs. However, a number of environmental impact problems are associated with this option, most importantly those of odour and nitrate contamination of surface and groundwater. Additionally, disposal in this manner is seen publicly as a waste of a resource. The introduction of Nitrate Sensitive Areas (NSAs) and borehole protection zones, and the presence of existing dedicated land in

areas of aquifer outcrop vulnerable to leaching, restrict expansion of this disposal route.

Land restoration

The restoration/reclamation of waste land enables sewage sludge and co-composted sludge and municipal waste to be used to build soils and improve existing soil fertility, their role being primarily as soil conditioners. An increasing amount of sludge has been used for this purpose in recent years but demand is insufficient and too variable to enable a realistic expansion into this disposal area. In addition to the problem of long-term security of the outlet, the metal content in sludge has to be addressed and the soil metal content tailored to comply with the final designated use of the land. The variability of sludge quality may present a problem in this instance. Restoration schemes often involve large-scale works (e.g. landscaping of former mining areas) and therefore large storage facilities are required for stockpiling sludge.

Forestry

Forestry is an increasing outlet for sludge disposal. Problems associated with this option are those of sludge storage facilities; insecurity of outlet through commercial changes; sludge quality and public health concerns; availability of forest land; acid soils either preventing or reducing sludge applications; and leaching of nitrogen as ammonium or nitrate into surface and groundwater. If carefully managed, the environmental impact and possible aesthetic problems can be overcome but availability and distance to forest land may restrict operations due to transport costs.

Agricultural use

The application of sewage sludge to agricultural land (both pastoral and arable) has been an active disposal route in the UK for over 30 years (Davis 1989) but sludge is currently only received by approximately 1% of farm land. In the UK, sludge to agriculture is seen as a practical, low-cost, low-energy alternative to most other forms of sludge disposal which, if carefully managed, has a low environmental impact. The main drawbacks of this option are those of security; a need for storage facilities since, on arable land at least, sludge can only be applied whilst the land is bare (improvements in application

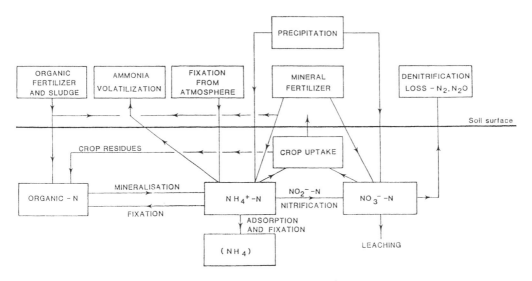

Fig. 1. The soil nitrogen cycle.

plant may alleviate this item although seasonal problems may continue, related to the rate of liquid loading of LUS and LDS); environmental restrictions; problems over sludge type and quality; and transport costs. Agricultural disposal suffers from a lack of security through an inability to control the outlet, which may be subject to a change in farming attitudes, land use (e.g. setaside) and the delineation of NSAs and borehole protection zones. It is also endangered by public health scares, the odour nuisance (with emission of greenhouse and acid gases) and potential water contamination from metals, pathogens, organic chemicals and, perhaps most significantly, nitrate.

Until recently the nitrate content of sludge was ignored and application rates were based on the sludge metal content. Concern over increasing nitrate concentrations in groundwater caused the implementation of a Code of Good Agricultural Practice for the Protection of Water (MAFF 1991) which recommended further limitations on the use of sludge by restricting the amount applied in any one year to 250 kg N/ha total N. It also recommended that the available nitrogen in sludge should be considered when calculating fertilizer dressings and that applications should be avoided next to water courses and boreholes supplying drinking water and under certain field conditions. The total N content, N-availability and mineralization behaviour of various sludge types is therefore of importance and has been the subject of ongoing research (Hall 1985, 1986; Coker *et al.* 1987*a,b*; Andrews 1993). The various elements of the soil nitrogen cycle under consideration are shown in Fig. 1. Fast nitrogen release sludge fertilizers such as LDS are suitable for grassland where nitrogen uptake is rapid throughout the growing season; however, under arable conditions such sludge types present a nitrate leaching hazard.

Current application techniques are restricted to surface spreading of liquid sludge or cake and injection of liquid sludge. Both techniques may be rapidly followed by cross-ploughing to mix the soil and sludge together and to reduce odour problems. At present, sludge cannot be applied to a standing arable crop so its use is restricted to between harvest and the sowing of the following crop. With the gradual shift from spring- to winter-sown cereals, this 'window' is diminishing. The limited period over which sludge can be applied to arable land has a number of repercussions. Firstly, sludge produced outside this period must be stored. Liquid sludge can be kept either at sewage treatment works or in purpose-built lagoons at farms, although these are subject to further regulation and present a risk to groundwater and may create an odour nuisance. Cake can be stored more easily at both treatment works and farms with less risk of water contamination and odour problems. In addition, the lower volume of cake compared to liquid sludge for a given total N content reduces transport costs.

Secondly, the type of sludge and the time at which it is applied in relation to the recharge period, and type and sowing date of the following crop, are of considerable importance in

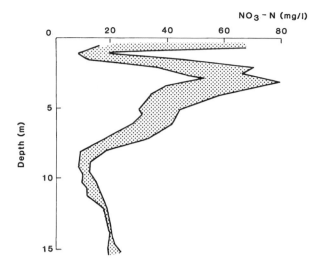

Profile beneath arable land from predominantly fertilizer application
(after Foster et al., 1982)

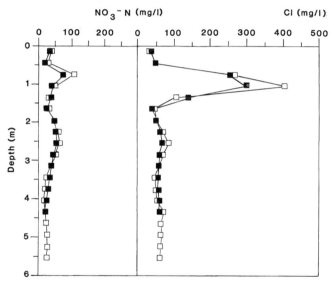

Profile beneath arable land for mixed fertilizer and periodic LUS
(3 yearly) application (Andrews, 1993)

Fig. 2. Unsaturated zone nitrate profiles in soil/chalk.

controlling nitrate leaching. Applying a fast-release, ammonium-rich sludge, such as LDS, after harvest and leaving the land bare over winter should be avoided. Application of a slow-release, ammonium-poor sludge, such as DSC, which is then followed as soon as possible by a high-uptake, deep-rooting winter crop (such as a winter cereal) will minimize nitrate leaching.

In this way the crop can be used to remediate the effect of sludge application whilst the farm benefits from the ensuing reduction in fertilizer costs. The application technique, itself, is also of significance since it affects the volatilization rate of ammonia. Surface spreading produces higher volatilization losses than injection so that less nitrogen remains in the soil, thus reducing the

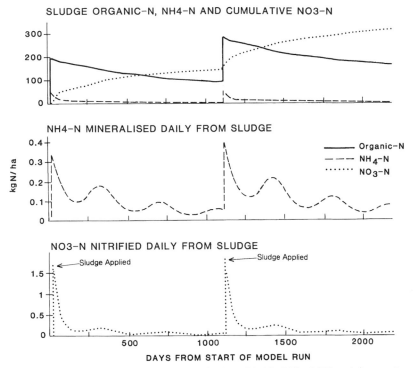

Fig. 3. Modelled sludge breakdown from an autumn injection of 250 kg N/ha LUS at three-year intervals.

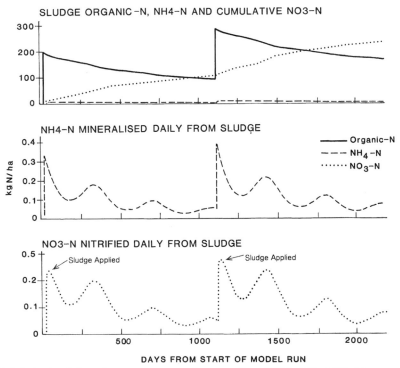

Fig. 4. Modelled sludge breakdown from an autumn surface spreading of 250 kg N/ha LUS at three-year intervals.

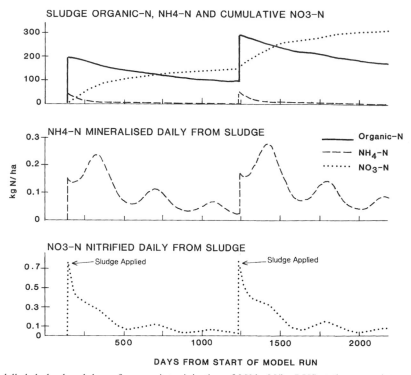

Fig. 5. Modelled sludge breakdown from a winter injection of 250 kg N/ha LUS at three-year intervals.

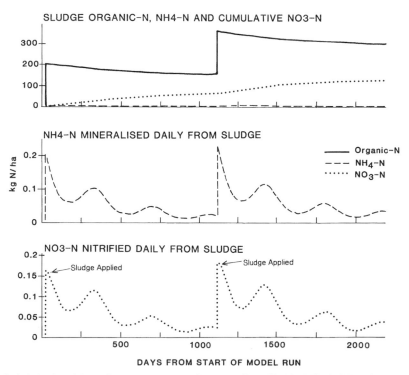

Fig. 6. Modelled sludge breakdown from an autumn surface spreading of 250 kg N/ha DSC at three-year intervals.

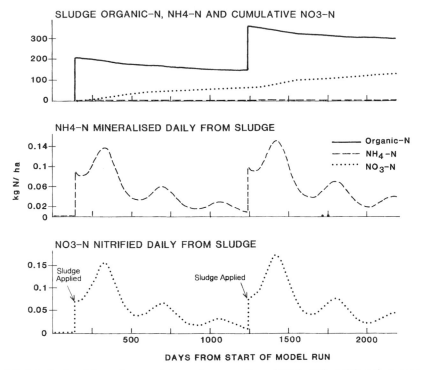

Fig. 7. Modelled sludge breakdown from a winter surface spreading of 250 kg N/ha DSC at three-year intervals.

potential amount of nitrate that can be leached. Nitrate input from sludge application can increase leaching concentrations in following years to levels above that for chemical fertilizer applications alone. In general, however, loading profiles may often be similar to those for fertilizer application (Fig. 2) although the agricultural regime and type of crops sown may have a significant control.

Finally, the small amount of time available for sludge applications to arable land, together with regulations on metal and nitrogen content and public sensitivity mean that increasingly well-controlled and co-ordinated management is required. Figures 3–7 show the availability of nitrogen species in the soil for certain sludge–time applications, based upon modelling of the nitrogen cycle given in Fig. 1 and calibrated against field and laboratory data (Andrews 1993).

Attention to the risk of nitrate contamination from sewage sludge applications mean that water supply companies must begin to supply farmers with information on the total N content of the sludge that they have received and also the proportional N-availability over several years, to enable allowance to be made in progressing fertilizer treatments. Good sludge/nitrogen manage-

ment entails making a well-timed application of sludge in a suitable quantity to enable selected crops to remove as much of the nitrate produced as possible. If badly managed then sludge application can considerably increase the mass and concentration of nitrate leached into aquifers.

Improvements in sludge quality control and screening methods with increased and aggressive marketing of the obvious natural benefits of sludge to agriculture may eventually enable water supply companies to charge farmers and other users for the product rather than giving it away, as is the current practice.

Conclusions

The disposal of sludge remains a significant problem to water supply companies as European Community environmental legislation continues to restrict existing activities. Most UK companies are attempting to safeguard and consolidate major outlets such as sludge to agriculture whilst beginning to embrace the strategic and practically appealing option of incineration. The short- to medium-term solution offered by landfill disposal is recognized and other, more innovative, options are being examined for the future. The small percentage of farm land currently receiving sludge means that there is considerable scope for expansion in this direction although potential future legislation and restrictions may limit the strategic viability of this option.

The authors would like to thank Anglian Water Services Limited for funding the research relating to sludge application to arable land undertaken by the Hydrogeological Research Group at the University of Birmingham.

References

ALCOCK, R. & JONES, G. 1993. *Cheriosphere*, **26**(12), 2199–2207.

ANDREWS R. J. 1993. *The Impact of Sewage Sludge Application on Nitrate Leaching from Arable Lland on the Unconfined Chalk Aquifer of East Anglia, England.* PhD Thesis, Birmingham University, UK.

ANON 1989. *Sludge (Use in Agriculture) Regulations 1989.* SI 1989 No. 1263. HMSO (ISBN 0 11 097263 5) as amended by the Sludge (Use in Agriculture) Regulations 1990 SI 1990 No. 880 (ISBN 0 11 0038800). HMSO, London

ANON 1992. *HAZNEWS. International Hazardous Waste Management Monthly.* **57**, December 1992.

COKER, E. G., HALL, J. E., CARLTON-SMITH, C. H. and DAVIS, R. D. 1987*a*. Field investigations into the manurial value of lagoon-matured digested sewage sludge. *Journal of Agricultural Science*, **109**, 467–478.

—— 1987*b*. Field investigations into the manurial value of liquid undigested sewage sludge when applied to grassland. *Journal of Agricultural Science*, **109**, 479–494.

DAVIS, R. D. 1989. Agricultural utilization of sewage sludge: a review. *J. IWEM.* **3**, 351–355.

DoE/NWC 1983. *Sewage sludge survey, 1980 data.* Department of the Environment/National Water Council Standing Committees on the Disposal of Sewage Sludge, August 1983. EC (1980), EC (1986), EC (1991).

EC 1980. *EC Directive 80/778/EEC.* Council Directive of 15 July 1980 relating to the quality of water intended for human consumption. Official Journal of the European Committees, 30 August 1980, OJ No. L229/11.

—— 1986. *EC Directive 86/278/EEC.* Council Directive of 12 June 1986 on the protection of the environment, and in particular the soil, when sewage sludge is used in agriculture. Official Journal of the European Communities, 4 August 1086, OJ No. L181/6.

—— 1991. *EC Directive 91/271/EEC OJ L135.* 30 May, 1991.

FOSTER, S. S. D., CRIPPS, A. C. & SMITH-CARRINGTON, A. 1982. Nitrate leaching in groundwater. *Philosophical Transactions of the Royal Society of London*, **296**, 477–489.

HALL, J. E. 1985. The cumulative and residual effects of sewage sludge nitrogen on crop growth. *In*: WILLIAMS, J. W., GUIDI, G. and L'HERMITE, P. (eds) *Long-term effects of sewage sludge and farm slurry applications.* Elsevier Applied Sciences, London, 73–83.

—— 1986. *The agricultural value of sewage sludge.* WRc Report ER 1220-M.

MAFF (Ministry Of Agriculture, Fisheries And Food) 1991. *Code of Good Agricultural Practice for the Protection of Water.* Ministry of Agriculture, Food and Fisheries. Welsh Office Agricultural Department, July 1991.

POWLESLAND, C. & FROST, R. 1990. *A methodology for undertaking BPEO studies of sewage sludge treatment and disposal.* Water Research Centre Report PRD 2305-M/1.

ROGERS. H. R. 1987. *Organic contaminants in sewage sludge. Occurrence and fate of synthetic organic compounds in sewage and sewage sludge – a review.* Water Research Centre Report PRD 1539-M.

ROYAL COMMISSION 1988. 12th Report of the Royal Commission 1988. HMSO, London.

SCHRODER, P. 1989. *The Water Share Offers: Pathfinder Prospectus.* Schroder Wagg.

WALKER, J. B. 1991. *Sludge disposal atrategy in Yorkshire.* Institute of Water Engineers and Manageers Seminar, Options for Sewage Sludge Disposal, 30 May 1991.

Investigation of contamination at a public supply borehole in Hertfordshire, UK

I. DAVEY, I. MOXON & D. HYBERT

Environment Agency Thames Region, Kings Meadow House, Kings Meadow Road, Reading RG1 8DQ, UK

Abstract: For several years a public supply borehole in Hertfordshire has suffered intermittent low level contamination by ammoniacal nitrogen. The situation worsened in about 1990 when the concentration began to rise steadily. Concentrations of other ions, including chloride, had also risen over a number of years. The site has limited treatment capacity and, because of the ammoniacal nitrogen contamination, it became necessary to take the source out of supply. Investigation of the problem began in 1994, concentrating on the potential contaminant sources nearby which were a pig slurry lagoon, a landfill and a river, which carries a loading of sewage effluent. Monitoring indicated the slurry lagoon to be the most likely cause of contamination to the public supply. The lagoon was emptied in 1995 which revealed that no liner existed and that there was apparent continuity with groundwater. Subsequent monitoring reinforced the evidence for a connection between the slurry lagoon and the public supply. However, contaminants were also found in association with the landfill and the river and contributions from these cannot be ruled out.

A public water supply borehole in Hertfordshire (PWS1) began to suffer intermittent contamination from ammoniacal nitrogen during the late 1970s. Concentrations of other determinands also rose, including chloride and sulphate. No bacteriological contamination was reported.

Ammoniacal nitrogen concentration increased sharply during the 1990s. The site has limited treatment capacity and by 1992 it proved necessary to take the source out of supply.

Figure 1 shows the geographical setting of PWS1. A joint field investigation with the public supply operator was begun in February of 1994 to assess the origin of the pollution. The objective was to propose remedial action to protect the public water supply source and groundwater resource for the future. This paper summarizes how the investigation was carried out and discusses the likely origin of pollution.

Potential pollution sources

There are several potential sources of pollution in the area. A landfill site, which accepts commercial and industrial waste, is located 300 m to the north of PWS1. A borehole abstracts water in the vicinity of this landfill for gravel washing and disposes of the effluent to a silt lagoon 250 m to the north of PWS1. A further, closed landfill site which took domestic waste lies some 2 km to the northwest of PWS1. Other landfills where inert wastes are deposited are present in the locality.

There is also a pig farm with slurry lagoon, which lies just to the south of the landfill. Between the slurry lagoon and PWS1 lies a river, which carries a loading of sewage effluent, with consequent elevated concentrations of nitrate and intermittent occurrences of raised concentrations of ammonia. The area is unsewered and sewage is disposed of to ground from septic tanks, although at some distance from PWS1.

The potential pollution sources are shown in Fig. 2.

Geology and hydrology

PWS1 is located in the shallow valley of the river. In the valley bottom, alluvial deposits and river gravels overlie the Upper Chalk. The top surface of chalk strata is commonly weathered and structureless, forming so-called 'putty chalk'. This has a relatively low permeability and can confine groundwater below it. Away from the river, glacial sands and gravels with boulder clay horizons overlie the Chalk. The glacial deposits have been quarried for many decades.

Regionally, groundwater flow is generally towards the southeast (British Geological Survey 1984). However, solution features in the Chalk are numerous in this area and groundwater flow is anomalous. Tracer tests (Harold 1938) have shown that surface water entering swallow holes some 8 km to the southwest of PWS1 appears about three days later in springs and boreholes 10 km to the west-northwest of

Davey, I., Moxon, I. & Hybert, D. 1998. Investigation of contamination at a public supply borehole in Hertfordshire, UK. *In*: Mather, J., Banks, D., Dumpleton, S. & Fermor, M. (eds) *Groundwater Contaminants and their Migration*. Geological Society, London, Special Publications, **128**, 75–92.

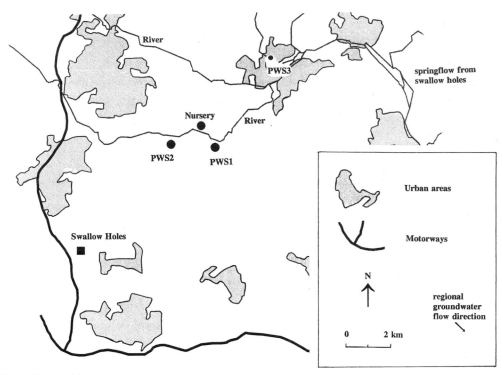

Fig. 1. Geographical setting.

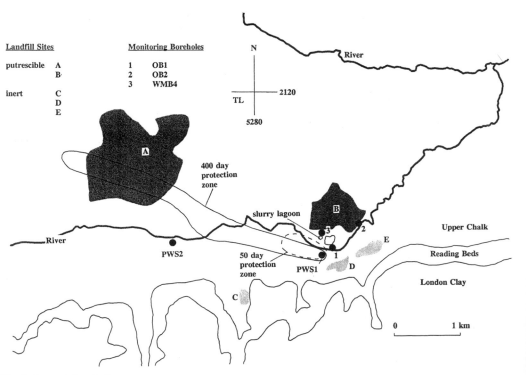

Fig. 2. Potential pollution sources.

PWS1. This is in conflict with the regional flow direction (see Fig. 1).

Contaminants entering the swallow holes affect the quality of groundwater at a further public supply borehole, PWS2, which lies 2 km to the west of PWS1 along the river valley (Fig. 1). However, the impact is not felt at PWS1, as evidenced by water company and Environment Agency records.

Flow in the river is, on average, around $20\,000\,\mathrm{m}^3\,\mathrm{day}^{-1}$, but varies seasonally between about $10\,000\,\mathrm{m}^3\,\mathrm{day}^{-1}$ and $30\,000\,\mathrm{m}^3\,\mathrm{day}^{-1}$. The degree of hydraulic continuity between the river and groundwater is not known.

PWS1 boreholes

PWS1 is located some 50 m to the south of the river and abstracts from the Chalk aquifer. It comprises two boreholes of 0.46 m diameter and 125 m depth and is licensed to abstract up to $1360\,\mathrm{m}^3\,\mathrm{day}^{-1}$. Although the abstraction rate is relatively small, PWS1 is of strategic importance to the water company.

The Upper Chalk is overlain by about 4 m of alluvium and gravels at PWS1 and rest water level is some 3 m below ground level. Plain casing extends to 26 m in depth.

The National Rivers Authority (NRA) modelled flow to PWS1 using the two-dimensional steady-state particle tracking model, Flowpath, as part of the programme of defining protection zones around boreholes (National Rivers Authority 1994a). The modelled zone and sampling point locations are shown in Fig. 2.

Background water quality and trends

As is to be expected, groundwater in the area is calcium bicarbonate type. Variations in concentrations of various ions can however be noted and Table 1 shows analyses from various locations in the area.

Figure 3 compares the trends in concentrations of ammoniacal nitrogen, total oxidized nitrogen (TON) and chloride for the river and PWS1 during the 1970s to 1995. Trends for ammoniacal nitrogen and chloride at PWS1 are clear but there appear to be few corresponding changes in the river data.

The river has its origins as baseflow from the Chalk aquifer but it has relatively high concentrations, for chalk type water, of chloride, sulphate, nitrate, sodium and potassium. This can be attributed to the discharge upstream of sewage effluent from works which serve two major conurbations.

At the nursery site, situated 1.5 km to the northwest of PWS1, groundwater has low concentrations of most ions. Similar comments apply to PWS3, 4.5 km to the northeast.

PWS2 has elevated concentrations of sodium, magnesium, potassium, chloride, sulphate and also of nitrate. Although apparent similarities can be seen with the quality at PWS1 during the 1980s, the provenances have distinctive differences as evidenced by the tracer studies at the nearby swallow holes and subsequent records.

A borehole, BH1, installed in 1977 to monitor the landfill, has shown little indication of any impact of leachate, although ammoniacal nitrogen has risen to $2\,\mathrm{mg}\,\mathrm{l}^{-1}$ on occasion. However, BH1 is located adjacent to the silt lagoon and may be influenced by the discharge of gravel washing water.

Further monitoring boreholes were installed around the landfill site in 1993. However, only WMB4, which lies near to the slurry and silt lagoons, shows contamination by ammoniacal nitrogen ($4.94\,\mathrm{mg}\,\mathrm{l}^{-1}$ in November 1993). Chloride, sodium and possibly nitrate appear elevated in some of the other boreholes.

Investigation

Shallow boreholes were installed to augment existing monitoring boreholes in the locality. A pumping exercise was carried out at PWS1 during March 1994, the nearby boreholes being monitored for groundwater level and quality. Periodic monitoring was continued during 1994 and 1995, during which time the slurry lagoon was emptied and relined. In March and April of 1996 a further pumping exercise was carried out at PWS1, with a view to returning the borehole to supply.

The modelled protection zone for PWS1 is narrow but elongated, the 400-day flow zone being 4 km long but only about 350 m wide. Groundwater flow direction has been taken as broadly west to east, although in reality recharge to the slurry and silt lagoons may influence local flow direction. The main parameters used are shown in Table 2.

The zone includes partly within or near to it two major landfill sites that have received putrescible or other degradable wastes, the pig slurry lagoon, the nearby river and several discharges of sewage effluent from septic tanks, as shown in Fig. 2.

In view of resource limitations it was decided to restrict investigation to assess likely contaminant sources nearest to PWS1. In effect,

Table 1. *Background water quality (all concentrations in mg/l)*

Site		PWS1	River	PWS2	PWS3	Nursery
No. of samples		20	5	10	9	3
Years		1964–1985	1994–1995	1985–1995	1985–1995	1994–1996
Ca	min.	104	113	103	100	95
	med.	120	118	111	114	99
	max.	127	129	125	121	104
Mg	min.	1	3.8	3.6	2.9	3.2
	med.	5.3	4.1	4.4	3.2	3.3
	max.	10.4	4.1	7	3.7	3.4
Na	min.	12.5	50	22	9	8
	med.	24	64	24.5	10.4	8
	max.	30	79	29	11	8
K	min.	2.4	7.7	2.6	1.4	1.1
	med.	3.3	9.3	3.2	1.7	1.4
	max.	5.6	11.4	4.4	2	1.5
NH_4N	min.	<0.05	<0.05	<0.05	<0.05	<0.05
	med.	0.15	0.3	<0.05	<0.05	
	max.	0.55	5	0.14	0.1	0.12
			(1201 samples 1970–1994)			(2 samples)
HCO_3	min.	245	231	139	216	194
	med.	250	264	222	220	198
	max.	270	300	239	226	203
Cl	min.	20	2.4	27	19	11
	med.	36	75	39	20	13
	max.	45	203	44	41	13
			(776 samples 1970–1994)			
SO_4	min.	26	8	32	15	7
	med.	54	80	62	20	
	max.	64	243	173	138	10
			(227 samples 1970–1994)			(2 samples)
NO_3	min.	22	19	28	20	7.5
	med.	27	54	31	21	10.2
	max.	34	100	37	22	11.5
			(993 samples 1970–1994)			

investigations were designed to assess the impact of the river in the vicinity of PWS1, the pig slurry lagoon and the nearest landfill site.

The first step was to augment existing monitoring boreholes in the area and carry out a pumping exercise at PWS1.

Monitoring points

Existing monitoring points were used where possible. Locations of all monitoring points are shown in Fig. 4. The bridge across the river at the entrance to the landfill site was chosen as a monitoring point, with further data available from an automatic quality monitoring station (AQMS), 300 m downstream of the bridge.

Borehole BH1, located south of the landfill site and adjacent to the silt lagoon, was monitored to identify contaminant flow from the landfill and assess impact from gravel washing water recirculated to the silt lagoon. Borehole BH2, located some 900 m to the northeast of and down hydraulic gradient from PWS1 and considered far enough down gradient to be unaffected by pumping at PWS1, was taken as a reference for groundwater levels.

At both these boreholes, groundwater was apparently confined beneath a layer of putty chalk, which was found to underlie the sand, gravel and clay drift at 5 m below ground level (bgl). BH1 was completed to 16 m bgl and BH2 to 14 m bgl. The drift deposits were cased out.

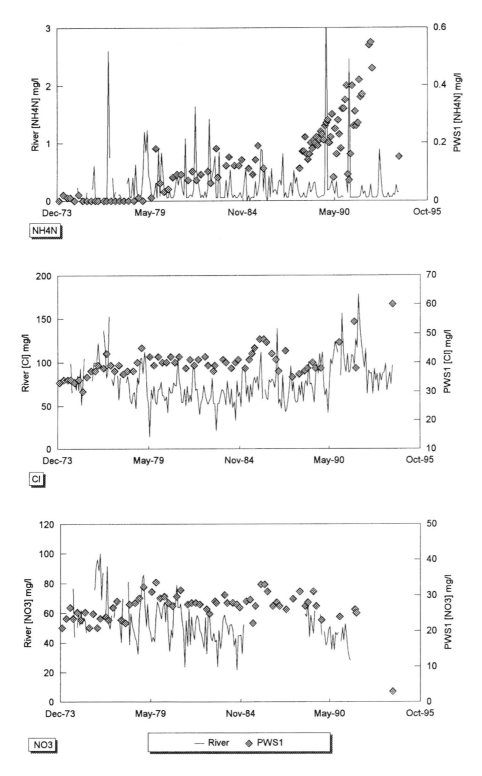

Fig. 3. River and PWS1 water quality trends.

Table 2. *Model parameters*

Aquifer thickness	70 m
Effective rainfall	209 mm a^{-1}
Groundwater gradient	0.003
Permeability	5–30 m day^{-1}
Transmissivity	1600 m^2 day^{-1}
Storage coefficient	0.01

Borehole WMB4 is a 50-mm-diameter piezometer completed to 16 m bgl. It is located to the north of the slurry lagoon but immediately adjacent to the southern perimeter of the oldest part of the landfill site and was chosen to assess contaminant flows from the landfill.

Three further boreholes were installed at strategic points in the zone of influence of PWS1. Borehole BHA was installed adjacent to the pig slurry lagoon, to assess the possible effects of leakage from the lagoon. Borehole BHB was installed between the river and PWS1, to assess whether any contaminated water may flow south beneath the river towards PWS1, and to help identify any flow contribution from the river itself. Borehole BHC was installed to the west of PWS1, up hydraulic gradient, to identify any contaminant contribution from this flow direction.

Chalk was found below clays, sands and gravels at each new borehole at around 5 m bgl. Groundwater was found at approximately 3 m bgl. Each borehole was completed to a total depth of 15 m with the upper 2 m plain cased. Although the boreholes were developed by bailing until the water cleared, suspended solids proved problematic for some weeks following borehole construction.

A schematic geological cross-section from the landfill site to PWS1 is shown in Fig. 4, with the line of the section indicated on Fig. 5.

Pumping exercise

Groundwater contours prior to pumping are shown in Fig. 5, produced partly from information presented in a Hertfordshire County Council report (Hertfordshire County Council 1995). The general west to east flow direction is evident, as is mounding associated with recharge from the silt and slurry lagoons. The presence of this recharge mound may help flow to reach PWS1 via BHB and potentially via BHC.

Pumping from PWS1 borehole number 1 was carried out at a rate of about 1000 m^3 day^{-1} between 24 February and the end of March 1994.

The water was pumped to waste into the river via a section of the public supply pipework, isolated from the rest of the network during the exercise. Chlorination was necessary as a precaution to protect the supply network, but to protect the river the NRA set a low allowable chlorine concentration (0.2 mg l^{-1}) for the discharge to the river. As a result automatic switches, triggered by excessive chlorine concentration in the treated groundwater, caused the pumps to trip out frequently during the initial stages of pumping.

Groundwater levels were measured at PWS1 and all the monitoring boreholes throughout pumping. Water level change during pumping and pumping rate are shown in Fig. 6. The slight fall in level at BH2 is taken to represent natural recession.

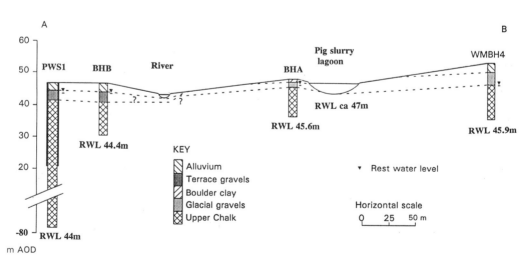

Fig. 4. Schematic cross-section at PWS1.

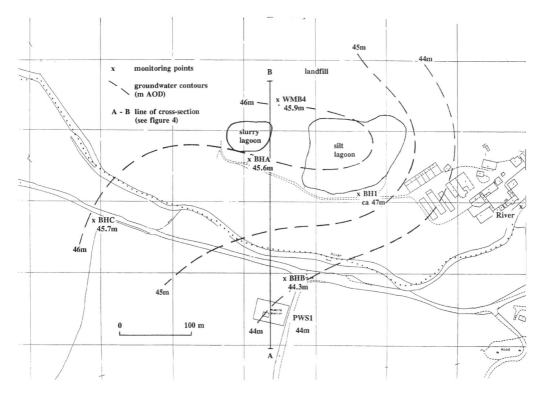

Fig. 5. Monitoring points.

Discussion of results: quantity/levels

Changes in water level at boreholes BHB and BHC are clearly related to pumping at PWS1, BHB responding to tripping out of the pump. This shows that some flow to PWS1 is derived from these directions. Water levels also changed at BHA, WMB4 and BH1, which although not pronounced, appeared greater than the apparent regional recession. This may indicate a flow contribution to PWS1 from this area. However, this cannot be taken as conclusive since these levels are also likely to be affected by the amount of recharge from the slurry and silt lagoons. This effect has not been quantified.

Analysis of pumping test results show the Chalk aquifer in this area to have moderately high transmissivity (c. $650\,m^2\,day^{-1}$ using the Theis method, National Rivers Authority 1994b). The storage value obtained is relatively low (0.002), indicating that the Chalk in the zone of influence may be confined. This could possibly be related to the presence of a putty layer at the surface of the Chalk. The pumping test data did not show identifiable flow to PWS1 from the river, although a flow contribution cannot be ruled out.

Assessment of water quality monitoring

Samples were taken, following purging, from all the monitoring points, including PWS1 and the river, before and during pumping. Figure 7 shows changes in concentrations of ammoniacal nitrogen, potassium and chloride at BHB, BHC, BH1, WMB4 and PWS1.

Prior to pumping, several concentrations were elevated at PWS1. Concentrations decreased straight after pumping, followed by a gradual rise. Broadly similar changes in concentrations were seen at BHB. The final concentration of ammoniacal nitrogen recorded at BHB was $3.7\,mg\,l^{-1}$, although this was the only sample with an elevated concentration.

At BHC, concentrations of potassium and chloride decreased steadily. The concentration of ammoniacal nitrogen varied, but with no trend maintained throughout pumping.

Ammoniacal nitrogen concentration in the river varied between less than $0.3\,mg\,l^{-1}$ to $1.16\,mg\,l^{-1}$. Potassium varied between $7.3\,mg\,l^{-1}$ and $9.3\,mg\,l^{-1}$. Chloride fell from $95\,mg\,l^{-1}$ to $65\,mg\,l^{-1}$.

Concentrations of several ions analysed for at BHA, nearest the lagoon, were higher than the

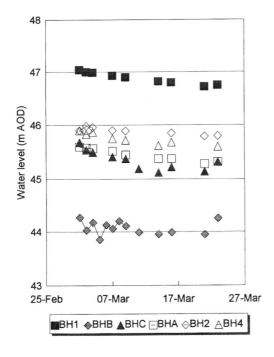

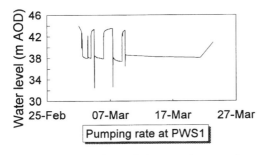

Fig. 6. Water level change during pumping excercise 1.

laboratory had been prepared for. Concentrations exceeded the reporting limits of $250\,\mathrm{mg\,l^{-1}}$ for sodium, $50\,\mathrm{mg\,l^{-1}}$ for potassium, $200\,\mathrm{mg\,l^{-1}}$ for chloride, $100\,\mathrm{mg\,l^{-1}}$ for nitrate and $1\,\mathrm{mg\,l^{-1}}$ for nitrite. This persisted throughout the pumping exercise and was not picked up until completion of pumping. Of note, however, is that nitrate and nitrite concentrations were high, but ammoniacal nitrogen concentration was negligible, despite the borehole's being located within a few metres of the slurry lagoon.

At WMB4, adjacent to the landfill, concentrations were elevated prior to pumping. During pumping, chloride showed an increase whereas potassium decreased. Ammoniacal nitrogen concentrations remained high, with no particular trend identifiable.

At BH1 ammoniacal nitrogen concentration increased overall. Trends for other ions were either absent or not possible to confirm.

At BH2 concentrations of most ions remained steady. The concentration of ammoniacal nitrogen did, however, vary.

Samples of slurry and leachate were also analysed, at this time and subsequently. Difficulties were experienced in sampling both. In the case of leachate, frequently none was present in the monitoring borehole. For the slurry, ensuring suspended solids were not included in the analysis was problematic, since filters quickly blocked up when sampling. Results are shown in Table 3.

The leachate is seen to be of fairly 'low strength', which may be expected given the relatively low input of putrescible wastes to the site. The slurry contains extreme concentrations of many of the determinands of interest. Ammoniacal nitrogen concentrations are very high and in contrast to nitrate, as expected. In addition to high concentrations of major ions, of interest are the concentrations of iron, fluoride and copper. Copper is a common dietary supplements for pigs.

Discussion

The changes in groundwater quality during pumping supported the evidence from water level change indicating that PWS1 draws water from the direction of BHB and BHC. Although changes in quality took place at the boreholes near to the slurry lagoon and landfill, these may be coincidental rather than conclusively due to pumping.

The final concentration of ammoniacal nitrogen recorded at BHB, if not spurious, was significant. The only known contaminant sources in the vicinity that were likely to be capable of providing such a concentration at this time were the landfill and the slurry lagoon. Overall the situation was evidently complicated and the following is put forward as a possible explanation of changes in quality at the monitoring boreholes.

During the time PWS1 was out of operation a contaminant plume may have developed within its zone of influence. Mounding caused by recharge to the slurry and silt lagoons may have encouraged radial flow of contaminants from this area, helping the plume to spread. Equilibrium conditions in the vicinity of the slurry lagoon during the period of non-pumping may have allowed nitrification of ammoniacal nitrogen to nitrate.

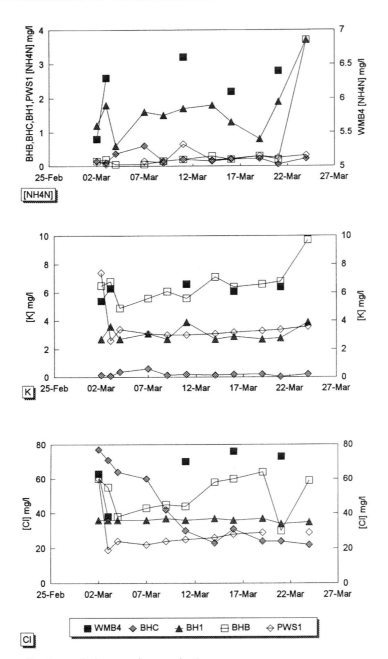

Fig. 7. Water quality change during pumping exercise 1.

Reaction to pumping was rapid at some of the boreholes and contaminant concentrations fell at PWS1 and BHC as cleaner water was drawn in. However, subsequently, pumping drew in water from nearer the pollutant source and contaminant concentrations at PWS1 and BHB rose.

This hypothesis does not, however, take account of other potential contaminant sources. Overall, although the exercise provided circumstantial evidence that contaminants might be drawn in from the vicinity of the landfill or the slurry lagoon, it was not conclusive. However, in view of the apparent high concentrations of some

Table 3. *Leachate and slurry quality*

	pH	EC (μs cm^{-1})	Ca (mg l^{-1})	Mg (mg l^{-1})	Na (mg l^{-1})	K (mg l^{-1})	Alkalinity	Cl (mg l^{-1})	SO$_4$ (mg l^{-1})	NO$_3$ (mg l^{-1})	NH$_4$N (mg l^{-1})
Leachate											
23 May 1995	7.1		47	12	53.3	7.6	3230	812	99	<0.1	71.9
Slurry											
18 Nov. 1993		13 000				1600	3380	747	5260	1.8	1250
06 Dec. 1994	7.7						6870	1110	29	20.8	2770
19 June 1996		24 800					<40	1320	1360	1.9	2460

	Fe dissolved (μg l^{-1})	Fe total (μg l^{-1})	Mn dissolved (μg l^{-1})	Mn total (μg l^{-1})	Cu dissolved (μg l^{-1})	Cu total (μg l^{-1})	Zn dissolved (μg l^{-1})	Zn total (μg l^{-1})	F (μg l^{-1})	TOC (μg l^{-1})	Fatty acids (mg l^{-1})
Slurry											
18 Nov. 1993	660	11 800	<50	1900	1430	4660	520	25 200	77 000	886	1389
06 Dec. 1994					95				911 000	5700	
19 June 1996											valeric iso-valeric iso-caproic n-caproic iso-butyric ethanoic propanoic

determinands in BHA, which seemed most likely to be attributable to the slurry lagoon, it was recommended that the lagoon be emptied and the liner inspected. Continued monitoring at all the monitoring points was also recommended.

Inspection of the slurry lagoon

During the autumn of 1994 the concentration of ammoniacal nitrogen at BHA rose dramatically from a few milligrams per litre to $250 \, \text{mg} \, l^{-1}$. This rise was generally not mirrored by other determinands. However, the concentration of fluoride rose significantly.

Of the other boreholes, only concentrations at BHB are noteworthy. Here concentrations of several ions fell significantly initially, including ammoniacal nitrogen, in contrast to BHA. Concentrations of sodium, magnesium, potassium, manganese and sulphate also fell. Fluoride concentration increased significantly initially. Changes in concentrations at BHA and BHB are shown in Fig. 8.

In view of the ammoniacal nitrogen concentrations at BHA, the need to assess the lining of the slurry lagoon became urgent and the lagoon was emptied during the summer of 1995. It was found to be some 5–6 m in depth and chalk strata were visible in parts. There was no evidence of a low permeability liner and it was apparent that an old sand and gravel excavation had been used without preparation for storage of slurry. Groundwater was present in the base of the site and it was clear that some degree of hydraulic continuity had been possible, although undoubtedly the lagoon base had silted up over the years.

The presence of the slurry lagoon at this location is not desirable, given the proximity of PWS1. However, in view of the presence of several hundred pigs at the farm a practical solution was needed with some urgency. To this end, the lagoon was lined with 2.5 m of local boulder clay by the operators of the adjacent landfill. This was carried out with equipment and methods used for capping the landfill for which a permeability of less than $1 \times 10^{-10} \, \text{m} \, \text{s}^{-1}$ had been recorded. To keep flow through the liner to a minimum, but to allow adequate volume for storage, it was agreed that the slurry in the lagoon should not exceed a depth of 1.5 m. This would allow a flow through the liner of around $0.035 \, \text{m}^3 \, \text{day}^{-1}$ or $12 \, \text{m}^3 \, \text{a}^{-1}$.

Subsequently, concentrations at BHA fell, including those of ammoniacal nitrogen, potassium, sodium and chloride. The concentration of calcium rose, as did that of fluoride.

Pumping exercise 2

The water company undertook a further pumping exercise at PWS1 between 23 February 1996 and 3 April 1996 to assess whether the source could be returned to supply. In view of the persistent high concentration of ammoniacal nitrogen at BHA, an attempt had been made prior to this to scavenge pump from this borehole. However, the borehole yield was low, with a 7 m drawdown recorded at a pumping rate of $4 \, \text{m}^3 \, \text{day}^{-1}$, and the attempt was abandoned.

PWS1 was pumped at a rate of around $1.3 \, \text{Ml} \, \text{day}^{-1}$, although difficulties with the pump were experienced and continuous pumping could not be maintained. Figure 9 shows the changes in quality (ammoniacal nitrogen, chloride and sulphate are plotted) at BHA, BHB, WMB4 and PWS1 along with the pumping rate.

There was a general rise in concentration of most ions at PWS1 as contaminants were drawn in. However, the ammoniacal nitrogen concentration did not exceed $0.15 \, \text{mg} \, l^{-1}$. Concentrations at BHA responded to pumping within a few days, with a significant rise in concentration of chloride. Increases were also recorded in sodium, fluoride, manganese and iron concentrations. Concentration of sulphate fell, possibly as reduction to sulphide took place. Since the lagoon had been lined these changes indicated the persistence of the pollution plume in the vicinity of the lagoon.

At BHB there was a large increase in ammoniacal nitrogen concentration from near detection limit to $11 \, \text{mg} \, l^{-1}$, with increases also in concentrations of sodium, potassium and chloride. Slight falls occurred part way through which corresponded with the period of non-pumping at PWS1. Sulphate concentration fell initially and gradually tailed off, which corresponded approximately with the situation at BHA.

Changes also took place at WMB4, mostly falls in concentration, as with sulphate (shown in Fig. 9), and also for nitrate, calcium, magnesium and potassium. In contrast, chloride concentration rose steadily, as did fluoride initially although this fell subsequently. Evidently WMB4 was affected by pumping at PWS1.

Water quality relationships

Historical data and that from the initial pumping exercise are plotted on the trilinear diagram in Fig. 10 for PWS1 and the most relevant monitoring points. Some trends are clear. The quality of water at PWS1 varies between a clean water, such as seen at the nursery, and

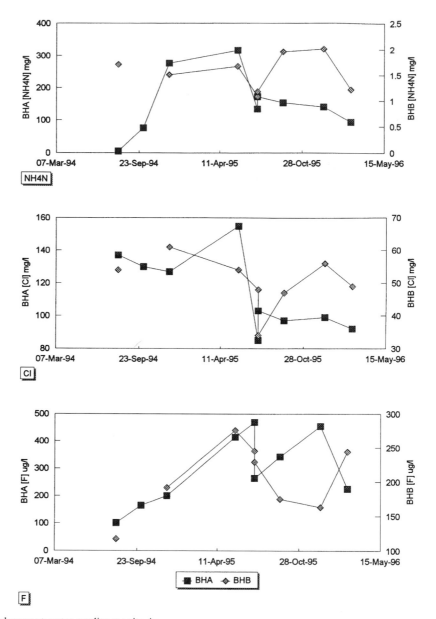

Fig. 8. Subsequent water quality monitoring.

one contaminated with sodium, chloride and sulphate. From the trilinear plot, water of both WMB4 and the river types could contribute to this contamination. Further attention is given to this below under 'Mass balance'.

The concentrations at BHA are considered to be due to leakage from the slurry lagoon. The water quality at WMB4 may result from mixing of chalk water with leachate or slurry.

Groundwater sampled at BHC may be a mixture of various waters, depending on hydraulic conditions. The variables are thought to be whether pumping is taking place at PWS1 or whether there is influence from recharge mounding below the lagoons during non-pumping periods, and whether there is a contribution from the river. A similar case may be put forward for BHB.

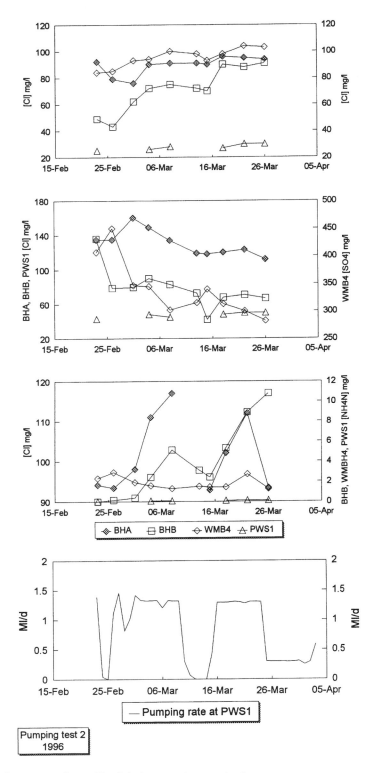

Fig. 9. Changes in water quality and level during pumping exercise 2.

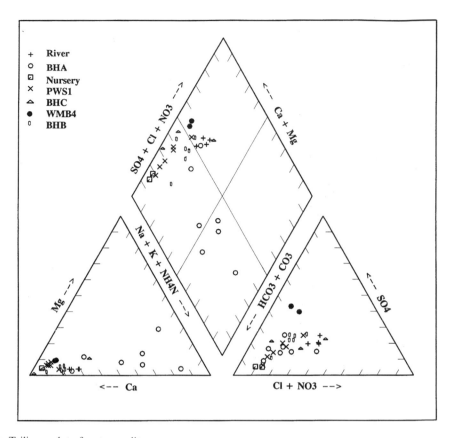

Fig. 10. Trilinear plot of water quality.

The geochemical processes taking place within the plume are evidently complex, although this study has mostly considered contaminant concentrations in terms of advection and dilution. Many other processes may be taking place, including ion exchange, sorption, oxidation–reduction processes, as well as changes arising from dilution, dispersion, fissuring and recharge to the silt lagoon.

Mass balance

Simple mass balances have been estimated to account for some of the concentrations seen at PWS1 and also at BHB. The starting point for the estimate has been taken from the trilinear plot (Fig. 10). The diamond and anion fields suggest that water of the river, nursery and WMB4 types may contribute to PWS1 in equal proportions. The cation field suggests that these water types may contribute in the ratio 2:2:1.

Various permutations of contributions were tested and some of the more reasonable solutions are shown in Table 4.

Concentrations for the contributing waters, the nursery, the river, WMB4 and also BHA, have been taken as 'representative', rather than absolute. For PWS1 and BHB, concentrations have been taken from the end of pumping exercise 1 and 2 respectively.

No exact solution has been found possible, which may be for various reasons. The assumption of the water types involved may be incorrect or there may be additional water types not accounted for. The mass balance has been calculated on the basis of dilution whereas more complex contaminant attenuation mechanisms are undoubtedly taking place.

However, if the estimates are taken at face value, water at PWS1 may comprise a large proportion of nursery type water (more than 70%); a large proportion of river type water (up to 55%); and a small proportion of WMB4 type water (up to 10%) or of BHA (1%, BHA

Table 4. *Mass balance*

Ion	Nursery	River	WMB4	BHA	PWS1 (end Ptest1)	PWS1a	PWS1b	PWS1 (end Ptest2)	PWS1c	BHB (end Ptest2)	BHB1	BHB2
Ca	100	120	300	110	100*	172	131			130	112	137
Mg	3.2	4	23	10	5	10	6			8	4	6
Na	8	60	45	40	48	37	40	15	22	47	40	32
K	1.2	8	10	60	8	6	6	3	4	11	11	5
NH_4N	0.04	0.2	4	100	0.7	1.4	0.5	0.2	1	10	10	1
Cl	3	90	75	90	60	55	58	31	30	95	60	44
SO_4	10	85	350	150	86	147	85	50	31	65	65	87
NO_3	11	45	15	2	13	23	30	27	18	15	29	24
Fe	46	<10	50	3500				<5	32	<10	366	31
Mn	29	<10	1500	500				<15	24	40	60	240

* Assumed value.
Calculations based on:
PWS1a = 33% nursery, 33% river, 33% WMB4; compares with PWS1 (end Ptest 1).
PWS1b = 35% nursery, 55% river, 10% WMB4; compares with PWS1 (end Ptest 1).
PWS1c = 70% nursery, 20% river, 10% BHB (end Ptest2); compares with PWS1 (end Ptest 2).
BHB1 = 35% nursery, 55% river, 10% BHA; compares with BHB (end Ptest 2).
BHB2 = 50% nursery, 30% river, 20% WMB4; compares with BHB (end Ptest 2).

being 10% of BHB, which itself is 10% of PWS1).

Although speculative, this may have implications for water quality at PWS1 when it is returned to supply. While the apparent contamination plume remains then the following concentrations of ammoniacal nitrogen may be calculated for PWS1 (assuming concentrations observed during or at the end of pumping exercise 2): assuming $100 \, \text{mg} \, l^{-1}$ at BHA, then $1 \, \text{mg} \, l^{-1}$ at PWS1; assuming $4 \, \text{mg} \, l^{-1}$ at WMB4, then up to $0.4 \, \text{mg} \, l^{-1}$ at PWS1; assuming $10 \, \text{mg} \, l^{-1}$ at BHB, then $1 \, \text{mg} \, l^{-1}$ at PWS1. However, it is anticipated that the plume should disperse in time.

If the river type water is actually derived from the river, for concentrations seen normally in the river ($<0.2 \, \text{mg} \, l^{-1}$), the 55% contribution suggested above should not be problematic. However, peak concentrations (in excess of $2 \, \text{mg} \, l^{-1}$), on this assumption, would be of concern. Attention to monitoring and pumping rates will be important when PWS1 is returned to supply.

Use of minor ions as tracers

Concentrations of several minor ions were monitored and these are shown in Table 5. However, attempts to use these as tracers were not conclusive.

Very high concentrations of iron and manganese have been recorded in BHA and WMB4. Elevated concentrations of these ions were found in soil pore water in a study into the effects of applying pig slurry to land (Joseph 1983). A study of cow slurry contamination in chalk (Withers 1995) found elevated concentrations of manganese and iron in the unsaturated zone. Iron and magnesium were detected in samples from the PWS and BHB, although in concentrations which might not be considered unusual for the Chalk aquifer (Environment Agency 1996). These ions were, however, not detected in the river.

A copper dietary supplement is commonly fed to pigs and copper concentrations are very high in the slurry. Copper concentrations above $20 \, \mu\text{g} \, l^{-1}$ were found in samples from some boreholes and may originate from the lagoon, although concentrations have generally been only a few micrograms per litre. A rising trend in concentration of copper was recorded at BHB during pumping test 2 which appeared to correlate with total organic carbon (TOC), although since the copper concentration was only a few micrograms per litre this would be of limited use without supporting evidence.

Zinc is another additive to pig food and was found in elevated concentration at BHA. However, attempts to trace zinc away from BHA are inconclusive.

Table 5. *Minor constituents*

Site		PWS1	River	BHA	BHB	BHC	WMB4	BH1	Nursery
No. of samples		14	4	8	11	12	12	10	
Cu	min.	6	5.8	3.4	1.8	<0.5	4.1	0.7	0.5
	med.	11	6.3	4.4	3.5	1.8	5.3	1.6	
	max.	15	16	7.9	5.2	28	8.2	12.7	
Zn	min.	17	16	16	3	3	10	2	4
	med.		30	35	8	11	32	5	
	max.		32	78	22	46	46	9	
Fe	min.	<15	<20	520	<30	<30	<30	<30	<20
	med.	<15		1830	<30	<30	40	30	
	max.	2420	28	2410	70	410	1420	50	
Mn	min.	<5	6	200	<10	<10	260	30	5.5
	med.	<5		240	<10	<10	1190	40	
F	min.	160	133	225	176	140	312	128	146
	med.	166	229	263	276	158	580	141	
	max.	194	266	290	306	177	621	179	169
		(4 samples)							
TOC	min.	1.1	3.4	7.2	1.1	<1	9.9	1.2	<1
($\text{mg} \, l^{-1}$)	med.		4.7	9.8	2.5	1.2	12.1	1.7	
	max.	1.9	5.1	10.1	3.8	14.2	13.3	2.1	1.2
^{2}H	min.				<33		<42	<33	
	med.				79.2		97.5	75	
	max.				95		120	87.5	
					(4 samples)		(4 samples)	(4 samples)	

Fluoride was measured in very high concentration in slurry ($77\,000\,\mu g\,l^{-1}$). This has impacted on groundwater in BHA and possibly in WMB4. BHB may also have been affected, although elevated concentrations have also been recorded in the river. The concentrations recorded for PWS1 would not be considered unusual for the Chalk aquifer.

Robinson has reported high concentrations of tritium in some landfill leachates (Department of the Environment 1996). Up to 120 tritium units (TU) were found in WMB4 and 95 TU in BH1 and BHB. A few tens of TU might be expected more normally for an aquifer such as the Chalk. Unfortunately no measurement of tritium concentration has been possible for the leachate since, at the time of writing, it had not been possible to obtain further leachate samples. Tritium was not found in a sample of slurry.

Taken together, the various ions help confirm the impact of the slurry lagoon on groundwater at BHA and suggest contaminant migration to BHB. They cannot themselves confirm the connection to PWS1 nor can they confirm the impact of leachate on groundwater. They appear to rule out the river from being the sole source of pollution.

Conclusions

PWS1 has suffered elevated concentrations of several ions, most significantly of ammonium and chloride, since the early to mid 1970s. In 1992, rising ammoniacal concentrations were such that PWS1 had to be taken out of supply. The investigation has identified a major source and route of contamination to PWS1 and remedial actions have been taken. It is not known what caused the onset of rising concentrations of contaminants.

The geochemical conditions are evidently complex, with processes likely to include cation exchange, sorption, oxidation–reduction as well as changes due to dilution, dispersion, fissuring and recharge to the silt lagoon. However, assessment of the contamination in terms of the more complex processes has not been necessary to achieve the main objectives of the investigation.

It has been shown that pig slurry has contaminated groundwater adjacent to the lagoon as shown at BHA and by the emptied lagoon itself. BHB, between the slurry lagoon and PWS1, shows the impact of contamination from the area of the slurry lagoon, which is emphasized by pumping at PWS1. Groundwater levels at BHB have also been shown to respond directly to pumping at PWS1. It is concluded

that contamination due to the slurry lagoon has affected quality at PWS1.

Leachate has been found in the landfill, but not consistently. The assumption is that it is lost from the landfill into groundwater, although there appears to be limited impact on quality. Of the boreholes around the landfill, only WMB4 and BH1 have shown any significant concentrations of relevant contaminants, which could in any case be influenced by the slurry lagoon or by recharge to the silt lagoon.

Groundwater abstracted at PWS1 may be a mix between clean 'background' water, such as seen at the nearby nursery, and water with enhanced concentrations of ions, particularly sodium, chloride and ammonium. Water of types seen at WMB4 and the river together may account for this. The low storage coefficient calculated from pumping test data suggests that inflow from the river may be very limited. On the other hand, a mass balance estimate suggests that water similar in quality to that in the river could provide a substantial proportion (50%) of flow to PWS1.

The water company understandably would like to return the source to supply as soon as possible. It is hoped that the remedial works to the lagoon will be sufficient to safeguard PWS1 in the future. It seems likely, though, that it will be some time before the existing pollution plume dissipates and continued monitoring is needed. Monitoring is also warranted in case of leakage from the lagoon and in view of the other potential pollution sources in the area.

The water company regularly monitors the source. During 1997 the company will commission plant to treat elevated concentrations of ammoniacal nitrogen, should these persist. The work is being undertaken to safeguard a valuable resource while ensuring there is no risk to the public supply.

The co-operation and involvement of Three Valleys Water Company and Hertfordshire County Council Waste Regulation Unit (now Environment Agency) are gratefully acknowledged. The views expressed in this paper are those of the authors and do not necessarily represent those of the Environment Agency.

References

BRITISH GEOLOGICAL SURVEY 1984. *Hydrogeological map of the area between Cambridge and Maidenhead.*

DEPARTMENT OF THE ENVIRONMENT 1996. *A review of the composition of leachates from domestic wastes in landfill sites.* Report by Aspinwall and Co.

ENVIRONMENT AGENCY 1996. *Groundwater quality in the Middle and Upper Lee catchments.* Report.

HAROLD, C. H. H. 1938. *The results of the chemical and bacteriological examination of the London waters for the twelve months ended 31st December 1937.* Metropolitan Water Board 32nd Annual Report.

HERTFORDSHIRE COUNTY COUNCIL 1995. *Waterhall landfill site assessment of the potential to pollute water.* Report by Aspinwall & Co for Hertfordshire County Council.

JOSEPH, J. B. 1983. *The effects of applying pig slurry to land over an unconfined aquifer.* International Conference on Groundwater and Man, Sydney.

NATIONAL RIVERS AUTHORITY 1994*a*. Internal report.

NATIONAL RIVERS AUTHORITY 1994*b*. Internal report.

WITHERS, P. 1985. Agricultural Development and Advisory Service, personal communication.

Pollution resulting from the abandonment and subsequent flooding of Wheal Jane Mine in Cornwall, UK

G. G. BOWEN[1], C. DUSSEK[1] & R. M. HAMILTON[2]

[1]*Marcus Hodges Environment, 14 Cathedral Close, Exeter, Devon EX1 1HA, UK*
[2]*Environment Agency South Western, Manley House, Kestrel Way, Exeter, Devon EX2 7LQ, UK*

Abstract. The closure of Wheal Jane Mine in Cornwall and the withdrawal of dewatering pumps resulted in the flooding of mine workings extending over a large area. The mine drainage was historically very acidic and high in metals. In order to predict the impact of the mine discharge, the National Rivers Authority commissioned a programme of investigations and monitoring. This paper describes the monitoring and assessment work undertaken at Wheal Jane during mine flooding in order to predict the discharge, location, time, quantity and quality, and the potential impact on groundwater sources used for potable supply. It also describes the subsequent mine water discharge and actual monitored levels of flow and water quality.

Investigations and monitoring of the flooding mine system included a detailed mining survey, the recording of water levels at suitable shafts, the collection and analysis of mine water samples and water quality depth profile surveys in a number of shafts. Surveys and monitoring of groundwater sources were also undertaken. Measurement of water levels and the mining survey allowed accurate prediction of the location and time of discharge. An estimate of the mine water discharge of between 5000 m^3 day^{-1} and 20 000 m^3 day^{-1} was prepared from catchment water balance calculations. This compares with the seasonally fluctuating, actual discharge of between 5000 m^3 day^{-1} and 40 000 m^3 day^{-1}. Mine water quality samples and water quality depth profiles taken from different shafts across the mine system identified large variations; in general, becoming more acidic with higher concentrations of metals with depth. Following flooding, the initial discharge quality was within the range predicted, with a pH of 2.8 and total metals of approximately 5000 mg l^{-1}, notably consisting of iron, zinc and cadmium. Surveys identified a number of groundwater wells and boreholes close to the mine system which required regular monitoring. None were identified to have been affected during the mine flooding or following discharge.

Wheal Jane is the collective name for a group of interconnected metalliferous mines situated in the Carnon Valley, near Truro in Cornwall (Fig. 1). Wheal Jane Mine itself was developed in the late 1960s using modern techniques to exploit the rich mineral lodes previously worked by shallow mines in the 17th, 18th and 19th centuries. In the 1970s, the mine extended beneath the Carnon Valley to connect with another working mine, Mount Wellington, and exploratory workings further to the west, connected into a large group of abandoned workings known as United Mines.

The workings associated with Wheal Jane are extensive, reaching to a depth of 450 m and for several kilometres laterally along a number of main mineral lodes. The mine was renowned for being very wet and substantial dewatering at up to 60 000 m^3 day^{-1} was required in winter months. The pumped water was typically very acidic, with high concentrations of dissolved metals resulting from the sulphide mineral deposits. Approximately half of the pumped water was treated by the mining company prior to discharge to the Carnon River.

In March 1991, Wheal Jane was closed, and shortly afterwards, mine dewatering ceased and the pumps were removed. The National Rivers Authority (NRA) identified the potential for a substantial impact on the water environment and immediately commissioned an integrated investigation to characterize the mine water discharge and assess the probable impact.

Environmental setting

Wheal Jane was the last of more than 50 mines to operate in the Gwennap Mining District, an area with a rich mining history dating back to at least the 17th century (Dines 1956). These mines were developed to exploit minerals associated with quartz porphyry dykes intruded into the Killas mudstones, during a period of regional metamorphism associated with the emplacement of the Cornish granite batholiths. The main mineral lodes exploited by Wheal Jane trend approximately SW–NE and dip at approximately 45° to the northwest. They typically include the following minerals: cassiterite

BOWEN, G. G., DUSSEK, C. & HAMILTON, R. M. 1998. Pollution resulting from the abandonment and subsequent flooding of Wheal Jane Mine in Cornwall, UK. *In*: MATHER, J., BANKS, D., DUMPLETON, S. & FERMOR, M. (eds) *Groundwater Contaminants and their Migration*. Geological Society, London, Special Publications, **128**, 93–99.

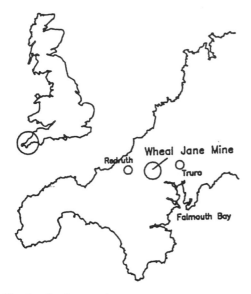

Fig. 1. Wheal Jane Mine location.

(tin), chalcopyrite (copper), pyrite (iron), wolframite (tungsten) and arsenopyrite (arsenic). Silver, galena (lead) and a number of alteration minerals also occur in lesser amounts.

The Killas rocks in the vicinity of Wheal Jane have low primary porosity. However, fractures,

faults (cross courses) and weathered zones provide storage and allow limited groundwater flow. In general, the groundwater table away from the mined areas forms a subdued replica of the surface topography with locally steep hydraulic gradients. These rocks provide small yields and support a number of wells, boreholes and springs which supply cottages and small farms (British Geological Survey 1990) They are classified as a minor aquifer by the NRA.

In mined areas the low permeability of the Killas rocks contrasts with the high conduit flow permeability engineered by mine workings. The mine workings therefore provide a preferential flow route for groundwater. Dewatering of Wheal Jane occurred to 450 m depth, resulting in a cone of water-table depression. The precise extent of the cone has never been identified; however, the low permeability of the Killas rocks would have resulted in steep hydraulic gradients towards the dewatered mine workings, thus restricting its extent.

The mine workings underlie the Carnon Valley, extending at shallow depths beneath the Carnon River (Fig. 2). The river and a number of its tributaries are historically of poor water quality, with high metal concentrations resulting from discharges associated with previously abandoned mines and mine spoil.

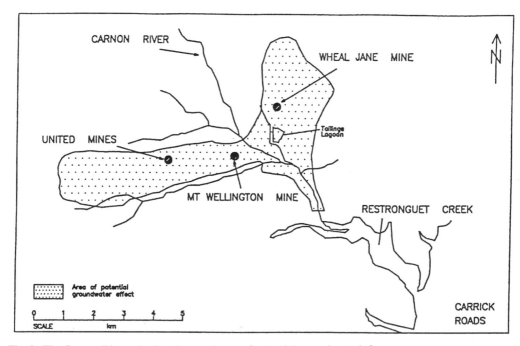

Fig. 2. The Carnon River mine locations and area of potential groundwater influence.

The Carnon River flows into Restronguet Creek, approximately 3 km downstream of Wheal Jane, and subsequently into Carrick Roads estuary. These tidal waters have shellfish and amenity value.

Mine flooding

Closure of the mine in March 1991 was effected rapidly, the mine dewatering ceased and the mine was stripped of equipment including all pumps. The extensive mine void began to fill, and a large surface area of mineral deposits exposed by mining was now exposed to groundwater leaching.

Investigations and monitoring

The NRA was concerned that once the mine void was flooded a significant discharge of poor quality water would drain from the mine, impacting upon the Carnon River and tidal waters in Restronguet Creek and the Carrick Roads. It was also concerned that flooding mine waters would pollute groundwater sources close to Wheal Jane. They commissioned a programme of investigations and monitoring which included those described below.

Mine system and mine waters
- A detailed mining survey to identify all mine workings, interconnections, adits, mineral variations, underground structures, etc., relevant to the future drainage of the flooded mine.
- A water level survey of mine shafts throughout the Wheal Jane mine system and adjacent workings to identify hydraulic connections and appropriate monitoring points.
- Measurement of the rise of mine water levels in six appropriate shafts across the mine system.
- Collection of mine water samples from varying depths (bottom, middle and top waters), in three main shafts across the mine system.
- Continuous depth profile measurement of water quality in top waters at the three main shafts.

Groundwater sources
- Survey and measurement of groundwater levels in boreholes, wells and shafts (not interconnected with Wheal Jane) to identify the zone of groundwater depression from mine dewatering.
- Survey of private and licensed groundwater sources of supply.

- Baseline sampling of water supplies at risk from rizing mine waters.

Predictions of mine water discharge

The information obtained from investigations and monitoring was regularly reviewed during the period of mine flooding and the programme modified as appropriate. The main objective of the data collection was to enable the following assessments to be made to predict the mine water discharge:

- location and time
- quantity and water quality
- impact on private groundwater sources

Assessment of the mine water discharge location and time Monitoring of water levels across the mine system showed that the mine workings were hydraulically well connected, flooding at the same rate. The possible mine system decant points were identified from the mining survey and are shown in Fig. 3. The lowest decant is through Jane's Adit estimated at a level of 14–15 m AOD. However, the main mineral lode had been worked to within a few metres of the bed of the River Carnon at an elevation of 15–16 m AOD and there remained a potential for direct discharge to the river by diffuse seepages. The next decant point at Nangiles Mine was estimated to be at 16—17 m AOD elevation. A graph of the mine water rise at one of the main shafts monitored is shown in Fig. 4.

Early recovery was rapid but soon slowed and became irregular, presumably due to the varied filling time for voids at different levels. The flooding rate was considered to be related to the volume of workings and recharge from groundwater. As the mine flooded, the hydraulic gradient into the mine reduced and, therefore, the rate of filling from groundwater decreased. It was also known that at Wheal Jane, the volume of interconnected workings increased higher in the mine due to the number of extensive shallow old mines. Towards the end of mine flooding the recovery was slow, as expected, and discharge occurred on 17 November 1991 from Jane's Adit into the Carnon River.

Assessment of the mine water discharge quantity and quality Investigation and monitoring of the mine system identified a highly complex hydraulic system (Dussek 1992). Accurate prediction of mine discharge flow and water quality was not possible. However, a best estimate was required to allow prediction of the potential

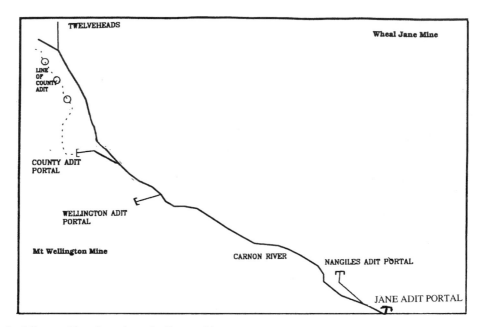

Fig. 3. Adit portal locations along the Carnon River.

impact on surface waters and also to enable appropriate treatment works to be designed and constructed.

Wheal Jane is interconnected with a large number of unsurveyed 18th century workings (Hamilton Jenkin 1963) and it was not possible to predict the discharge flow from rate of rise and workings volume calculations. Therefore, a preliminary water balance was undertaken projecting the known workings to the surface to define a recharge catchment.

Water balance calculations using long-term average recharge data predicted a mine water discharge of between $5000 \, m^3 \, day^{-1}$ and

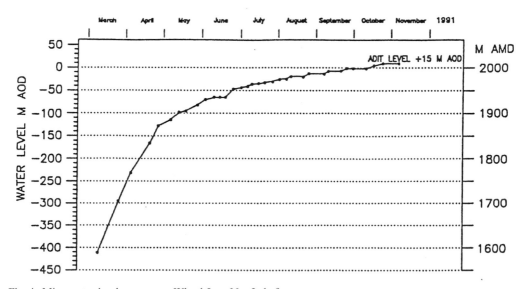

Fig. 4. Mine water level recovery – Wheal Jane No. 2 shaft.

Table 1. *Water quality in Wheal Jane No. 2 shaft when mine water was 45 m below adit level*

Sample depth (m bwl)	pH	Sp. Elec. Cond. ($\mu s\,cm^{-1}$)	Cadmium ($mg\,l^{-1}$)	Iron ($mg\,l^{-1}$)	Zinc ($mg\,l^{-1}$)	Copper ($mg\,l^{-1}$)
Shallow (10 m)	3.7	1640	0.13	94	98	4.5
Middle (90 m)	2.6	7390	2.6	1846	1379	40
Deep (180 m)	2.5	8440	4.8	2162	1541	44

$20\,000\,m^3\,day^{-1}$. The actual discharge since flooding has been recorded is between $5000\,m^3$ day^{-1} and $40\,000\,m^3\,day^{-1}$. Surface samples collected regularly from the rising mine waters in six shafts across the mine system identified widely fluctuating levels of acidity and dissolved metal concentrations. These were considered to be due both to variations in mineral deposits and exposure, and to complex patterns of mine water flow.

To provide a better understanding of the variation in mine water quality, regular depth sampling and depth profile monitoring was also undertaken. Samples were taken using a flow-through bailer to a depth of up to 180 m below water level (bwl) at three main shafts across the mine system. A multi-probe water quality meter modified to allow monitoring to depths of up to 10 m was also used to identify water quality variations near surface. Table 1 identifies selected parameters indicative of the water quality at Wheal Jane No. 2 shaft on 5 July 1991 when mine waters were 45 m below adit level.

In general, water quality was found to deteriorate with depth within Wheal Jane, with samples from depth being very acidic with high concentrations of dissolved metals. Better water quality was observed close to the surface and waters sampled in United Mines were of much better quality. Mine water circulation was considered to be complex, possibly controlled by a combination of recharge, temperature gradient, density variation, and hydraulic mechanisms.

Accurate prediction of the discharge water quality was not possible, although a likely range was identified. The actual quality of the discharge was better than the worst water quality monitored, but it was still very acidic with high concentrations of dissolved metals.

Assessment of the impact of mine water discharge on private groundwater sources During mine flooding, groundwater drained into the dewatered zone, reducing the lateral extent of the zone. The groundwater gradient into the mine prevented sources of supply outside the dewatered area from becoming contaminated. However, on the boundary of this area, as groundwater recovered,

it was possible that sources of supply could become affected. Groundwater sources identified to be within this boundary area were sampled and hydrogeological assessments undertaken to identify any impact by rising contaminated mine waters. None of the sources were considered to be directly at risk. However, seven sources were also identified adjacent to the Carnon River downstream of the discharge point. These were all shallow wells supported by alluvial deposits and therefore potentially at risk from a deterioration in river water quality. Following mine water discharge, these supplies were regularly monitored by Carrick District Council Environmental Health Department. Fortunately, none were identified as being contaminated by the river.

Mine water discharge

On 17 November 1991, the mine system flooded to 14.5 m AOD and mine water decanted through Jane's Adit into the Carnon River. A treatment lagoon constructed by the mine owners was rapidly overwhelmed by a flow of approximately $5000\,m^3\,day^{-1}$ of mine water with pH 2.8 and total dissolved metals of approximately $5000\,mg\,l^{-1}$.

The quality and quantity of the discharge was similar to that predicted. Back-up contingency treatment measures, funded by the NRA, were brought into operation immediately. These involved adding lime into the head of Jane's Adit to lower the acidity and precipitate metals, and pumping of mine water from behind a constructed plug near the adit portal into the Wheal Jane Tailings Dam.

In late December 1991, operational problems with the tailings dam halted pumping from the adit, which resulted in a gradual backup of water within the mine system. The mine water level increased by approximately 4 m resulting in seepages through the bed of the Carnon River and forcing a blockage within the second lowest known decant point, Nangiles Adit, to clear. Between $25\,000\,m^3$ and $50\,000\,m^3$ of poor quality mine water was suddenly released over

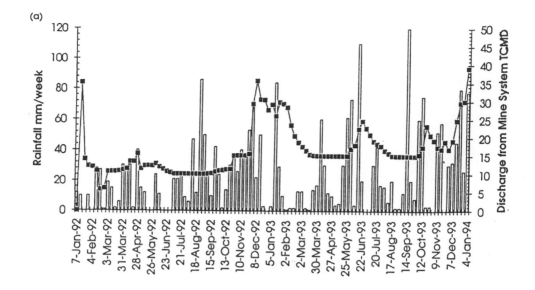

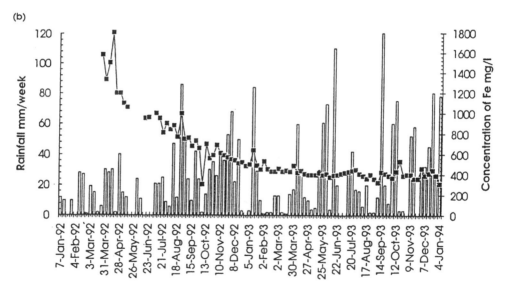

Fig. 5. (a) Mine water discharge flow and rainfall January 1992 to January 1994. (b) Mine water discharge quality monitored at No. 2 Shaft and rainfall January 1992 to January 1994.

a period of 24 h. The resultant pollution extended down the Carnon River into Restronguet Creek, the Carrick Roads and the ochre plume of contamination was visible into Falmouth Bay.

Following this incident, additional treatment measures were implemented. Pumps were installed into one of the main shafts at Wheal Jane and the treatment process control was optimized to improve metal precipitation within the tailings dam.

Intensive surface water and tidal water monitoring and assessment work was undertaken following the uncontrolled mine water discharge. In addition, monitoring of the mine system and mine water discharge was continued. This information was used to construct simple models to predict future mine water discharge quality and quantity.

Figure 5 graphically shows the discharge flow and water quality (Fe concentration only) compared with rainfall (bar chart) between January 1992 and January 1994. The flow is closely related to rainfall events although a lag exists of up to one month between a rainfall event and corresponding peak flow. The quality of the discharge in terms of metal concentrations has greatly improved, but the levels are still substantial and the water remains very acidic.

Long-term mine water discharge

The monitoring and investigations undertaken during the flooding of the mine and since the mine waters overflowed have provided information essential in mitigating the short-term pollution of the aquatic environment. However, the pollution of groundwater from metalliferous mining at Wheal Jane will continue for many tens of years or even centuries. Mines abandoned for more than a century close to Wheal Jane are still discharging acidic groundwaters, high in metals concentrations which impact upon streams and rivers.

The majority of drainage from Wheal Jane is currently being treated by an expensive short-term operation. Investigations are under way to identify a long-term, low-cost treatment system that will mitigate the pollution. In order to design such a treatment system effectively it is essential to characterize fully the future mine water discharge flow and water quality. The investigations and monitoring undertaken to date have provided a good understanding of the characteristics, and this will be developed by further monitoring over the next few years.

Conclusions

Metaliferous mining at Wheal Jane resulted in an extensive void, exposing metaliferous minerals to oxidation and weathering. Following the cessation of dewatering and with a knowledge of the mine system, it was possible to monitor the mine water rebound rate and water quality to accurately predict the location, quantity and quality of the discharge.

The eventual discharge required a substantial treatment operation involving co-operation between the National Rivers Authority and the mining company. This treatment operation is continuing and will need to be continued for many tens of years into the future if the water quality and aquatic environment of the region is to be protected.

References

BRITISH GEOLOGICAL SURVEY 1990. *The Carnmenellis Granite – Hydrogeological, Hydrochemical and Geothermal Characteristics*. British Geological Survey, Keyworth.

DINES, H. G. 1956. *The Metalifererous Mining Region of South-West England*. 2 vols, Memoirs of the Geological Survey of Great Britain. HMSO, London.

DUSSEK, C. 1992. *The Hydrogeology of Wheal Jane Mine*. MSc Thesis, University of Birmingham.

HAMILTON JENKIN, A. K., 1963. *Mines and Miners of Cornwall. VI Around Gwennap*. Truro Bookshop, Truro.

Minewater remediation at a French zinc mine: sources of acid mine drainage and contaminant flushing characteristics

PIERS J. K. SADLER

Steffen, Robertson and Kirsten (UK) Ltd, Summit House, 9/10 Windsor Place, Cardiff CF1 3BX, UK.

Abstract: Recent closure of a zinc mine in southwestern France has resulted in flooding of workings and discharge of contaminated minewater from the workings, polluting a small stream. The minewater discharge is of near neutral pH and is contaminated with high levels of zinc, cadmium, manganese, iron and sulphate. With the exception of iron, concentrations of these contaminants have fallen dramatically since the discharge arose.

The main source of recharge to the mine is likely to be overlying aquifers associated with near-surface weathering. A classification is proposed to explain the hydrogeochemical sources of contaminants and the changes in discharge water quality with time. Primary sources related to active sulphide oxidation dominate the minewater chemistry in the unflooded workings, whilst secondary sources related to remobilization of secondary minerals dominate in the flooded workings. Prediction of contaminant flushing rates has been carried out on the basis of this classification. Zinc and cadmium levels are expected to reach stable medium-term levels within 2–5 years of the discharge first occurring.

Potential remediation strategies include plugging of the discharging mine portal and flooding of upper workings, coupled with conventional chemical and wetland treatment.

This paper concerns a recently closed zinc mining complex in southwestern France. The site details are not disclosed for reasons of confidentiality, but the hydrogeochemical concepts illustrated, the classification of minewater sources and the remediation options discussed are widely applicable.

The complex involves two mines which will be referred to as the West Mine and the East Mine. There are also a number of waste rock dumps and a tailings impoundment. These areas are shown schematically in Fig. 1. Following closure of both mines in December 1993 and cessation of pumping of water, flooding of the abandoned workings began.

Minewater recovery in the East Mine was complete by July 1994 when minewater began to discharge at the surface. Minewater recovery in the West Mine is continuing. This paper is concerned with the discharge from the East Mine. The minewater has been contaminated by acid mine drainage (AMD) processes resulting from sulphide mineral oxidation, and is characterized by high concentrations of zinc, manganese, cadmium, iron and sulphate, with near neutral pH (initial acid production being buffered by reaction with neutralizing minerals). The discharge has caused severe pollution of a small stream (Fig. 1).

An investigation was carried out to identify appropriate remediation strategies based on the prognosis of minewater discharge quality. The approach was to interpret water level, pumping and discharge information and water quality data, in the context of the known geological environment, to develop a conceptual model of the hydrogeological and hydrogeochemical mechanisms of contaminated discharge development. Prediction of future water quality and the response of the system to potential remediation options was based on the conceptual model and a classification of AMD source types.

Sources of information

The main sources of information on which this paper is based are as follows:

(a) background information on the geology, mineralogy, mining history and mine geometry provided by the mine operator;

(b) minewater levels as monitored in several parts of the mine during recovery;

(c) minewater quality data obtained during minewater recovery and following discharge;

(d) average monthly minewater pumping rates during mine operation;

(e) minewater discharge rates following recovery.

Minewater samples were taken regularly by the mine operator between January 1994 and January 1995 and analysed for total zinc, iron, manganese, cadmium, lead, nickel, copper and cobalt in the mine laboratory by graphite furnace flame atomic adsorption spectrometry (GFAAS). Measurements of pH, electrical conductivity (EC) and temperature were taken using portable HANNA meters. Occasionally

SADLER, P. J. K. 1998. Minewater remediation at a French zinc mine: sources of acid mine drainage and contaminant flushing characteristics. *In*: MATHER, J., BANKS, D., DUMPLETON, S. & FERMOR, M. (eds) *Groundwater Contaminants and their Migration*. Geological Society, London, Special Publications, **128**, 101–120.

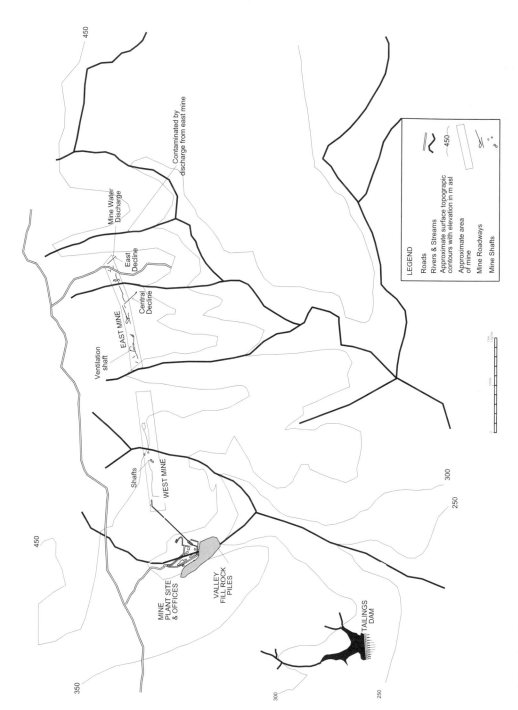

Fig. 1. Schematic location plan.

samples were sent off site for analysis for a wider suite including the major ions: sulphate, chloride, bicarbonate, sodium, potassium, calcium, magnesium and silicate.

In February 1995 the sampling scheme was revised to include laboratory determinations of aluminium, arsenic, cadmium and lead by GFAAS; calcium, iron, magnesium, nickel, sodium, zinc and manganese by flame atomic adsorption spectrometry; silica, ammonium and sulphide by colorimetric analysis; and chloride, nitrate and sulphate by ion chromatography. Samples were filtered at $45\,\mu m$ on site at the time of sampling and preserved with a few drops of concentrated nitric acid for determination of dissolved and total metals. Site measurements of pH, EC, dissolved oxygen and temperature were taken at the time of sampling, again using HANNA meters. Alkalinity was determined by titration in the laboratory immediately after sampling, following confirmation that alkalinity did not change appreciably between sampling and returning to the laboratory.

Samples were taken from the mine discharge and various other parts of the mine accessed by pipes submerged in the flooded descending mine roadways or 'declines'.

Physiography and hydrology

The mine is located in southwestern France in the Department of Tarn, and is aligned WSW–ENE on the southern margin of an area of granitic high ground. The main geographic features are shown in Fig. 1. The area is one of variable topography between approximately 600 m above sea level (asl) 1 km to the north of the mine, to approximately 220 masl 5 km to the south. This southward-sloping country is incised by steep-sided stream valleys aligned approximately N–S. The minewater discharge from the East Mine enters the stream that passes the eastern end of the complex. The streams are tributaries of a small trout river which passes approximately 1 km to the south of the East Mine.

The mean annual rainfall and effective rainfall for the site are approximately 1200 mm and 250 mm respectively. In general, the rainfall in the period June to September is markedly lower than in the remainder of the year.

Geology

The mine workings are located within a sulphide orebody aligned along a regional fault zone, along which movement was originally horizontal and at a later stage vertical. The zone of disturbance in which the mineralization occurs is approximately 30 m in width, and dips at approximately 80° to the south.

The fault zone lies between 150 m and 200 m to the south of a granitic intrusion and follows the margin of the granite. The surrounding country rock consists of Cambrian schists which vary in lithology from arenaceous to argillaceous black schists, with minor calcareous schists and quartzites. The fault zone truncates the metamorphic aureole of the granite with thermal metamorphism only to the north of the structure, suggesting that movement occurred along the fault after granite emplacement. The emplacement of the granite occurred during the latter part of the Hercynian Orogeny in the Carboniferous.

The rocks in the fault zone are brecciated and silicified. The ore occurs as a massive sulphide emplaced by hydrothermal activity along the structure, sphalerite (ZnS) being the main ore mineral. The mineralization is associated with aplite intrusion. Two parallel mineralized veins occur. One is adjacent to the metamorphic aureole and one is adjacent to the weaker schists. Hydrothermal activity occurred in a single phase along planes of enhanced palaeo-permeability on the margins of the disturbed zone, in the later stages of granite emplacement. A schematic N–S cross-section is shown in Fig. 2.

The average zinc content of the ore was approximately 11.7% by weight (17.5% sphalerite) and the cut-off grade for economic ore was 7% by weight. The mean silver and germanium grades were $100\,g\,ton^{-1}$ and $160\,g\,ton^{-1}$ respectively. Siderite ($FeCO_3$) containing 3–5% manganese substituting for iron and pyrite (FeS_2) were the main gangue minerals present in the ore. Disseminated pyrite is also present in the black schists, which host much of the orebody.

Mining history and geometry

Exploitation of the orebody in the East Mine began in 1981. A simplified vertical section through the workings is shown in Fig. 3. The section represents workings over a width of approximately 30 m perpendicular to the trend of the ore zone.

The East Mine was worked from the surface via two declines (Central and East), as shown in Fig. 3. The workings are also linked to the surface via a ventilation shaft.

The elevation of the workings is between 275 masl and 440 masl. The elevation of the mine accesses are as follows:

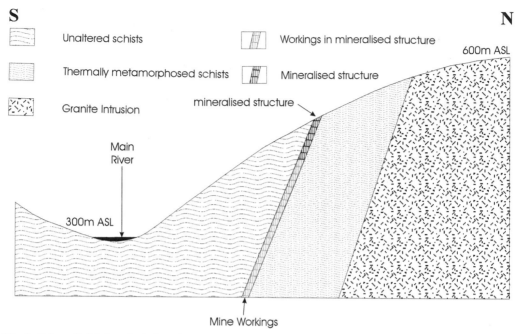

Fig. 2. Schematic N–S geological section.

Central Decline	465 masl
East Decline	430 masl
Ventilation shaft	515 masl

Both the declines are located on the edge of valleys which cross the ore bearing structure.

The ore was exploited by cut-and-fill methods, with waste rock generally used as the fill although some concrete was used in parts of the mine. Following closure of the mine the declines were backfilled with limestone and waste rock as shown in Fig. 3. The limestone was intended to encourage precipitation of metal carbonates.

Hydrogeology

Detailed analysis of the hydrogeology has been outside the scope of this project. The mine is clearly the major flow path for groundwater in the area, with background hydraulic conductivities several orders of magnitude less than in the mine. Background hydrogeological information has been obtained from observations in the mine and the surrounding area, some water level and pressure information from exploratory boreholes and from minewater pumping and discharge data. The focus has been identification of sources of water, as this will determine the

response of the discharge to changes in the system associated with remediation.

Groundwater occurrence and flow

The schists are likely to be of low permeability with minor flows occurring in joints and fractures. Significant permeability will only be developed where regional faults have created zones of disturbance, such as along the line of the mineralized structure.

Observations in the mine suggest that the metamorphosed schists are of very low permeability, since there is no visible evidence of water flowing in these strata. By contrast, the granite to the north has been the source of spring and well water supplies and is expected to be of relatively high permeability. Significant hydraulic connection between the granite and the structure can only be expected where there is connection via faults. Two such faults run along the valley which forms the boundary between the West and East mines.

The overall regional groundwater gradient is likely to follow the regional topographic gradient southwards. Locally the direction of flow will be controlled by the relative position of recharge and discharge areas and by the lithology and

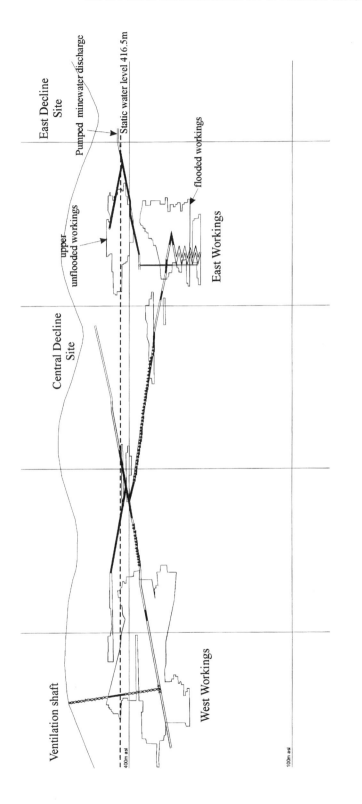

Fig. 3. Simplified vertical section through the East Mine.

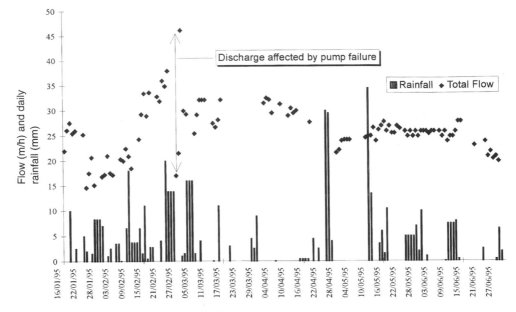

Fig. 4. Pumped minewater discharge and rainfall.

geometry of the aquifers. Consequently local flow directions are likely to be sub-parallel to the main structure from interfluves to streams and springs.

Impact of mine workings

The backfilled mine workings are likely to act as a drain, providing preferential flow paths for groundwater. The minewater discharge from the East Decline occurs at an elevation of 416.5 masl. This controls the minewater level throughout the East Mine to approximately this elevation as the permeability of the backfilled workings is very high. Groundwater gradients would be expected towards the workings except at the western end where there may be flow towards the West Mine, although this has not been confirmed.

Minewater pumping, recovery and discharge

Pumping records show that during mine operation a monthly average of between $17\,m^3\,h^{-1}$ and $42\,m^3\,h^{-1}$ was pumped from the mine between June 1992 and December 1993.

Following closure of the mine in December 1993 the water level began to rise. The water level stabilized below the lip of the East Decline at an elevation of approximately 415 masl due

to leakage through weathered strata, to the adjacent stream. The leakage area was sealed with concrete and a pumping station established to control the water level at 416.5 masl. The pumped water is discharged to the stream and the pumping rate monitored.

The total amount of water pumped from the mine from January to June 1995 is plotted in Fig. 4. The plot includes daily rainfall as measured at the mine site for comparison. The plot shows that the rate of flow to the mine when flooded varied between $15\,m^3\,h^{-1}$ and $38\,m^3\,h^{-1}$.

Source of minewater

The flow rate and variation shown in Fig. 4 are similar to the amount of water pumped when the mine was operational. This suggests that the flow rate to the mine is independent of minewater level, which implies that the main source of the water was from above, as, in order to maintain constant flow rates, the hydraulic gradient towards the mine would have had to remain constant with the rise in water in the mine. This could occur if the deep workings were associated with very low permeability strata and significant groundwater flow was limited to more permeable strata at relatively shallow depths. Since the main permeability development is likely to be associated with fracturing, and fracture permeability is commonly very much reduced at depths of greater than 60–80 m below the surface, this explanation would appear highly plausible.

Analysis of the data has provided an indication of the amount of baseflow in the discharge. In this context, baseflow is used to describe flow that varies according to changes in regional groundwater head, and is not affected rapidly by rainfall events. The data suggest that approximately 60% of the discharge is derived from baseflow. The main source of water to the mine is therefore likely to be a relatively shallow aquifer intercepting the upper workings or permeable regions above the workings. Direct recharge to the mine workings is, however, likely to be a significant additional source.

For equilibrium conditions to exist, the inflow to the aquifer, through regional recharge, must be equal to the outflow either through mine discharge, or through direct discharge to surface waters. The surface catchment of the mine has been estimated from a 1:10 000 scale topographical map using an electronic planimeter, and is approximately 0.99 km². Assuming that 100% of effective rainfall recharges the aquifer, the total recharge in the catchment area is equivalent to a flow of approximately 28 m³ h⁻¹. This is probably greater than the mean discharge rate, but it is likely that there is some spring flow to surface water courses within the mine catchment. The preliminary conclusion is therefore that there are no major hydraulic sinks other than the mine discharge, and that there is no other significant groundwater outflow from the workings.

Hydrogeochemistry

Groundwater quality

Groundwater samples were taken from a number of sites, including underground and surface exploratory boreholes and springs, in February 1994. These samples are taken to be representative of baseline groundwater quality.

The results show that background levels of metals in the groundwater are relatively high, with zinc concentrations up to $3.55 \, \mathrm{mg \, l^{-1}}$, iron up to $3.8 \, \mathrm{mg \, l^{-1}}$ and cadmium up to $40 \, \mu\mathrm{g \, l^{-1}}$. Significant variations have been observed below these levels. Sulphate concentrations are relatively consistent at approximately $100 \, \mathrm{mg \, l^{-1}}$. These results should be viewed with caution, however, as they represent some waters which could have been affected by mining.

Minewater quality

Minewater samples have been taken from a number of sites in the East Mine, at inflow points and in flowing channels above the flooding level, at rising minewater surfaces, from pipes emplaced below the flooded level and from the discharge to surface at the East Decline. The following discussion focuses on quality of water sampled from the free water surface in the East Decline prior to and after discharge to the surface in July 1994, since this discharge is the only significant outflow to the environment.

Typical results of minewater discharge analyses are presented in Table 1. The table also presents the percentage error in the ionic balance of each sample. The error range is approximately 3–30%. Although this puts the accuracy of the data in doubt, it is clear that the concentrations given, particularly for metals (analysed independently at two laboratories), and the trends observed are broadly correct. Waters of this type present a complex chemical matrix for analysis and errors are common, particularly where laboratories are not accustomed to analysing waters with such high levels of dissolved metals (Morin 1990).

The concentrations of total metals are plotted on a logarithmic scale versus time (April 1994 to

Table 1. *Typical minewater discharge chemistry*

Sample date	pH	SO_4^{2-}	HCO_3^-	Cl^-	SiO_2	Ca^{2+}	Na^+	Mg^{2+}	Zn^{2+} (total)	Fe (total)	Mn	Cd^{2+}	Ionic balance error (%)
11 Aug. 1994	–	2229	163	9	16	453	18	68	262	0.07	29	1.06	20
19 Aug. 1994	6.6	1602	149	7	17	416	17	58	219	0.09	28	1.16	2.8
1 Dec. 1995	6.7	967	243	–	12	485	17	33	82	7.8	19	0.29	11
3 Apr. 1995	6.8	915	151	11	10	112	15	27	65	8.5	18	0.34	30
26 June 1995	6.6	487	140	9	14	200	13	21	36	2.9	14	0.21	6

*All values are quoted in milligrams per litre with the exception of pH.

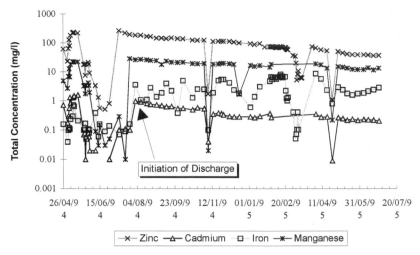

Fig. 5. Total metals concentration in the minewater discharge.

July 1995) in Fig. 5. The data presented in Table 1 and Fig. 5 show that the minewater was highly contaminated with metals and sulphate (2229 mg l^{-1} initially) when the discharge first occurred, but that water quality rapidly improved with time. The pH of the rising minewater and discharge waters has remained near constant, generally in the range 6.8–7.0, with no trends observed with time. In general, levels of dissolved oxygen were low, although sampling procedures often produced erroneous results initially. Typical dissolved oxygen results show approximately 10–20% saturation. Occasional Eh measurements suggest that a value of −50 mV is approximately representative. The degree of variation is not known.

The waters are consistently rich in sulphate, although levels have fallen from over 2000 mg l^{-1} to less than 500 mg l^{-1}. The main cations are zinc and calcium, both of which have shown significant decreases with time. Calcium levels were originally between 400 and 500 mg l^{-1}, but have fallen to between 100 and 200 mg l^{-1}. Sodium and chloride levels are comparatively low, with sodium concentrations between 10 and 20 mg l^{-1} and chloride concentrations between 7 and 11 mg l^{-1}. Bicarbonate ion concentrations have generally been in the region of 150 mg l^{-1}. All the major ions, with the exceptions of chloride and bicarbonate, have shown a decreasing trend in concentration with time.

During the monitoring period the pattern of variation of zinc was followed by manganese, cadmium, nickel and cobalt. The general form of the trend is exponential, plotting approximately

as a straight line on the semi-logarithmic graph. Before the initial discharge occurrence the zinc level fell from a maximum of approximately 230 mg l^{-1} in May 1994 to less than 1 mg l^{-1} in June 1994. The lower value is likely to be a result of recharge water collecting above the minewater body in the decline as the water level was rising. Following commencement of the discharge and stabilization of flows in July 1994, the zinc concentration fell from 262 to 35 mg l^{-1} in June 1995.

The iron concentration did not conform to this pattern, but fluctuated at levels generally between 1 and 10 mg l^{-1}.

Surface water quality

Samples were taken from the receiving water course on several occasions during 1994. Samples taken from the receiving water course downstream of the discharge entry point show a marked deterioration in water quality compared to upstream samples. Downstream zinc concentrations ranged between 14 and 41 mg l^{-1}, with cadmium at 0.07 to 0.16 mg l^{-1} and sulphate at 300–800 mg l^{-1}. There was not a major increase in iron level except for immediately downstream of the discharge in December 1994 when a level of 0.39 mg l^{-1} was recorded.

AMD source characterization

It is not the purpose of this paper to present a review of AMD processes, but a broad understanding of these processes and the controls on solubility of certain metals is essential in characterizing the source.

To aid interpretation and prediction, sources of AMD have been classified into primary and secondary sources (Connelly *et al.* 1994), relating to active oxidation of sulphide minerals and dissolution of secondary minerals respectively. A third very important source of contaminants is dissolution of buffering minerals such as carbonates and aluminium silicates.

Primary sources Primary AMD sources occur from active oxidation of sulphide minerals. Sphalerite was the main economic sulphide present in the ore zone and the primary oxidation reaction can be represented as

$$ZnS + 2O_2 \rightarrow Zn^{2+} + SO_4^{2-}. \quad (1)$$

This reaction is not acid-producing, the products being zinc and sulphate in solution. Cadmium is commonly contained as a trace element in sphalerite in solid solution with zinc, and therefore this reaction is also likely to release trace amounts of cadmium (Bowell *et al.* 1996).

Pyrite is also present as a primary sulphide although information on the concentration is not available. The oxidation of pyrite is generally bacterially mediated by species such as *Thiobacillus ferrooxidans* and is net acid-producing (Nordstrom 1982):

$$2FeS_2 + 2H_2O + 7O_2 \rightarrow 2Fe^{2+} + 4SO_4^{2-} + 4H^+. \quad (2)$$

It is likely that the proton acidity produced by this reaction is also a driving force for sphalerite oxidation, since bacterial activity is enhanced under more acid conditions (Robertson & Barton-Bridges 1990).

Sulphide oxidation can also occur in the presence of ferric iron as follows (McKibben & Barnes 1986):

$$FeS_2 + 14Fe^{3+} + 8H_2O \rightarrow 15Fe^{2+} + 2SO_4^{2-}$$
$$+ 16H^+ \quad (3)$$

This reaction is rapid under acidic conditions, but likely to be of little importance at higher pH, as ferric iron, if present, rapidly hydrolyses to ferric hydroxide.

Primary sources occurring in oxygenated systems such as unflooded workings are likely to result in temporally persistent contamination, due to the relatively slow rate of sulphate depletion, with high sulphate concentrations. Drainage water quality is likely to fluctuate according to flow regime. Oxygen entry to the unflooded workings is likely to occur by diffusion through the fractured bedrock and barometric pumping, the process by which differential pressures in the mine and atmosphere cause movement of air into the mined void (Robertson & Barton-Bridges 1990).

Secondary sources Secondary sources of contamination occur by dissolution of secondary minerals which are products of the primary AMD reactions described above.

Secondary mineral precipitation occurs when solutions associated with primary AMD become super-saturated with respect to certain minerals. The main processes which cause secondary mineral precipitation are likely to be concentration of solutions with oxidation products and changes in pH and Eh.

Concentration occurs on or close to surfaces of sulphide mineral grains, by evaporation, or when oxidation occurs in the presence of humid air. Changes in pH and Eh can occur by reaction of AMD with the solid phase, mixing with other waters and dissolution or exsolution of gases. Complex interactions between the above processes make identification of the dominant processes of secondary mineral precipitation very difficult, and it is likely that all the above processes operate to some degree.

The mineralogy of secondary minerals is partly dependent on which of these processes causes its formation. In general, concentration results in the formation of highly soluble sulphate salts. Where pyrite is the sulphide, hydrated ferrous sulphates such as melanterite ($FeSO_4.7H_2O$) may be precipitated on the surface of mineral grains, and over time, in the presence of air, these will oxidize to meta-stable iron-sulphate hydrate minerals such as, römerite ($Fe^{II}Fe_2^{III}(SO_4)_4.14H_2O$) and copiapite ($Fe^{II}Fe_4^{III}(SO_4)_6(OH)_2.20H_2O$) (Nordstrom 1982; Cravotta 1994). Removal of zinc from solution can occur by substitution into the structure of some of these salts; for example, zinc melanterite where zinc is substituted for iron, or by precipitation of equivalent zinc salts such as goslarite ($ZnSO_4.2H_2O$) (Nordstrom *et al.* 1990) and bianchite (($Zn, Fe)SO_4.6H_2O$).

In contrast, increases in pH and Eh generally result in precipitation of more stable minerals such as jarosites ($KFe_3(SO_4)_2(OH)_6$), and metal hydroxides, carbonates and silicates. Iron is most likely to be removed from solution by precipitation of jarosites at low pH, or ferric hydroxides such as ferrihydrite ($Fe(OH)_3$) at higher pH. Removal of zinc and cadmium from solution is more likely to be by precipitation of carbonates and silicates (Hem 1972), and manganese will only precipitate very slowly, probably as an oxide, at the prevalent pH levels (Hedin *et al.* 1994).

Reaction with certain neutralizing minerals such as siderite and rhodochrosite ($MnCO_3$) is likely to result in the release of further iron and manganese into solution. Where this occurs in close proximity to oxidizing sulphides and little water is present, more sulphate salts such as mallardite (manganese melanterite – $MnSO_4.7H_2O$) may form. At the pH values observed, even if alumino-silicates reacted with acid waters, the mobility of aluminium would be minimal.

For secondary minerals to become significant sources of AMD, as opposed to sinks for AMD, there must be a change in conditions. Flooding of the workings will change conditions in the following ways:

(a) highly soluble secondary minerals will come into contact with dilute waters;

(b) in general, waters will come into contact with more mineral surface for longer time periods, and the opportunity for buffering of AMD will be increased;

(c) oxygen will be prevented from entering the workings so pyrite oxidation will be limited by the solubility of oxygen in water, and Eh will be reduced.

On the basis of the above discussion, two types of secondary AMD source which could contribute to contamination of rising minewaters can be defined:

● *type 1*: highly soluble secondary minerals which dissolve rapidly into the rising minewaters;

● *type 2*: less soluble secondary minerals which either dissolve slowly due to kinetic controls or dissolve in small amounts due to thermodynamic controls.

Type 1 sources may be converted to type 2 sources if solubility of type 2 minerals is exceeded. An example of such a process is the conversion of iron sulphate hydrates to jarosites or ferrihydrite on contact with significant amounts of dilute waters (Nordstrom 1982).

Source identification Under the flooded conditions with low oxygen and near neutral pH, in most of the mine, primary sources of AMD related to sulphide oxidation are likely to be low. The main contaminant source is therefore likely to be secondary.

The concentrations of zinc, manganese and cadmium have decreased exponentially since the discharge first occurred. This demonstrates that the source of these metals is being depleted as recharge water flows through the system, and if continuing mineral dissolution is resulting in production of these metals in solution, this process is minor compared to the process which resulted in the initial high discharge concentrations. In contrast, the temporal persistence of iron suggests that there is a continuous source of iron to the water.

The implication is that the main sources of zinc, manganese and cadmium are of secondary type 1, whilst the main sources of iron are of secondary type 2.

Geochemical modelling with the USGS geochemical speciation code MINTEQA2 (Allison *et al.* 1991) was undertaken to evaluate the thermodynamics of the system and to aid identification of likely controlling mineral phases and sources of types 1 and 2. It should be noted that the model will identify the thermodynamic tendency for a mineral to precipitate or dissolve, but does not give information on kinetics, i.e. the rate at which the reaction will occur, or whether it will occur within the relevant time frame.

The model was run for various samples with a sweep of Eh values between $-200\,mV$ and $+50\,mV$. The pH was not varied because the sampling has demonstrated that there is little pH variation. Calculated saturation indices for selected minerals for the discharge sample taken on 26 June 1995 (as shown in Table 1) are presented in Table 2. The main observations are as follows:

(a) the discharge is close to saturation with respect to carbonates of calcium, iron, manganese, zinc and cadmium, particularly under higher redox conditions (Eh $-50\,mV$ to $+50\,mV$);

(b) the water is under-saturated with respect to all metal sulphates through the Eh range evaluated;

(c) sulphides of the main metals are thermodynamically stable at Eh $-150\,mV$ and $-200\,mV$;

(d) the water is super-saturated with quartz throughout the Eh range;

(e) the water is under-saturated with respect to ferrihydrite throughout the range, but becomes super-saturated with respect to goethite (FeOOH) between Eh $-150\,mV$ and Eh $-50\,mV$.

The preliminary indication is that sulphates could potentially act as type 1 secondary sources, and that goethite, ferrihydrite and the carbonates, siderite, rhodochrosite, smithsonite ($ZnCO_3$) and otavite ($CdCO_3$) could act as type 2 secondary sources, depending on the Eh of the waters. Since the source of zinc, manganese and cadmium is likely to be of type 1, it is

Table 2. *Calculated saturation indices of selected minerals for water sampled on 26 June 1995*

Mineral (stoichiometry)	Redox potential (Eh) of sweep (mV)			
	200	150	50	+50
Anhydrite ($CaSO_4$)	−4.601	−0.815	−0.744	−0.744
Calcite ($CaCO_3$)	−0.720	−0.817	−0.852	−0.852
Dolomite ($CaMg(CO_3)_2$)	−2.272	−2.458	−2.526	−2.526
Ferrihydrite ($Fe(OH)_2$)	−9.033	−5.464	−3.726	−1.967
Goethite (FeO_2H)	−5.041	−1.472	0.266	2.026
Gypsum ($CaSO_4.2H_2O$)	−4.279	−0.494	−0.422	−0.422
Hematite (Fe_2O_3)	−5.125	2.012	5.488	9.008
Na jarosite ($NaFe_3(SO_4)_2(OH)_6$)	−38.704	−20.244	−14.842	−9.562
Melanterite ($FeSO_4.7H_2O$)	−11.337	−4.767	−4.693	−4.693
Pyrite (FeS_2)	11.772	9.904	−14.566	−39.206
Quartz (SiO_2)	0.530	0.530	0.530	0.530
Siderite ($FeCO_3$)	−3.488	−0.801	−0.833	−0.833
Rhodochrosite ($MnCO_3$)	−0.097	−0.184	−0.217	−0.217
$MnSO_4$	−13.539	−9.743	−9.670	−9.670
Smithsonite ($ZnCO_3$)	−8.922	−2.602	−0.393	−0.393
$ZnCO_3.1H_2O$	−8.542	−2.222	−0.013	−0.013
$Zn(OH)_2$	−10.377	−4.056	−1.836	−1.836
Sphalerite (ZnS)	3.310	6.473	−5.291	−19.371
Willemite ($ZnSiO_3$)	−17.651	−5.008	−0.569	−0.569
Bianchite ($ZnSO_4.6H_2O$)	−17.037	−6.833	−4.518	−4.518
Goslarite ($ZnSO_4.7H_2O$)	−16.746	−6.543	−4.227	−4.227
Otavite ($CdCO_3$)	−9.221	−3.069	0.932	0.932
$CdSiO_3$	−16.185	−10.032	−6.021	−6.021
$CdSO_4.1H_2O$	−21.489	−11.454	−7.348	−7.348
Greenockite (CdS)	3.703	6.698	−3.275	−17.355
Sulphur	−1.400	−2.798	−13.262	−23.822

likely that the main sources are sulphates of these metals.

The processes controlling the concentration of iron in the discharge are clearly different. The iron concentration has remained steady, generally at between 1 and $10\,mg\,l^{-1}$ since the discharge first occurred. Fluctuations cannot be attributed to variations in pH, but there is a seasonal control, with high levels during summer and low levels during winter. This form of variation indicates that mineral dissolution is the most likely source of iron, because low flow rates in summer are accompanied by long contact times between potential source minerals and the minewater. Potential type 2 sources of iron are ferric hydroxides (ferrihydrite and goethite) and siderite (Younger 1995).

MINTEQA2 was run to simulate dissolution of goethite and then siderite, which were included in the model as finite solid phases at concentrations of $0.1\,mole\,l^{-1}$. Ferrihydrite was not included in the assessment because sufficient information on ferrihydrite is not present in the MINTEQA2 database, but it is likely that ferrihydrite would show similar trends to the more structured mineral goethite, with slightly higher dissolved equilibrium concentrations.

The dissolved iron concentrations predicted for various Eh values for dissolution of goethite and siderite are presented in Table 3. The results

Table 3. *Calculated equilibrium dissolved ferrous iron concentration in the presence of goethite and siderite under various redox conditions*

Redox Potential (Eh mV)	−100	−75	−50	−25
$[Fe^{2+}]$ dissolved ($mg\,l^{-1}$)				
Goethite dissolution	12.15	4.41	1.60	0.58
Siderite dissolution	17.5	17.5	17.5	17.5

show that siderite solubility is not a function of Eh in the range evaluated, but that goethite solubilities are sensitive to Eh. The solubilities of both minerals are sensitive to pH, but this is not reported in detail as the pH conditions in the minewaters are relatively constant and well constrained. The calculated range of values for goethite was $0.58-12.15 \, \mathrm{mg} \, l^{-1}$, whilst for siderite a value of $17.5 \, \mathrm{mg} \, l^{-1}$ was calculated. These values compare reasonably well with the observed iron concentrations. The results indicate that these minerals could be a source of iron in solution.

The sensitivity of goethite and ferrihydrite dissolution to Eh and oxygen concentration may provide a clue to the main source of iron in discharge waters, since it is the flooding of workings which has resulted in high iron concentrations. Drainage from waste rock piles on the surface consistently has low iron concentrations (less than $1 \, \mathrm{mg} \, l^{-1}$) even at pH as low as 5.5, due to the controlling mechanism of ferric iron precipitation. Dissolution of ferric iron under anoxic conditions is also invoked to explain increases in the iron loading in discharge waters from the waste rock piles at the Bersbo mine in Sweden following covering with oxygen exclusion barriers (Lundgren 1990), and for increases in dissolved ferrous iron concentration with depth in submerged tailings (Pederson et al. 1990). Such increases in iron loading would not be expected if siderite was the main source of iron, since siderite solubility is not sensitive to Eh.

Furthermore, it is more plausible that the relatively unstructured secondary mineral ferrihydrite would dissolve at a rate sufficient to produce the observed iron concentrations, than would the well-structured primary mineral siderite.

Finally, siderite is a common mineral in the area in contact with groundwaters, but iron concentrations of the order observed in the minewater discharge do not occur in local groundwaters.

It is therefore postulated that dissolution of ferric hydroxides, particularly ferrihydrite is the main source of iron in the discharge.

Dissolution of the relatively stable mineral goethite can be achieved by protonation or reduction (Stumm & Sulzberger 1992). Under the current circumstances, reductive dissolution is most likely. The requirement is for dissolved ferrous iron (or other reduced metallic ion) and a complex-forming ligand such as oxalate, or for reductants such as ascorbate or sulphide. In view of its less well ordered crystal structure, ferrihydrite may dissolve more readily than goethite by similar processes.

Other possible sources of iron are discussed below:

(1) Dissolution of type 1 sulphate salts results in super-saturation of minewaters in flooded workings with respect to ferric hydroxides, which precipitate from the water and control the iron concentration to a maximum level. This model does not take account of the flushing of the system by relatively clean recharge waters which would result in continuous improvement in water quality. The flushing process would actually be enhanced by mineral precipitation.

(2) Active oxidation of pyrite above the water table results in ongoing production of dissolved iron in solution. Although this may be occurring it is unlikely that this would contribute significant amounts of iron to the discharge, because iron is unlikely to be mobile in the drainage from the backfilled workings due to oxidation of ferrous to ferric and precipitation of ferric hydroxides. In addition, samples from deep in the mine which could not have come into contact with drainage from unflooded workings show very similar quality to the discharging waters.

(3) Sulphide oxidation below the water level by ferric iron in solution could be contributing to dissolved iron concentrations. It is unlikely that ferric iron is present in dissolved form in the minewaters however.

It is concluded therefore that dissolution of ferric hydroxides such as ferrihydrite and goethite is likely to be the main source of iron in the discharge. These minerals represent type 2 sources.

Implications for future water quality   The exponential form of the decrease in zinc, cadmium and manganese concentrations is commonly observed in discharges from abandoned workings (Fernandez-Rubio et al. 1987; Robins & Younger 1996). The form of the curve suggests that these metals are being flushed from the mine by replacement of contaminated water with relatively clean recharge. This process has been modelled assuming that the mechanism is similar to that of a completely stirred tank reactor (CSTR), and that the flushing water is of a quality represented by relatively clean recharge water flowing into the upper workings. This type of methodology has been used successfully for prediction of contaminant removal from groundwaters by flushing (Brusseau 1996), although the solution is more applicable to minewaters where the mined void approximates more closely a mixing tank.

Table 4. *Flushing model input data*

Parameter	Source	Zinc	Cadmium
C_0	Calculated by curve matching, but of the same order as the discharge concentration in July 1994	$210\,\mathrm{mg\,l^{-1}}$	$0.97\,\mathrm{mg\,l^{-1}}$
C_i	Estimated from recharge water quality in Combe Maurel	$1.5\,\mathrm{mg\,l^{-1}}1$	$0.005\,\mathrm{mg\,l^{-1}}$
Q	Representative figure for flow from monitoring data	$25\,\mathrm{m^3\,h^{-1}}$	$25\,\mathrm{m^3\,h^{-1}}$
V	Calculated by curve matching	$110\,000\,\mathrm{m^3}$	$70\,000\,\mathrm{m^3}$

The CSTR process can be represented by the following equation:

$$C_t = \{C_0 \cdot \exp(-(Q/V)t)\}$$
$$+ \{C_i \cdot (1 - \exp(-(Q/V)t))\} \quad (4)$$

where C_t is the concentration at time t; C_0 is the initial concentration ($t = 0$); C_i is the concentration in inflow water; Q is the flow rate; and V is the volume of reactor

In this case all the parameters are well known except the volume of the reactor (i.e. the void in the workings in which the mixing occurs or the effective volume). In order to test that this mechanism could be the main control, the discharge concentration was plotted for zinc and cadmium for the period July 1994 to January 1995. Simulations were carried out using the equation above, and varying the effective volume to achieve a match. The model is not highly sensitive

to the volume, but the volumes used to achieve the best match are reasonable estimates of the volume that would have to be flushed. The input data are presented in Table 4 and results of the modelling are shown in Figs 6 and 7.

Figures 6 and 7 show discharge water concentrations to June 1995 (i.e. later than the model calibration data). This includes a period during which water was pumped from the Central Decline (as indicated in Figs 6 and 7) which has affected the water chemistry in the East Decline. After the pumping ceased, changes to the flushing pattern occurred, particularly in the case of cadmium where higher concentrations than those modelled were observed. Reduced removal of zinc during pumping resulted in a relative high when pumping ceased, followed by an increased rate of flushing. The effect on cadmium was different in that there was a hiatus in the flushing process and no increase in the flushing rate after

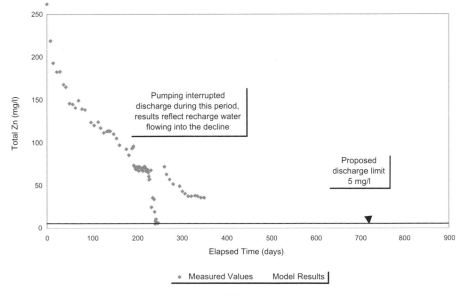

Fig. 6. Discharge zinc concentration and CSTR modelling results.

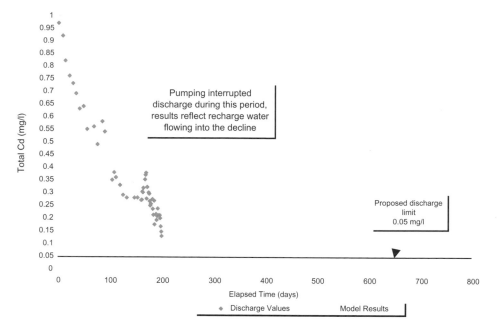

Fig. 7. Discharge cadmium concentration and CSTR modelling results.

pumping was discontinued. The reasons for these differences are unclear.

Given that the model calibration is not very sensitive to the chemistry of the recharge water C_i,, the tail of the plots is not intended to be a prediction of the medium- to long-term water quality. It is likely that long-term metals levels will be above those modelled for the following reasons:

(a) at lower metals concentrations the effects of processes such as type 2 mineral dissolution, which are initially minor compared to the flushing process as a whole, will become more significant;

(b) the model has assumed thorough mixing, but there will be parts of the mine which are flushed relatively slowly, and will contribute to the tail concentration for longer than the model suggests;

(c) the model does not account for the potential contribution of metals from upper, unflooded workings.

The model results do serve to indicate the likely flushing rate to medium-term levels, however, and on the basis of the calculated curves and taking into account the comments above, it is expected that the discharge water will have attained levels of less than $10 \text{ mg} \text{l}^{-1}$ (zinc) and $0.05 \text{ mg} \text{l}^{-1}$ (cadmium) within a period of 2–5 years after discharge initiation.

Concentrations of most other contaminants are likely to decrease according to similar form. The main exception is iron, future concentrations of which are likely to remain steady until such a time as the quantity of secondary source material has diminished so that the dissolution rates cannot be maintained. The long-term prognosis is for a very gradual reduction in iron concentration.

Prediction of the time for iron levels to decrease to background conditions would be based on estimates of the amount of pyrite oxidation that occurred during operation. Given the lack of information, it is highly likely that estimates of this would be wide-ranging. Consequently, this has not been attempted at this stage.

Conceptual model

The conceptual model is a summary of the hydraulic and hydrogeochemical mechanisms of discharge development which provides an assessment of the present understanding of the systems that give rise to the minewater discharge. The model forms the basis for predictions of the impacts of potential remedial measures, and will be developed as the investigation proceeds and our understanding of the system improves.

Figure 8 is a schematic illustration of the model incorporating and summarizing the interpretation of flow and geochemistry.

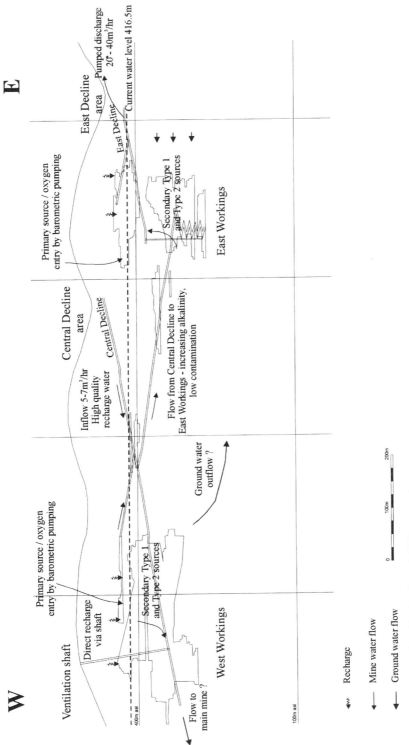

Fig. 8. Conceptual model of hydraulic and AMD generation processes.

Table 5. *Proposed discharge water quality limits*

EC	pH	O_2 (dis)	Zn	Mn	Cd	Fe	Pb	Ni	Cu
$750\,\mu S\,cm^{-1}$	6.5–8.5	70–90%	$5\,mg\,l^{-1}$	$1\,mg\,l^{-1}$	$0.05\,mg\,l^{-1}$	1–$2\,mg\,l^{-1}$	$0.05\,mg\,l^{-1}$	$0.5\,mg\,l^{-1}$	$0.05\,mg\,l^{-1}$

Minewater remediation

Remediation objectives

In France the compliance point for discharge water quality is not the discharge itself, but the point where water leaves the site boundary. Consequently, some benefit may be obtained in mixing the water with the receiving water before the water crosses the site boundary. However, for the purpose of this study it was assumed that the compliance conditions will apply directly to the discharge water and not to a mixture of discharge and stream water.

The compliance conditions have not been set but levels have been proposed by the mine operator. These levels are based on various EC standards for drinking water (CEC 1980), industrial discharge (CEC 1983) and salmonid fisheries (CEC 1976, 1978), and local limits set by French authorities for discharges. The proposed discharge concentrations are listed in Table 5.

These values take account of the local geographical circumstances which give rise to high background concentrations of metals in groundwater and surface water.

Remediation options

A wide range of options relating to source control, migration control, treatment and dilution have been considered. Those options which are worthy of further consideration and detailed examination are described below.

Sealing of the East Decline Sealing of the East Decline would have the effect of preventing the current discharge and causing the minewater level to rise. Since the principal water source is above the workings it would be expected that the water level would continue to rise until a discharge occurred. Some leakage from the workings could be expected since water in the void would be at elevations above surrounding country, but in view of the low permeability nature of the host rocks and the large distance between the workings and the ground surface, it is likely that the majority of the existing

discharge would also surface at the Central Decline at an elevation of about 465 masl. This is about 25 m above the highest worked levels and therefore all the workings would have to flood before the discharge occurred.

The benefits associated with this option are listed below:

(1) Unflooded parts of the mine where active oxidation is currently occurring would be flooded and the active oxidation process reduced (Fernandez-Rubio *et al.* 1987; Banks 1994). This would have the effect of reducing the ultimate long-term discharge concentrations after the flushing process was complete. The long-term discharge water quality would then be close to the background groundwater quality, although some elevated metals concentrations could be expected.

(2) The rise in water level may have the effect of reducing the total rate of minewater discharge either by increasing circulation of groundwater in the near-surface zone and reducing the underdrain effect of the mine void, or by increasing seepage from the mine to the environment. It is expected that the latter process would be very slow, and acceptable dispersion and dilution in the environment could be achieved. If minewater levels were to stabilize below the elevation of the Central Decline portal, the problem could be solved without further action. This is unlikely however.

(3) There is limited space for treatment at the current discharge site and the land is not owned by the operator, whilst at the Central Decline site there is significantly more space for treatment and the land is owned by the operator.

There are a number of risks and disbenefits associated with flooding of upper workings, as follows:

(1) There would be an increase in the volume of void and reduction in the flow rate so that the flushing time would increase.

(2) Dissolution of oxidation products would result in the upper workings in a new secondary type 1 source which would cause a temporary increase in contaminant levels. Given the volume of void to be flooded, the impacts are

expected to be short term and therefore unlikely to offset the benefits of long-term improvement in discharge water quality.

(3) There may be uncontrolled and unacceptable leakage to the environment at discrete points. If this occurs there may be benefit in flooding the workings to a level at which there is no unacceptable leakage. In addition, the plug could be used to control the discharge at the East Decline portal so that water would only be released during high flows.

(4) High water pressures may result in structural instability. Geotechnical stability assessments would have to be carried out prior to plug installation, but in view of the large distances between mined void and the ground surface, problems are not anticipated. To some extent the stability will be improved as hydrostatic pressures counter gravitational forces in the void.

(5) There would be a risk of catastrophic plug failure (Banks 1994). Although plug failure occurred at the Wheal Jane mine in Cornwall, UK, causing a notorious pollution event in the Fal Estuary in January 1992 (Hamilton et al. 1990; Bowen et al. 1998) the circumstances of that plug failure were very different. The Nangiles plug which failed at Wheal Jane was constructed to prevent ingress of water to the adit and was not designed to withstand high pressures. Design and construction of plugs to prevent flow of water through mined and unmined conduits is a routine part of groundwater control in mining operations. Such a plug would only be installed if the conditions were such that risk could be reduced to acceptable levels.

Sealing of the adit would be achieved by placing a plug in the decline some distance from the entrance, where the rock is sufficiently competent to withstand high pressure, and permeability of the rock should be sufficiently low to allow control of by-pass flows. The plug would be designed to withstand a head of at least 75 m of water and fitted with a valve so that the water pressure could be measured and water could be released if problems arose.

A similar seal would also be installed in the Central Decline at depth so that leakage to the surface through fractures and weathered rock would not occur as the water level rose. The plug would have a drainage pipe through it so that water could flow up the decline to the surface in the pipe.

Water treatment at the Central Decline site Consideration has been given to both long- and short-term treatment requirements. In the short term

there will be a need to remove high levels of zinc, manganese, cadmium and iron from the water, whilst in the longer term it is expected that only iron will be present at levels requiring treatment, since the secondary type 1 sources of other metals will have been flushed from the system.

Passive removal of zinc, manganese and cadmium in wetlands is not well established. Aerobic manganese removal, although achievable, requires very long aeration times, whilst removal of zinc and cadmium in anaerobic systems has only been achieved with limited success.

For this reason, a system of lime dosing to raise the pH and settlement of metal precipitates in tanks and ponds has been proposed. The system should be designed to minimize active operation. The main elements are an automatic lime dosing system, manual pH control to maintain pH at about 9.5 in a mixing vessel, primary settlement tanks which will remove most of the precipitated sludge and a polishing pond.

With adequate treatability testing, a system of this type will provide a high level of control, whilst minimizing capital and operational costs. At the expected rates of sludge production this will be more cost-effective than construction and operation of a fully active system involving aeration cells and settlement in a conventional circular thickener.

The polishing pond would provide a buffer between the main treatment system and the receiving water course and a pH equilibration cell in which the pH would fall by aeration and CO_2 dissolution. The polishing pond would also be the long-term treatment cell designed for aerobic iron precipitation and settlement. This may be designed as an aerobic wetland or an open water body, but the area should be sufficient for iron removal and the storage volume should be sufficient to enable long-term treatment without regular sludge disposal. In either case, the cell should be blended into the natural environment to minimize the visual impact and maximize the ecological benefit of the long-term treatment system. The area of the pond will probably be of the order of 1000 m^2.

Other potential remediation strategies A number of other strategies were considered, but were rejected on the basis of predictions of the short-term nature of the problem. Some of these are described briefly below.

In situ sulphate reduction has been considered as a means of treating minewaters in flooded abandoned mines (Kuyucak et al. 1990) but to the knowledge of this author has never been implemented as a full-scale treatment operation in an underground mine. Metals removal can be achieved by inducing reduction of sulphate in the workings. The sulphate is reduced to hydrogen sulphide which reacts with metals and precipitates them as insoluble sulphides. This process is mediated by sulphate-reducing bacteria in the presence of a utilizable organic substrate and can be induced by addition of such a substrate to the workings.

Due to the novel nature of the method and the potential planning problems associated with adding organic waste to what is essentially groundwater, and taking into account the relatively short time period during which the discharge is likely to be highly contaminating, this strategy was considered inappropriate.

Another option that was considered was diversion of the stream at the Central Decline site into the workings to accelerate the flushing process. Temporary treatment plant could be installed at the East Decline site to receive this water. This was rejected because it is likely that the unflooded upper workings would continue as a primary AMD source after flushing was complete.

Conclusions

Selection of remediation strategies should be based on a sound conceptual model of the hydraulic and hydrogeochemical mechanisms of minewater discharge development, so that future developments in water quality and the response of the system to engineered manipulation can be predicted.

Probably the most important aspect of the conceptual model is interpretation of the AMD source, as this will dictate future discharge water quality and the response to remediation options. For this reason, a classification system has been proposed which facilitates prediction of discharge water quality and contaminant flushing. Sources of AMD have been divided into two types; namely primary sources and secondary sources. Primary sources occur mainly in unflooded workings and relate to active oxidation of sulphides in the presence of oxygen, although under acid conditions significant sulphide oxidation in the presence of ferric iron could be expected even in the absence of oxygen. Secondary sources occur under most circumstances in the presence of flowing water but are the dominating AMD source in flooded workings. Secondary sources can be subdivided into two types: type 1 second-

ary sources relate to dissolution of highly soluble sulphate salts; type 2 secondary sources relate to equilibrium dissolution of less soluble secondary minerals.

Primary sources of AMD can be considered open systems (Cravotta 1994) which will continue to contribute contaminants long after mine closure. Secondary AMD sources are closed systems with a finite stored source of contamination which will be flushed from the mine after flooding, provided that there is a discharge. Contaminants dominated by type 1 secondary sources are likely to show exponential decreases in concentration with time, as the initially contaminated minewater is flushed from the system by relatively clean influent waters. This flushing process can be modelled using a CSTR type equation provided the minewater mixes well with recharge in the workings. Type 2 secondary sources are likely to result in temporally persistent contamination, with seasonal fluctuations due to changes in contact time with source minerals as flow rate varies.

In this case the upper workings are likely to be a primary source of iron, zinc and cadmium. The main source of manganese is probably dissolution of siderite or rhodochrosite in neutralizing reactions. The flooded workings are a secondary type 1 source of zinc, manganese and cadmium, and a secondary type 2 source of iron.

Numerous other geochemical complexities are expected to contribute to the discharge water quality, but the processes outlined above and described in this paper serve as a useful means of classifying source and predicting future developments.

In this case the source of water appears to be aquifers at relatively shallow depth, and the host rock permeability is known to be very low. The system therefore lends itself to hydraulic manipulation such as plugging the East Decline and allowing the water level to rise until a new discharge occurs at the Central Decline. The elevation of the new discharge will be above the highest worked levels and therefore all the workings will become permanently flooded. This will all but eliminate primary AMD sources and improve the long-term discharge water quality.

The new discharge will occur on land owned by the operator and where there is space for treatment, which could be achieved by a system of lime dosing and settlement. A polishing pond designed as a buffer between the treatment system and the receiving water course and as a pH equilibration pond would also provide the site for long-term aerobic precipitation and settlement of iron.

Thanks are due to R. Campi for his input on mining history and geology and his continued support and co-operation during this work, F. Lopez for carrying out most of the sampling and analytical work, and G. Garcia for setting up pumping and monitoring systems in the mine.

References

ALLISON, J. D., BROWN, D. S. & NOVO-GRADAC, K. J. 1991. *MINTEQA2/PRODEFA2: A Geochemical Assessment Model for Environmental Systems. Version 3.2 User's Manual.* US Environmental Protection Agency, EPA/600/3-91-021, Athens, Georgia.

BANKS, D. 1994. The abandonment of the Killingdal Mine, Norway: a saga of acid mine drainage and radioactive waste disposal. *Mine Water and the Environment*, **13**, 35–47.

BOWELL, R. J., FUGE, R., CONNELLY, R. J. & SADLER, P. J. K. 1996. Controls on ochre chemistry and precipitation from coal and metal mines. *In: Minerals, Metals and the Environment II.* Institute of Mining and Metallurgy, London, 293–321.

BOWEN, G. G., DUSSEK, C. & HAMILTON, R. M. 1997. Pollution resulting from the abandonment and subsequent flooding of Wheal Jane Mine in Cornwall, UK. *This volume.*

BRUSSEAU, M. L. 1996. Evaluation of simple methods for estimating contaminant removal by flushing. *Ground Water*, **34/1**, 19-22.

CONNELLY, R. J., HARCOURT, K. J., CHAPMAN, J. T. & WILLIAMS, D. 1994. Approach to remediation of ferruginous discharges in the South Wales Coalfield. *In: REDDISH, D. J. (ed) Fifth International Mine Water Congress,* Proceedings Volume 2, International Mine Water Association and University of Nottingham, 521–531.

COUNCIL OF THE EUROPEAN COMMUNITIES (CEC) 1976. *Directive of 4 May 1976 on pollution caused by certain dangerous substances discharged into the aquatic environment of the Community.* 76/464/EEC; OJ L **129**.

—— 1978. *Directive of 18 July 1978. On the quality of fresh waters needing protection or improvement in order to support fish life.* 78/659/EEC; OJ L **221**.

—— 1980. *Directive of 15 July 1980 relating to the quality of water intended for human consumption.* 80/778/EEC; OJ L **229**.

—— 1983. *Directive of 26 September 1983 on limit values and quality objectives for cadmium discharges.* 83/513/EEC; OJ L **291**.

CRAVOTTA III, C. A. 1994. Secondary iron-sulfate minerals as sources of sulfate and acidity. *In: ALPERS, C. N. & BLOWES, D. W. (eds) Environmental Geochemistry of Sulfide Oxidation.* American Chemical Society Symposium Series 550, 345–364.

FERNANDEZ-RUBIO, R., FERNANDEZ-LORCA, S. & ESTABAN-ARLEGUI, J. 1987. Preventive techniques for controlling acid water in underground mines by flooding. *International Journal of Mine Water*, **6**(3), 39–52.

HAMILTON, R. M., TABERHAM, J., WAITE, R. R. J., CAMBRIDGE, M., COULTON, R. H. & HALLEWELL, M. P., 1994. The development of a temporary treatment solution for the acid mine water discharge at Wheal Jane. *In: REDDISH, D. J. (ed) Proceedings of the 5th International Mine Water Congress.* Nottingham, UK.

HEDIN, R. S., NAIRN, R. W. & KLEINMANN, L. P. 1994. *Passive treatment of coal mine drainage.* Bureau of Mines Information Circular 9389. United States Department of the Interior, Pittsburgh, PA.

HEM, J. D. 1972. Chemistry and occurrence of cadmium and zinc in surface water and groundwater. *Water Resources Research*, **8**(3), 661–679.

KUYUCAK, N., LYEW, D., ST-GERMAIN, P. & WHEELAND, K. G. 1990. *In situ* bacterial treatment of AMD in open pits. *In: SKOUSEN, J., SENCINDIVER, D. & SAMUEL, D. 1990 (eds) Proceedings of the 1990 Mining and Reclamation Conference and Exhibition.* West Virginia University, Morgantown, WV.

LUNDGREN, T. A. 1990. The first full scale project in Sweden to abate acid mine drainage from old mining activities. *In: GADSBY, J. W., MALICK, J. A. & DAY, S. J (eds) Acid Mine Drainage Designing for Closure.* BiTech, Vancouver, BC.

MCKIBBEN, M. A. & BARNES, H. L. 1986. Oxidation: pyrite in low temperature acidic solutions: rates laws and surface textures. *Geochimica et Cosmochimica Acta*, **50**, 1509–1520.

MORIN, K. A. 1990. A case study of data quality in routine chemical analyses of acid mine drainage. *In: GADSBY, J. W., MALICK, J. A. & DAY, S. J (eds). Acid Mine Drainage Designing for Closure.* BiTech, Vancouver, BC.

NORDSTROM, D. K. 1982. Aqueous pyrite oxidation and the consequent formation of secondary iron minerals. *In: KRAL, D. M. & HAWKINS, S. (eds) Acid Sulfate Weathering.* Soil Science Society of America, Special Publication No. 10, Madison, Wisconscin.

—— BURCHARD J. M. & ALPERS, C. N. 1990. The production and seasonal variability of acid mine drainage from iron mountain, California: a superfund site undergoing rehabilitation. *In: GADSBY, J. W., MALICK, J. A. & DAY, S. J (eds) Acid Mine Drainage Designing for Closure.* BiTech, Vancouver, BC.

PEDERSEN, T. F., MUELLER, B. & PELLETIER, C. A. 1990. On the reactivity of submerged tailings in Fjords and a lake in British Columbia. *In: GADSBY, J. W., MALICK, J. A. & DAY, S. J (eds). Acid Mine Drainage Designing for Closure.* BiTech, Vancouver, BC.

ROBERTSON, A. MACG. & BARTON-BRIDGES, J. 1990. Cost effective methods of long term acid mine drainage control from waste rock piles. *In: GADSBY, J. W., MALICK, J. A. & DAY, S. J (eds). Acid Mine Drainage Designing for Closure.* BiTech, Vancouver, BC.

ROBINS, N. S. & YOUNGER, P. L. 1996. Coal abandonment – mine water in surface and near surface environment: some historical evidence from the United Kingdom. *In: Minerals, Metals and the Environment II.* Institute of Mining and Metallurgy, London, 253–262.

STUMM, W. & SULZBERGER, B. 1992. The cycling of iron in natural environments: considerations based on laboratory studies of heterogeneous redox processes. *Geochimica et Cosmochimica Acta*, **56**, 3233–3257.

YOUNGER, P. L. 1995. Hydrogeochemistry of mine-waters flowing from abandoned coal workings in County Durham. *Quarterly Journal of Engineering Geology*, **28**, S101–S114.

Section 4: Groundwater pollution by hydrocarbons

Hydrocarbon fuels probably represent the commonest discrete sources of soil and groundwater contamination in the UK. The reader can find one of the most comprehensive studies of the hydrogeological and geochemical processes taking place in the vicinity of a hydrocarbon spill in papers by Bennett *et al.* (1993), Eganhouse *et al.* (1993) and Baedecker *et al.* (1993). Hydrocarbon contamination may be derived from fuel stations (**Banks; Clark**), fuel depots (**Holden & Tunstall-Pedoe**), rail or road accidents, leaking pipelines or, perhaps most notoriously, from airports (**Clark & Sims**). It is often accepted that most major airports are underlain by a lake of spilled kerosene.

In the former Soviet Union, for example, anecdotal evidence suggests that military pilots were often paid by air-time flown, estimated by fuel consumption: a practice leading to pilots 'dumping' fuel in the ground. At Siauliai airport in Lithuania, so much fuel was lost into the ground that locals are abstracting fuel from wells around the perimeter for their own use or for resale: a highly innovative cottage industry (Paukstys *et al.* 1996). The situation may not be so different beneath NATO bases, for all we know, and there are certainly reports of serious contamination existing beneath USAF airbases in the UK (Pearce 1996). In contrast to the former Soviet Union, however, this information seldom enters the public domain.

Investigation of hydrocarbon spills is made awkward by a number of factors. Hydrocarbons are generally light, non-aqueous phase liquids (LNAPLs) and float on the water table. Careful borehole design is thus required to ensure representative monitoring. Even so, buoyancy effects mean that monitoring wells tend to exaggerate the thickness of free product present, albeit in a manner that is to some extent predictable (Abdul *et al.* 1989). Despite these difficulties, many hydrocarbons are susceptible to either natural degradation (Barker *et al.* 1987; Cozzarelli *et al.* 1990; Bennett *et al.* 1993; Wiedemeier *et al.* 1996) or to clean up by techniques such as vapour extraction, air sparging, bioremediation or pump-and-treat techniques (Clark 1995). Indeed, some authors feel that the threat to regional groundwater quality from petroleum hydrocarbons may be exaggerated (Hadley & Armstrong 1991). The papers presented here, however, do indicate that hydrocarbon contamination can cause problems on a more local scale, particularly where abstraction boreholes are located in the vicinity of spillages.

ABDUL, A. S., KIA, S. F. & GIBSON, T. L. 1989. Limitations of monitoring wells for the detection and quantification of petroleum products in soils and aquifers. *Ground Water Monitoring Review*, **Spring 1989**, 90–99.

BAEDECKE, M. J., COZZARELLI, I. M., EGANHOUSE, R. P., SIEGEL, D. I. & BENNETT, P. C. 1993. Crude oil in a shallow sand and gravel aquifer III. Biogeochemical reactions and mass balance modeling in anoxic groundwater. *Applied Geochemistry*, **8**, 569–586.

BARKER, J. F., PATRICK, G. C. & MAJOR, D. 1987. Natural attenuation of aromatic hydrocarbons in a shallow sand aquifer. *Ground Water Monitoring Review*, **Winter 1987**, 64–71.

BENNETT, P. C., SIEGEL, D. E., BAEDECKER, M. J. & HULT, M. F. 1993. Crude oil in a shallow sand and gravel aquifer I. Hydrogeology and inorganic geochemistry. *Applied Geochemistry*, **8**, 529–549.

CLARK, L. 1995. Hydrocarbon pollution control and remediation of groundwater: a brief review. *Quarterly Journal of Engineering Geology*, **28**, S93-S100.

COZZARELLI, I. M., EGANHOUSE, R. P. & BAEDECKER, M. J. 1990. Transformation of monoaromatic hydrocarbons to organic acids in anoxic groundwater environment. *Environmental Geology and Water Science*, **16**, 135–141.

EGANHOUSE, R. P., BAEDECKER, M. J., COZZARELLI, I. M., AIKEN, G. R., THORN, K. A. & DORSEY, T. F. 1993. Crude oil in a shallow sand and gravel aquifer II. Organic geochemistry. *Applied Geochemistry*, **8**, 551–567.

HADLEY, P. W. & ARMSTRONG, R. 1991. Where's the benzene? Examining California ground-water quality surveys. *Ground Water*, **29**, 35–39.

PAUKSTYS, B., TUCKER, C., MISUND, A., BANKS, D., KADUNAS, K. & TØRNES, J. A. (eds.) 1996. Assessment methodologies for soil/groundwater contamination at former military bases in Lithuania. Part 1: Chemical Investigation. *Norges Geologiske Undersøkelse Report* **96.146,** Trondheim, Norway.

PEARCE, F. 1996. Dirty groundwater runs deep. *New Scientist*, **2048** (21/9/96), 16–17.

WIEDEMEIER, T. H., SWANSON, M. A., WILSON, J. T., KAMPBELL, D. H., MILLER, R. N. & HANSEN, J. E. 1996. Approximation of biodegradation rate constants for monoaromatic hydrocarbons (BTEX) in ground water. *Ground Water Monitoring Review*, **Summer 1996**, 186–194.

Migration of dissolved petroleum hydrocarbons, MTBE and chlorinated solvents in a karstified limestone aquifer, Stamford, UK

DAVID BANKS

Associated Consultant with Scott Wilson Kirkpatrick, Bayheath House, Rose Hill West, Chesterfield, Derbyshire S40 1JF, UK
Present address: Norges Geologiske Undersøkelse, Postboks 3006 – Lade, N7002 Trondheim, Norway

Abstract: Two incidents of hydrocarbon contamination to the Lincolnshire Limestone in east Stamford, UK, have been investigated. No evidence of LNAPL contamination of groundwater was observed, suggesting that the spills may largely have been retained in the unsaturated zone. Some groundwater contamination by dissolved hydrocarbons occurred, apparently especially at times of high recharge. Rapid flow paths were proven to nearby springs in the River Welland (with groundwater flow velocities of up to $240 \, \mathrm{m \, day^{-1}}$), and dissolved hydrocarbon and MTBE contamination appears to have been flushed rapidly from these systems. MTBE contamination at Tallington Pumping Station (5 km east of the site) is not clearly linked to these incidents. Of potentially more concern was the discovery of dissolved chlorinated solvent contamination in the groundwater at the spill sites, possibly related to a landfilled quarry and/or a nearby engineering works. No direct evidence of DNAPL was observed. A conceptual model of solvent distribution suggests independent sources of TCE, PCE and TCA.

The area of Stamford is vulnerable to groundwater contamination. The built-up area is largely situated upon outcropping Lincolnshire Limestone, dipping gently eastward. The limestone is among the UK's most transmissive aquifers and is partially karstified, with considerable fissure and swallow hole development.

East of Stamford, water is abstracted for public supply via several boreholes operated by the regional water company. The Rivers Gwash and Welland, running through Stamford on limestone outcrop are generally of high quality and amenity value and are partially fed by baseflow from the limestone aquifer.

Two petroleum contamination incidents in the east of Stamford, at a filling station and an agricultural storage depot (Fig. 1), in 1988 and 1992 respectively, have given cause for concern over potential contamination of the rivers and groundwater supply boreholes by hydrocarbons, including the fuel additive methyl tertiary butyl ether (MTBE), which is widely used in unleaded petrol. During the course of investigation of these incidents, significant levels of chlorinated solvent contamination were discovered in the groundwater near these spill sites.

The nearest groundwater abstraction to the spillage sites is at the adjacent Stamford Hospital (Fig. 1: the borehole is still in occasional use, although the laundry which used most of the water is now closed). The water company have public supply boreholes at Pilsgate Pumping Station (PS), 3 km to the southeast, and Tallington PS, 5 km to the east (Fig. 2).

Investigations have been carried out by independent drillers and consultants, the University of East Anglia, and by the National Rivers Authority (NRA; now the Environment Agency, EA). All these sources of information were collated during 1995–1996 by Scott Wilson Kirkpatrick (1996), on behalf of the NRA, to attempt to construct a coherent conceptual model of contaminant migration from the incidents. This paper aims to provide a case study of multi-source organic contamination incidents in a rather unindustrialized part of the UK, with particular emphasis on the complexity of contaminant transport in fissured, heterogeous limestone aquifers. The complete data set is rather complex, with varying sample frequencies for different parameters at different locations (Scott Wilson Kirkpatrick 1996). No attempt is made in this paper to summarize the entire data set. Rather, attention is focused on a selection of parameters exhibiting the highest concentrations in groundwater and posing the greatest risk of exceedence of drinking water and environmental quality standards.

Incident 1: petrol filling station

The filling station is part of a retail development comprising a supermarket and car park (Fig. 3).

BANKS, D. 1998. Migration of dissolved petroleum hydrocarbons, MTBE and chlorinated solvents in a karstified limestone aquifer, Stamford, UK. *In*: MATHER, J., BANKS, D., DUMPLETON, S. & FERMOR, M. (eds) *Groundwater Contaminants and their Migration*. Geological Society, London, Special Publications, **128**, 123–145.

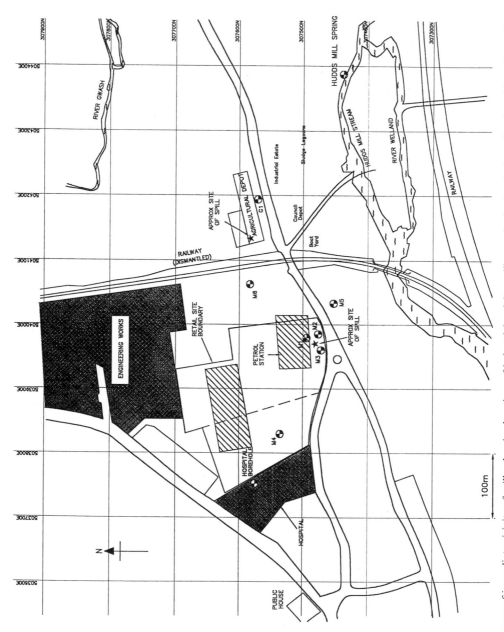

Fig. 1. Location map of immediate vicinity of spills, showing locations of industrial activity, spills, monitoring boreholes and springs. Note: the southern edge of the landfilled quarry coincides approximately with the location of the petrol station spill.

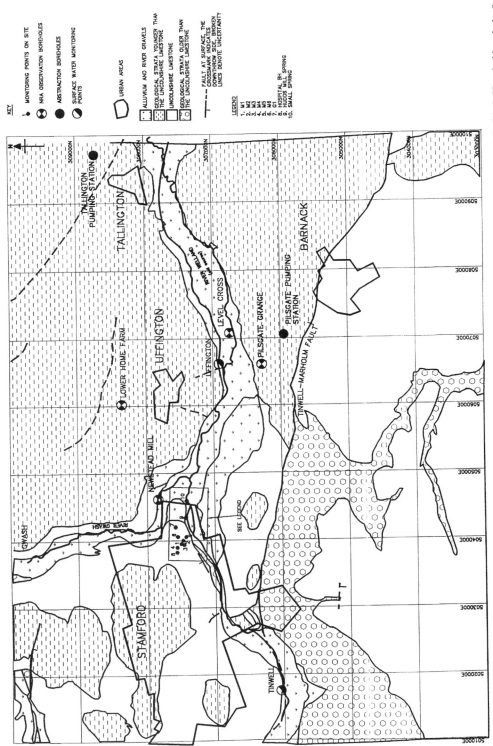

Fig. 2. Simplified geological map of the study area, showing locations of named boreholes. Key; 1, M1; 2, M2; 3, M3; 4, M4; 5, M5; 6, M6; 7, G1; 8, Hospital borehole; 9, Hudds Mill Spring; 10, Small Spring. The map covers the area between eastings TF [5]010 and [5]100 and northings TF [3]030 and [3]100.

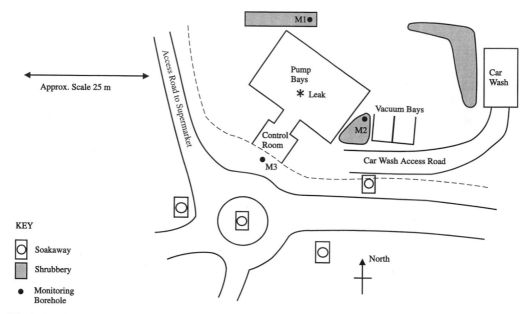

Fig. 3. Detailed sketch map of the petrol station, showing the locations of boreholes relative to the petroleum leak. The soakaways drain road runoff to the Lincolnshire Limestone. Scale is approximate.

The surrounding land is occupied by industrial and residential land uses. To the north of the site is a heavy engineering works. Part of the retail site area was originally used for quarrying Lincolnshire Limestone (probably since the 1840s) and for lime burning. The quarry was worked up until 1964. Between c. 1967 and 1977, the quarry was used as a landfill for the heavy engineering works, dominantly receiving waste foundry moulding sand.

The retail development and petrol station were constructed, partially on the landfilled quarry, in the autumn of 1988, following a detailed geotechnical investigation undertaken by Geotechnical Engineering (1987). The filling station area lies on the edge of the old quarry. The recorded fuel spill occurred at the time of commissioning of the retail complex on, or slightly before, 8 October 1988, from a fuel line adjacent to the pumps in the middle of the forecourt. The quantities spilt were 2800 l two-star leaded and 2000 l unleaded petrol (GTI 1988). Immediately following this incident, Groundwater Technology Services conducted a preliminary investigation involving the construction of three boreholes at the filling station area (M1, M2 and M3). The boreholes were completed open hole beneath the water-table, penetrating almost the full thickness of the limestone. No evidence of perched water-tables was noted. No light non-aqueous phase liquid (LNAPL)

hydrocarbon was observed at any time in the boreholes, suggesting that the bulk of the spill was effectively retained in the unsaturated zone. No attempt was made to effect remediation of the spill source area (e.g. by excavation).

At the time of the spill, fuel vapours from the spill became overpowering in a sewage tunnel (beneath the road in the vicinity of borehole M5; see Fig. 1) being constructed under a nearby road and tunnelling work had to be temporarily abandoned. Fuel vapours were also noted at the time of the spill in the cellars of the adjacent hospital and of a nearby public house. These fuel odours have not been reported since the spill. Migration of fuel towards the hospital was thus suspected and borehole M4 was drilled to investigate this possibility. The hospital borehole was sampled at the time of the spill but no significant contamination was detected.

Additionally, borehole M5 was drilled in a small redeveloped quarry site topographically downgradient of the spill, and M6 in front of the heavy engineering works, at the works' former solvent storage area. In all boreholes except M5, steel casing terminated at least 1 m above the water-table. In M5, steel casing terminated slightly below the water-table. Boreholes M1–6 were CCTV and gamma logged. The limestone in all was observed to be heavily fissured, with fissures up to 150 mm wide (GTI 1988).

During this investigation phase, the first detection of chlorinated solvents was recorded. GTI (1989) suggested that the solvents may have been deposited within the landfilled quarry area.

GTI (1989) concluded that most of the petroleum spill had been effectively retained within the c. 10 m thick unsaturated zone. Pilsgate PS and Tallington PS had been shut down following the spill as a precautionary measure, but were reinstated on 2 November 1988.

Incident 2: agricultural storage depot

The depot is located about 250 m to the east of the retail site, adjacent to the heavy engineering works (Fig. 1). The spill on 5 August 1992 occurred from an un-bunded above-ground unleaded fuel storage tank at the western part of the site. The quantity of unleaded (and MTBE-containing) fuel spilt is believed to be about 1500 l.

An observation borehole (G1) was installed c. 100 m east of the tank in order to investigate the spillage. At the spill site and at the borehole, the immediate substratum is limestone, no substantial thickness of fill being present. No LNAPL was observed in the borehole as a result of the spill, although the borehole is not ideally placed to detect any such contamination.

Geological and hydrogeological setting

Local hydrology

The study sites lie to the north of the River Welland (Fig. 1). The topographical gradient is towards this river and surface water and groundwater drainage is in this direction (i.e. towards the south). The minimum mean daily flow in the Welland in the period October 1992–August 1995 at Tinwell (TF ^{5}0180 ^{3}0605) was 2441 s^{-1} and the maximum 33 850 l s^{-1}.

The River Gwash runs to the east of the heavy engineering works and enters the Welland c. 1 km east of the retail site. A significant part of the engineering works area appears to fall within the surface (and presumably groundwater) catchment to this river. The minimum mean daily flow in the Gwash in the period October 1992–August 1995 at Belmersthorpe (TF ^{5}0380 ^{3}0970) was 218 l s^{-1} and the maximum 2543 l s^{-1}.

Regional geology

The Stamford area is underlain by strata of the Inferior and Great Oolite Series of Jurassic age (Fig. 2). The stratigraphic column is summarized in Table 1. In the northern part of central Stamford, strata as stratigraphically high as the Cornbrash crop out. The valleys of the Gwash and Welland cut down as far as the Lower Lincolnshire Limestone.

The strata dip gently eastward such that, at Tallington, some 5 km east of Stamford, the immediately subcropping solid strata are Oxford Clay and Kellaways Beds. The Lincolnshire Limestone is overlain by the Upper Estuarine Series immediately east of the Gwash.

South of the Welland, the strata are upthrown to the south by the E–W Tinwell–Marholm fault. Thus, immediately south of Stamford, Upper Lias Clay crops out immediately south of the fault. By easting TF ^{5}060, Lower Lincolnshire Limestone crops out adjacent to the south side of

Table 1. *Simplified stratigraphy of the Stamford area, with indicative thicknesses of strata*

Drift	Alluvium	
	Fenland deposits	
	Glacial sands and gravels	<6 m
	Glacial Boulder Clay	<15 m
	Oxford Clay	
	Kellaways Beds	
'Overlying Beds'	Cornbrash	0.1–3 m
	Blisworth Clay	2–12 m
	Blisworth Limestone	2.5–8 m
	Upper Estuarine Series	4–14 m
Lincolnshire	Lincolnshire Limestone	18 m
Limestone Aquifer	Lower Estuarine Series	6 m
Unit	Northampton Sands	6 m
Upper Lias	Upper Lias Clay	46–65 m

the fault, allowing the possibility of hydraulic continuity across the fault.

In the valleys of the Gwash and Welland, deposits of alluvium have been laid down. These are associated, along some reaches, with deposits of First Terrace gravels. To the east, the flat, lower lying areas underlain by the Oxford Clay and Kellaways Beds have largely been covered by a blanket of Quaternary Fenland Gravel.

Regional hydrogeology

Bradbury et al. (1994) consider the Lincolnshire Limestone Aquifer Unit to consist of the Northampton Sands, the Lower Estuarine Series of sands, silts, clays and shales, and the Lincolnshire Limestone itself. The bulk of the aquifer transmissivity is accounted for by the limestone, except in some parts of the unconfined region where the saturated thickness of the limestone is very low and where the transmissivity of the Northampton Sands may become significant.

The aquifer unit is underlain by the thick Upper Lias Clay aquitard and overlain by the so-called 'Overlying Beds' (see Table 1), of which the earliest is the Upper Estuarine Series, which consists of clays, shales and marls, with subordinate sand and limestone bands (Bradbury et al. 1994).

Rushton (1981) and Rushton et al. (1982) summarize the most important features of the regional hydrogeology of the southern part of the Lincolnshire Limestone. Both the Lincolnshire Limestone and the Blisworth Limestone are partially karstified, exhibiting swallow holes, particularly in river valleys and along the edges of the Boulder Clay-covered areas which occur elsewhere in the area (Bradbury et al. 1994).

Transmissivities in the confined aquifer have been estimated, from pumping tests, to be in the range $2000-10\,000\,m^2\,day^{-1}$ (NRA unpublished data), with storage coefficients of 10^{-4} to 10^{-3}. This is generally explained by the karstification of flow paths to areas of natural artesian discharge. At Tallington PS, Pilsgate PS and east of Stamford, transmissivities as high as $10\,000\,m^2\,day^{-1}$ have been recorded. This high transmissivity is associated with numerous post-Jurassic E–W faults.

In the unconfined aquifer, transmissivities are lower, tending to vary substantially with saturated depth, and unconfined storage is observed ($T = 50-500\,m^2\,day^{-1}$, $S = c.\ 0.05$; Rushton 1981). Water-table fluctuations are typically of the order of 2–10 m. Solution activity has been more intense in the zone of water-table fluctuation. Thus, one could regard the unconfined

limestone aquifer as consisting of an upper zone of solution-enhanced permeability, with very rapid groundwater flow, and a lower zone of slower flow.

Mean annual recharge to the aquifer is estimated by the NRA as 190 mm (unpublished data used for determination of Tallington PS groundwater protection zone). The main sources of recharge are thought to be as follows:

- direct infiltration of precipitation at outcrop;
- runoff from impermeable cover strata, often via fissure systems and swallow holes;
- soakaways (of which there is a high density in Stamford);
- possible infiltration from the Rivers Gwash and Welland.

The Tinwell–Marholm fault was originally regarded as a no-flow boundary, but evidence of flow and contaminant transport across the fault has been provided by studies of the Helpston waste disposal site (south of the fault). Contamination from Helpston has been found north of the fault (Rushton & Tomlinson 1994; Sweeney et al. 1998).

Local hydrogeology

The water-table underlying the retail site generally lies either at the fill/limestone interface or within the upper 4 m of weathered limestone bedrock beneath the fill. The average groundwater depths across the site in October 1988 ranged from 4.93 m below ground level (bgl) in M5 to 10.72 m bgl in M4.

Contoured, reduced groundwater levels for September 1989 are shown in Fig. 4. It can be seen that the water-table gradient across the site is rather small and groundwater flow is south towards the Welland. The small gradient may indicate that the site lies close to the groundwater divide between the Gwash and Welland catchments. The water-table must steepen towards the south, assuming that it is in hydraulic continuity with the River Welland.

The groundwater level in M4 is higher than that in the immediate vicinity of the spill (M1, M2 and M3), but sometimes (e.g. 20 March 1989) by a margin of as little as 2 cm. Water level elevations in the three boreholes around the fuel leak (M1, M2 and M3) are almost identical. At some times during the year, M3 has a very slightly higher groundwater level than M1, while at other times, the reverse is true. The possibility of northward flow of contaminants from the petrol station site, as well as the dominant southward flow, cannot thus be discounted.

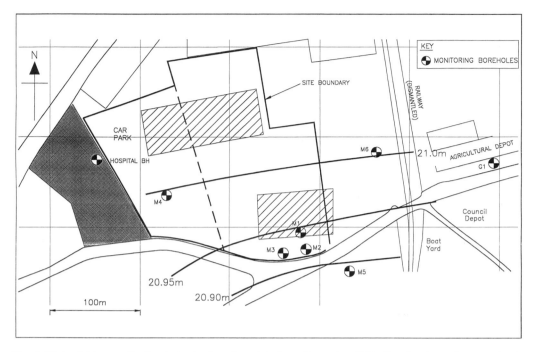

Fig. 4. Water-table map for the study site, based on data from September 1989. Contours in m above sea level.

Groundwater levels in G1 are at a depth of c. 7.7 m bgl. Figure 5 presents groundwater level hydrographs for M1–M6, and the EA's Newstead observation well. The water levels in all boreholes have almost identical seasonal range (a little under 1 m in 1989) and synchronized variations, including the observation well at Newstead, although the absolute water-table elevation here is somewhat below that of the other boreholes.

The contaminants

The most significant of the observed contaminants (in terms of concentrations and potential to exceed drinking water and environmental quality norms) in the groundwater of the study area are trichloroethene (TCE), tetrachloroethene (or perchloroethene PCE), 1,1,1-trichloroethane (TCA), the BTEX (benzene, toluene, ethylbenzene, xylenes) compounds and methyl tertiary butyl ether (MTBE).

Chlorinated solvents

The UK Drinking Water limits for PCE and TCE correspond to the WHO guidelines given in Table 2. The European Community recommends a guideline value of $<1\,\mu g\,l^{-1}$ for any chlorinated organic compound (Rivett et al. 1990)

TCE and PCE are on the EC Black List (List I Dangerous Substances Directive 76/464/EEC). This list is implemented in the UK via environmental quality standards (EQS). The specified EQS for both TCE and PCE in all surface waters is $10\,\mu g\,l^{-1}$.

Chlorinated solvents are volatile organics, typically used for the purposes of cleaning and degreasing. They are employed in metals, textiles and engineering works for degreasing metal components and fabrics and they are widely used in dry cleaners. They are also dense, non-aqueous phase liquids (DNAPLs), which are denser than water and migrate (in their free phase) in aquifers under gravitational rather than hydraulic gradients. Specifically, TCE is a high-volume industrial chemical, used as a solvent for degreasing metal and also as a septic tank cleaner and dry cleaning agent (Driscoll 1986; Rivett et al. 1990). Large-scale production of TCE began in 1928. Production of PCE began later (c. 1950) but consumption of both PCE and TCE peaked around 1972 in the UK, since when it has declined. PCE is used in dry cleaning (replacing TCE for this purpose) and as a degreasing agent.

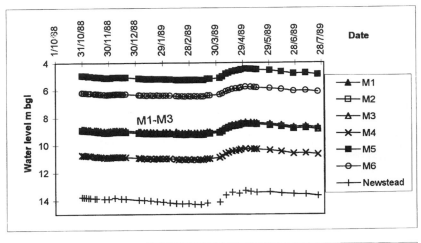

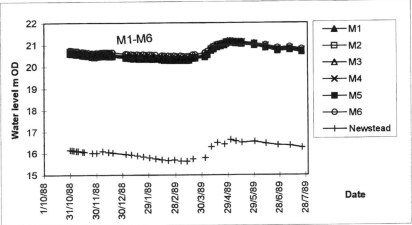

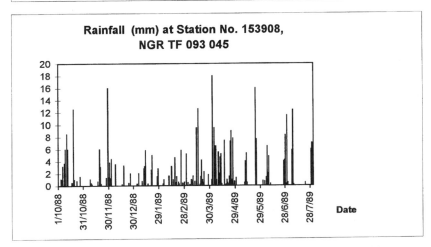

Fig. 5. Groundwater levels and daily rainfall at the study site, 1988–1989.

Table 2. *Tabulated values of solubility and K_{oc} (soil water partition coefficient) for selected contaminants (Fetter 1994), in order of decreasing groundwater mobility*

	Solubility (ppm)	K_{oc} (ml g^{-1})	Formula	UK drinking water MAC
Phenol	82 000	27	C_6H_5OH	$0.5\,\mu g\,l^{-1}$
MTBE	48 000		$(CH_3)_3COCH_3$	None set
1,1,2-TCA	4420	49	$CHCl_2CH_2Cl$	None set
Benzene	1780	97	C_6H_6	[†]
TCE	1100	152	CCl_2CHCl	$30\,\mu g\,l^{-1}$
1,1,1-TCA	700	155	CCl_3CH_3	None set
Toluene	500	242	$C_6H_5CH_3$	[†]
PCE	200	303	CCl_2CCl_2	$10\,\mu g\,l^{-1}$
o-Xylene	170	363	$C_6H_4(CH_3)_2$	[†]
p-Xylene	156	552	$C_6H_4(CH_3)_2$	[†]
m-Xylene	146	588	$C_6H_4(CH_3)_2$	[†]
Ethyl benzene	150	622	$C_6H_5C_2H_5$	[†]

[*] UK (Water Supply [Water Quality] Regulations 1989) drinking water maximum admissible concentrations (MAC) are also given. For TCE and PCE these correspond with WHO guidelines.
[†] The MAC for dissolved or emulsified hydrocarbons/mineral oils (total) is 10 µg/l.

1,1,1-TCA is used as an industrial cleaner and degreaser for metals and plastics, a resin adhesive, in inks and as a vapour-pressure depressant in aerosols. It was introduced in the UK in 1965, due to its lower toxicity than either TCE or PCE. Its consumption has been rising since then (Rivett *et al.* 1990).

Chlorinated solvents are not readily biodegraded in groundwater and, once present in DNAPL or dissolved form, are likely to remain for a long time (Rivett *et al.* 1990).

Petroleum hydrocarbons/BTEX

Petroleum is an LNAPL (light, non-aqueous phase liquid). The unsaturated zone has a significant capability to retain spilt hydrocarbons within pore spaces, particularly in granular aquifers. Generally, the finer grained the material, the higher its retention capacity. When spilt in sufficient quantities, LNAPLs can penetrate through the unsaturated zone and can accumulate as a pool 'floating' on the capillary fringe of the water-table.

Rainfall flushing through the unsaturated zone can leach out soluble components from the residual hydrocarbon body, carrying dissolved phase contamination to the water-table.

Petroleum is a mixture of hydrocarbon compounds. A large proportion consist of branching, straight- and cyclo-alkanes of carbon numbers 5–10. The solubility and vapour pressure of these compounds decrease as molecular weight increases (Nyer & Skladany 1989).

Petroleum also contains aromatic compounds. These are often of more concern in a dissolved hydrocarbon plume than the alkanes due to their higher solubility and higher toxicity. The most problematic are known as the BTEX components: benzene (C_6H_6), toluene ($C_6H_5CH_3$), ethyl benzene ($C_6H_5C_2H_5$), and xylenes ($C_6H_4(CH_3)_2$). These compounds, although toxic, are considerably more readily biodegraded than chlorinated hydrocarbons under most conditions. Barker *et al.* (1987) document almost complete biodegradation under saturated conditions of BTEX in a sand aquifer within 434 days, benzene being the only component persisting beyond 270 days.

MTBE

MTBE (methyl tertiary butyl ether) is an oxygenate additive used mostly, but not exclusively, in unleaded fuels, introduced to the UK in the mid 1980s. It is permitted to comprise up to 10% by volume of unleaded petrol. It is not especially toxic in water but imparts a taste or odour above levels of 2–$3\,\mu g\,l^{-1}$ (Symington *et al.* 1994). It is highly soluble in water ($48\,000\,mg\,l^{-1}$), has a low vapour pressure and is not readily retarded by sorption. This means that it can migrate very rapidly in groundwater, can reach substantial concentrations and is not readily remediated by vapour extraction. It is also persistent in groundwater, not being readily biodegraded.

There is no drinking water limit for this parameter in the UK, but levels of between 50 and $200\,\mu g\,l^{-1}$ have been suggested in the USA from the point of view of toxicity (Symington *et al.* 1994).

Hydrogeological framework for contaminant transport

Introduction

The Lincolnshire Limestone displays several features in common with other major UK limestone aquifers such as the Chalk, the Permian Magnesian Limestone and the Jurassic Limestones of the Cotswolds:

(i)   swallow holes, particularly draining Boulder Clay-covered areas (Bradbury *et al.* 1994);
(ii) very high apparent transmissivities;
(iii) very rapid response to recharge;
(iv) very rapid contaminant transport.

To illustrate this point, consider a value of transmissivity determined at Tallington PS ($10\,000\,\mathrm{m^2\,day^{-1}}$), which, with an aquifer thickness of 18.6 m, corresponds to a hydraulic conductivity (K) of $538\,\mathrm{m\,day^{-1}}$ ($6.2 \times 10^{-3}\,\mathrm{m\,s^{-1}}$). For a hydraulic gradient (i) of 0.01 and an assumed effective porosity (n_e) of 1%, the mean groundwater flow velocity can be estimated as:

$$v = K.i/n_e = 538\,\mathrm{m\,day^{-1}} \times 0.01/0.01$$

$$= 538\,\mathrm{m\,day^{-1}}$$

In the light of this calculation, the rapid migration of contaminants from Stamford to Tallington PS seems a possibility that needs to be considered.

Tracer tests

Tracer tests were carried out by Barnes (1993) of the University of East Anglia in 1993, by injection of rhodamine into borehole M1 and fluorescein into G1, and monitoring their breakthrough in the Hudd's Mill Spring (HMS), the Hudd's Mill Stream at point R, the Small Spring (SS) and the other M-series boreholes. The pathways proven by the tests (Fig. 6a), together with transport velocities and estimated hydraulic conductivities, are detailed in Table 3. The tests proved a connection between M1 and unknown springs feeding the Hudds Mill Stream, but failed to prove a connection between M1 and M5, Hudd's Mill Spring and Small Spring. The tests also demonstrated clear paths between G1 and Hudd's Mill Spring, Small Spring and the springs feeding the Hudd's Mill Stream. The proven pathways are in agreement with the dominant directions of subvertical fracture sets as mapped in quarry outcrops (Fig. 6b) and hydraulic gradients.

Barnes (1993) used a simple mixing tank model to estimate effective porosity, and a parallel plane fracture model to estimate fissure aperture, spacing and hydraulic conductivity. Despite these rather simplistic assumptions, the results obtained (Table 4) are within the range of values derived from pumping tests.

The recovery of tracer from the fluorescein tracer test during the monitoring period was around 50%, indicating predominant fissure flow mechanisms and little adsorption or pollution attenuation during transport.

Observed contaminant migration and fate

Hydrocarbon contamination from the filling station

Following the filling station spill, Groundwater Technology Services monitored water quality in boreholes M1–M6, the hospital borehole, Hudd's Mill Spring and Newstead observation well.

Initially, from 10 October 1988 to 2 November 1988, the monitoring points were monitored approximately daily for petroleum odour only. None was detected at Newstead, M5, Hudd's Mill Spring or the hospital. Faint odours were detected at M6 at the end of October. Odours were detected in M1–M4 during almost the entire period (GTI 1988).

From November 1988 to July 1989, samples were taken approximately every 3–4 days for analysis by gas chromatography, focussing on determinations of total unleaded petroleum (GTI 1989). No petroleum was detected at Hudd's Mill Spring or borehole M6 during this period. Only on 17 April 1989 were low levels of petroleum hydrocarbons detected in the hospital and Newstead boreholes (10 and $7\,\mathrm{\mu g\,l^{-1}}$ respectively). Hydrocarbon was only detected in M5 borehole at a level of $27\,\mathrm{\mu g\,l^{-1}}$ on 12 May 1989.

In the other boreholes, detection of contaminants tended to occur as brief 'episodes' of elevated concentrations, separated by periods of low concentrations. The highest concentrations of petroleum were detected in borehole M3 in several episodes. For example, petroleum hydrocarbons were detected at up to $240\,\mathrm{\mu g\,l^{-1}}$ on 9 December 1988 and during mid April 1989 at up to $10\,200\,\mathrm{\mu g\,l^{-1}}$. M2 also showed petroleum components at these times, with 39 and $2270\,\mathrm{\mu g\,l^{-1}}$ respectively. The petroleum detection episodes, particularly that in April, tended to follow rainfall events.

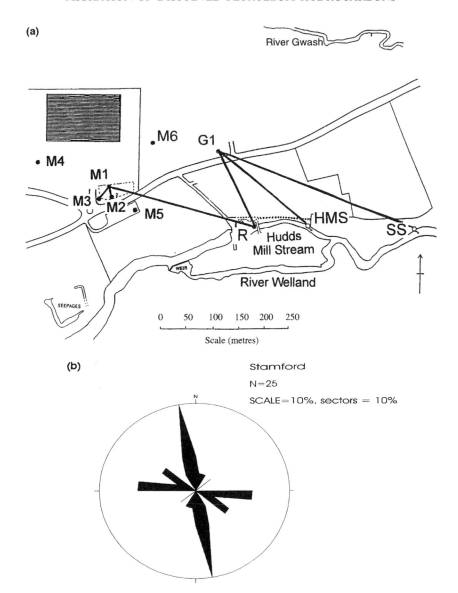

Fig. 6. (a) Summary of groundwater flow paths proven by tracer tests (reproduced, with permission, after Barnes 1993).(b) Rose diagram showing the preferred orientation of joint sets in the Lincolnshire Limestone of the study area, based on 25 dominantly subvertical fractures, after data presented by Barnes (1993). HMS, Hudds Mill Spring; SS, Small Spring.

In M1, hydrocarbons were detected in December 1988, January/February 1989 and April 1989. The concentrations were less than those during the same episodes in M3, but greater than those in M2 in the two earlier episodes. In April 1989, the maximum concentration was only $89\,\mu g\,l^{-1}$. In M4, the December and April episodes were also detected, at similar levels to M1.

The main detected hydrocarbon contamination episode occurred following heavy rainfall in early to mid April, which caused a significant rise of around 0.8 m in the water-table. On 17 April 1989, hydrocarbons were detected at the Hospital ($10\,\mu g\,l^{-1}$), Newstead ($7\,\mu g\,l^{-1}$), M1 ($89\,\mu g\,l^{-1}$), M2 ($2270\,\mu g\,l^{-1}$), M3 ($9920\,\mu g\,l^{-1}$) and M4 ($89\,\mu g\,l^{-1}$). By 20 April 1989, only M2 and M3 showed detectable hydrocarbons. Petroleum hydrocarbons were not detected at M5, M6 or Hudd's Mill Spring.

Table 3. *Summary of flow paths confirmed by tracer tests (summarized, with permission, from Barnes 1993)*

Pathway	Travel time (h)	Velocity (m day^{-1})	Estimated hydraulic conductivity (m day^{-1})
M1 to M2	27	22	290
M1 to M3	37	20	410
M1 to R	98	68	
G1 to HMS	22	242	370
G1 to R	67	56	
(3 peaks)	123	31	
	208	18	
G1 to SS	120	81	255

M1, M2, M3 and G1 are boreholes, HMS, Hudd's Mill Spring; SS, Small Spring; R, monitoring point in Hudd's Mill Stream (River Welland).

Analysis by gas chromatography of samples taken on 21 April 1989 for specific hydrocarbon components, revealed BTEX compounds and naphthalene at significant concentrations in M2 and M3, south of the filling station. Small concentrations of xylene were detected in M1, north of the filling station; other hydrocarbon parameters were below detection limits. The hydrocarbon components are thought to be derived from leaching of the spilt fuel from the unsaturated zone and were dominantly migrating south towards the Welland.

These results indicate that the main detected migration of the hydrocarbon contamination took place following a rise in water-table related to heavy rainfall in mid April 1989. The main migration took place in a southerly direction. The hydrocarbons appear either:

- to be effectively attenuated before they reach M5 or Hudd's Mill Spring; or
- not to be significantly detected at Hudd's Mill Spring or M5 because these points are not on the main flow path from the filling station.

The latter interpretation appears to be supported by the results of Barnes' tracer test. This does not preclude the possibility of a very diffuse flow connection between the points. It is likely, however, that the hydrocarbon contamination detected at M2 and M3 followed a flow path to unknown springs feeding the Welland.

Taken at face value, the results imply the possibility of limited northwestward and eastward migration, as hydrocarbons were detected at boreholes M1 and M4, Newstead and the hospital. Another possibility is that the hydrocarbons detected here may be derived from other sources, e.g. road/car park runoff following heavy rain. The fact that the hydrocarbon detection episode in December 1988 persisted for several days in M1 and M4 and yielded similar concentrations as in M2, suggests that some northwards migration from the spill really did occur, possibly due to a temporary reversal in water-table gradient.

NRA recommenced monitoring of boreholes M1–M6 from November 1992 to December

Table 4. *Summarized results and interpretation of tracer test: G1 to Hudd's Mill Spring and Small Spring (after Barnes 1993)*

	G1 to Hudd's Mill Spring	G1 to Small Spring
Distance	222 m	405 m
Travel time (peak)	22 h	120 h
Groundwater flow velocity	242 m day^{-1}	81 m day^{-1}
	2.80×10^{-3} m s^{-1}	9.38×10^{-4} m s^{-1}
Spring flow	9.15 l s^{-1}	2.04 l s^{-1}
Volume of equivalent mixing tank	3052 m^3	2163 m^3
Mass recovered	214 g (214 h)	3.87 g (214 h)
	243 g (840 h[*])	11.08 g (840 h[*])
Effective porosity	0.0172	0.0194
Head (G1)	20.5 m OD	20.5 m OD
Head (Spring)	18.0 m OD	18.0 m OD
Hydraulic gradient (i)	0.0112	0.00617
Hydraulic conductivity	372 m day^{-1}	255 m day^{-1}
Saturated thickness	7.25 m	8.5 m
Transmissivity	2697 m^2 day^{-1}	2168 m^2 day^{-1}
Hydraulic aperture	0.77 mm	0.60 mm
Fracture spacing	$N = 0.135$ m	$N = 0.093$ m
Reynolds Number	1.65	0.43
(<1000 for Darcy flow to be valid)		

[*] Extrapolated to 840 h.

1994. During this period, no persistent contamination from MTBE, toluene or benzene has been detected in these boreholes.

Summary The results support the interpretation of GTI (1989); namely that the spilt petrol was largely effectively retained in the unsaturated zone, without any LNAPL being discovered at the water-table. It is possible, however, that LNAPL contamination may have existed but was not detected in the boreholes due to the heterogenous, fissured nature of the aquifer. It is also possible that LNAPL may have been flushed away from the immediate spill vicinity by rapid groundwater flow in the aquifer before adequate monitoring commenced. During rainfall and during rises in water-table, soluble hydrocarbons were mobilized into groundwater in a dominantly southwards direction, causing hydrocarbon contamination episodes in M2 and M3. Occurrence of hydrocarbons in M1 and M4 indicates northward and westward migration of contaminants under some circumstances.

The majority of dissolved contamination did not appear to follow a flow path via M5 or Hudd's Mill Spring. Contaminants are likely to have followed an alternative flow path to unknown springs in the base of the Welland or Hudd's Mill Stream.

Leaching and biodegradation will have reduced the levels of hydrocarbon contamination in the unsaturated zone since the leakage, although the current situation is unknown. The lack of any significant MTBE, benzene or toluene contamination in M2 or M3 between 1992 and 1994 suggests that the hydrocarbon spill from the filling station may no longer represent a threat to groundwater quality.

Hydrocarbon contamination from the agricultural depot spill

No LNAPL is reported to have been observed in groundwater at the time of this spill. It is possible that a significant proportion of the petroleum may have been retained in the unsaturated zone. Unfortunately, borehole G1 is not situated immediately down gradient of the spill. In fact, higher concentrations of MTBE and benzene have been observed at Hudd's Mill spring than at G1, suggesting that the spring lies on a more direct flow pathway from the spill area.

Neither benzene nor MTBE contamination appears to have migrated westwards to the M-series boreholes. Rather it appears to have

migrated south, to the Hudd's Mill Spring, and east, to G1 and possibly beyond.

At Hudd's Mill Spring, concentrations of MTBE decreased from $157 \, \mu g \, l^{-1}$ on 5 October 1992 to $c. 5 \, \mu g \, l^{-1}$ in October 1993. Since then, MTBE has only been detected on one occasion. Benzene in Hudd's Mill spring shows a similar trend, declining rapidly from $75 \, \mu g \, l^{-1}$ in October 1992 to below detection limit by December 1992 (Fig. 7a). Xylenes (total) were detected at $29 \, \mu g \, l^{-1}$ in the Hudd's Mill Spring in October 1992, but after December 1992 have only been occasionally detected at low levels.

In G1, MTBE increased from $8.4 \, \mu g \, l^{-1}$ in January 1993 to a peak of $23 \, \mu g \, l^{-1}$ in summer 1993, since when concentrations have declined (Fig. 7b). After July 1994, MTBE was not detected in samples taken monthly. Benzene has only been detected on two occasions at G1, in early 1994, at levels of up to $1.5 \, \mu g \, l^{-1}$.

Further east, MTBE was detected during 1992 and 1993 at boreholes between the depot and Tallington PS (Table 5). It is thus possible to contour a continuous plume from the depot to Tallington PS based on the very sparse data points. The question then arises, is the apparent plume of MTBE a single plume, with MTBE contamination observed at Tallington PS derived from the depot? An examination of the plots in Fig. 8 demonstrates that it is not (or, at least, may be only partially) related to the depot.

Tallington PS has registered low levels of MTBE contamination. The first appearance of MTBE is unknown, but at the commencement of monitoring for this parameter in August 1990, levels of some $0.2 \, \mu g \, l^{-1}$ were reported, i.e. some two years before the spillage at the agricultural depot. Concentrations began to rise in October 1990 to levels of almost $1 \, \mu g \, l^{-1}$ by March 1991. Unfortunately, between March 1991 and October 1992, no records are available, but the highest recorded MTBE levels were observed in October 1992 ($1.25 \, \mu g \, l^{-1}$ in borehole A; $0.96 \, \mu g \, l^{-1}$ in borehole B). Similar levels of around $1 \, \mu g \, l^{-1}$ were also recorded by the NRA in Tallington observation boreholes 1, 2 and 3 in late 1992. MTBE levels at Tallington declined throughout early 1993 until they had returned to around the initial level of $0.3 \, \mu g \, l^{-1}$. The MTBE levels recorded are consistently below the taste and odour threshold of 2–$3 \, \mu g \, l^{-1}$. Several explanations are possible for the observed MTBE concentrations:

(i) The initial concentration of $c. 0.2 \, \mu g \, l^{-1}$ represents a fairly persistent 'background' of contamination, possibly derived from road runoff or some unknown source, and the increase

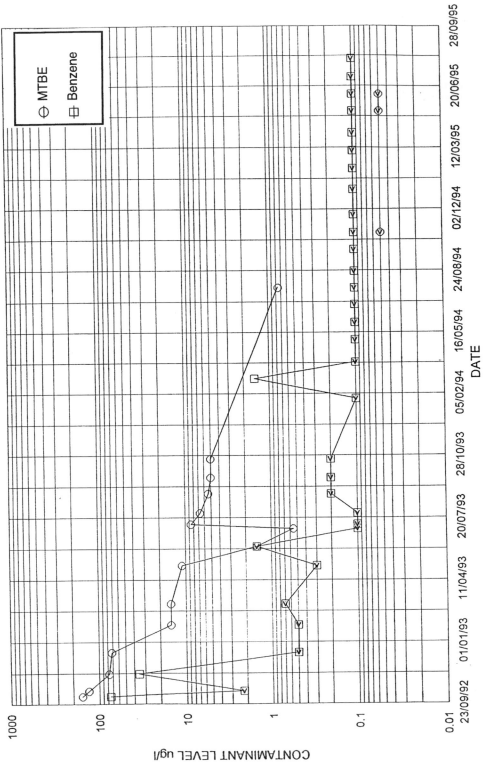

Fig. 7. (a) MTBE and benzene concentrations in the Hudd's Mill Spring (in micrograms per litre). Note that for benzene, only three samples returned values above the detection limit. The '<' sign implies a concentration below the analytical detection limit, which is represented by the plotted concentration of the point.

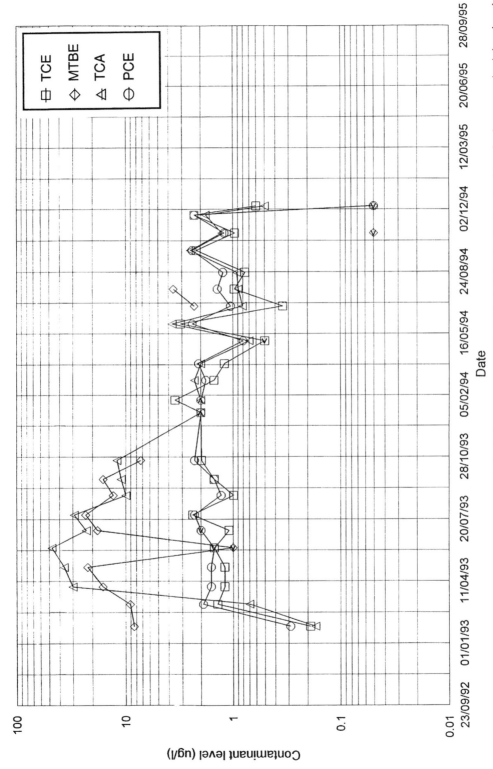

Fig. 7. (b) MTBE, TCE, TCA and PCE concentrations in borehole G1 (in micrograms per litre). The ' < ' sign implies that the determined concentration is less than the detection limit, represented by the value of the point.

Table 5. *Summary of measured chlorinated solvent and MTBE concentrations ($\mu g\, l^{-1}$) at monitoring wells between Stamford and Tallington/Pilsgate*

Location/date	TCE	TCA	PCE	MTBE
Lower Home Farm, Uffington				
16 Dec. 1992	<0.2	<0.16	<0.3	2.7
29 Jan. 1993	0.16	<0.10	<0.15	1.5
19 Mar. 1993	<0.15	<0.10	<0.20	1.5
Pilsgate Level Crossing				
16 Dec. 1992	<0.2	<0.16	<0.3	1.1
5 Feb. 1993	<0.15	<0.10	<0.15	<0.2
19 Mar. 1993	<0.15	<0.10	<0.20	<0.10
Pilsgate Grange (The Dingle)				
16 Dec. 1992	<0.2	<0.16	<0.3	<0.8

Note: <, Below detection limit. For locations, see Fig. 2.

in concentrations in October 1990 represents the arrival of a plume of MTBE from the petrol station spill (implying a travel time of around two years). This is regarded as unlikely, however, as one would expect a second similar plume two years after the depot spill (i.e. in autumn 1994), which is not observed.

(ii) The initial concentrations may represent a background or residual from the petrol station spill and the rise in October 1990 may represent the arrival of a MTBE plume from an unknown source. Declining MTBE concentrations measured in late 1992 and early 1993 might conceivably represent a component of MTBE from the depot spill (i.e. a double peak) or the decay of the peak from the unknown source.

There is thus no clear evidence to suggest that any of the MTBE observed at Tallington PS is derived from the depot spillage. It is possible that some of it may be derived from the filling station spill although this cannot be proven. It is also possible that there may be a third source of MTBE in the vicinity (road runoff or other petrol stations between Stamford and Tallington), which is causing the apparent arrival of a MTBE plume in late 1990. Scott Wilson Kirkpatrick (1996) identified other potential sources but, as yet, no investigations have been carried out to confirm whether MTBE has entered groundwater at these sites.

Summary The degree to which spilt fuel was retained in the unsaturated zone in the limestone at the agricultural depot is unknown. The lack of observed LNAPL contamination suggests that some retention may have taken place. Alternatively, LNAPL phase contamination may

have occurred at the water-table but not been detected at the available monitoring points. Finally, LNAPL contamination may have occurred but may have been flushed rapidly out of the aquifer before systematic monitoring was established.

Dissolved phase contaminants, notably benzene and MTBE, were released to groundwater from the spill at the agricultural depot. A large part of the dissolved phase contamination migrated very rapidly to Hudd's Mill Spring, where it was flushed out of the aquifer within just over 1 year.

There is some component of contaminant migration eastward, as evidenced by concentrations of benzene and MTBE in borehole G1. Whether contamination has migrated further eastward remains an open question, although there is no clear evidence to link the spillage at the depot to observed MTBE contamination at Tallington PS. This MTBE seems, at least in part, to have another, as yet unknown, source. No benzene or toluene contamination has been observed at Tallington.

Chlorinated solvent contamination

GTI (1989) encountered unexplained peaks during GC analysis of waters sampled from Hudd's Mill Spring, the hospital borehole and boreholes M1–6. More detailed analysis indicated the presence of TCE and PCE at substantial concentrations (TCE > drinking water limit of $30\,\mu g\, l^{-1}$) in M1–4 (Table 6). Concentrations declined with distance from the retail site. No TCE was detected at Newstead or the hospital borehole. The greatest TCE contamination in July 1989 was at M1 and M2 (Fig. 9).

No PCE was detected at Newstead observation well, while the greatest concentration was at M2. Traces of PCE in the hospital borehole, coupled with the absence of TCE, might possibly have been ascribable to dry-cleaning activities. The fact that levels of PCE at M4 are distinctly lower than those at M2 and M3 suggests, however, that the hospital is not the major source of PCE contamination.

The solvent concentrations suggest one or more sources of contamination to be located in the vicinity of the current retail site, with some component of migration and attenuation of concentrations towards the Hudd's Mill Spring.

To clarify the possible sources of chlorinated solvent contamination, concentrations of TCE, TCA and PCE have been plotted for M1–6, G1 and the Hudd's Mill spring for several months in 1993 (Fig. 10). During 1993, there

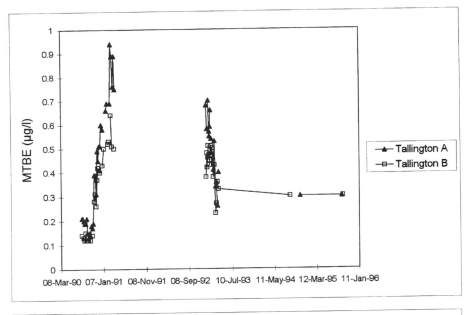

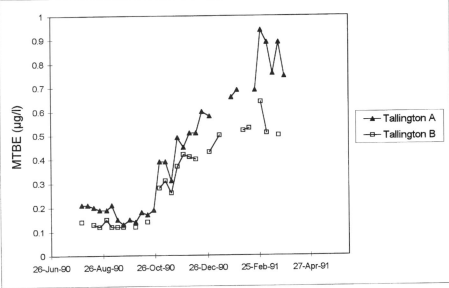

Fig. 8. Time series plots of MTBE concentrations at Tallington production boreholes A and B. The lower diagram is an expanded version of the rizing limb of the upper diagram.

are fluctuations, which may be seasonally influenced, in the concentrations of chlorinated solvents. In spring 1993, maxima occur in most boreholes for all three solvents. These peaks are documented in Table 7, for TCE, TCA, PCE and also MTBE for comparison. This table, together with Fig. 10, provides evidence that the chlorinated solvents do not appear to be

derived from the same source. The timing and value of the peaks for different parameters in different boreholes do not coincide. For example, the values of TCE in Hudd's Mill Spring do not approach those in M1 and M2, despite comparably high values of TCA.

G1 and Hudd's Mill Spring show closely related solvent fingerprints, with consistent

Table 6. *Summarized chlorinated solvent and BTEX concentrations ($\mu g\,l^{-1}$) measured by GTI in 1989*

Parameter	21 April 1989					25 July 1989*	
	TCE	PCE	Xylene	Benzene	Toluene	TCE	PCE
M1	44	7	2	<0.11	ND	109	3.9
M2	69	28	166	2.5	1.6	92.4	22.3
M3	41	12	230	3.3	1.5	54.5	7.3
M4						84.8	4.4
M5						13.0	2.6
M6						11.7	3.9
Hospital Borehole						<0.1	0.19
Hudd's Mill Spring						4.5	2.1
Newstead Obs. Well						<1.0	<1.0

* GTI (1989).

TCE and PCE values of a few micrograms per litre. A 'flush' of TCA contamination appears at these points during April 1993, resulting in concentrations of several tens of micrograms per litre. This indicates that there is a separate source of TCA to the sources of TCE and PCE. It also confirms the close hydraulic connection between Hudd's Mill and the agricultural depot. Borehole G1's solvent concentrations are typically less than those at Hudd's Mill Spring, indicating that the spring represents a more

direct flowpath from the solvent source than does the borehole.

The 'flush' of TCA also appears in April 1993 in M2 and M5. It then appears in M1 after a slight delay.

TCE does not reach particularly high concentrations during 1993 in the G1 or Hudd's Mill Spring. It is, however, present at very high concentrations (up to several hundred micrograms per litre) in M1 and at lesser concentrations in M2 and M5. PCE, on the other hand, is not

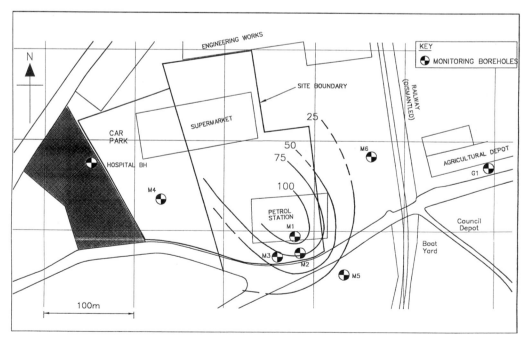

Fig. 9. Contours on TCE concentrations (in micrograms per litre) in groundwater at the study site, as measured on 27 July 1989 (after GTI 1989). The contours must be regarded as very tentative, given the sparse data points and the heterogeneous nature of the aquifer.

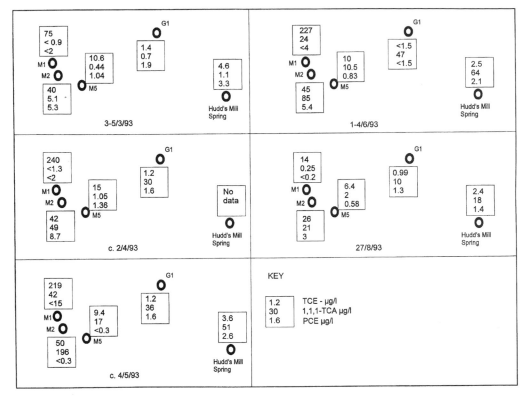

Fig. 10. Distribution of selected chlorinated solvents in groundwater throughout the spring and summer of 1993 (not to scale).

detected in M1, thus indicating that PCE has a different source to TCE. PCE is present in its highest concentrations in M2.

Figure 11 provides a model summarizing these observations. There appears to be a sporadic source of TCA, probably located somewhere up-gradient and between G1 and the retail site (i.e. possibly in the vicinity of the engineering works). From this source are direct transport pathways to Hudd's Mill Spring, G1, M2 and M5. There is also a delayed pathway to M1.

There appears to be a source of TCE up-gradient of M1, possibly within the landfilled quarry. The flow pathway from this source feeds M1, M2 and M5. It may discharge to the River Welland or Hudd's Mill Stream via

Table 7. *Dates and values of spring 1993 peaks in concentrations (µg/l) of selected solvents and MTBE at selected monitoring points.*

	TCE		TCA		PCE		MTBE	
	Date	Value (μg l^{-1})	Date	Value (μg l^{-1})	Date	Value (μg l^{-1})	Date	Value (μg l^{-1})
G1	(5/3/93)	(1.4)	4/6/93	47	(5/3/93)	(1.9)	4/5/93	22
M1	2/4/93	240	4/5/93	42	None		28/1/93	1.8
M2	4/5/93	50	4/5/93	196	2/4/93	8.7	28/1/93	1.3
M5	2/4/93	15	4/5/93	17	2/4/93	1.36	Consistently declining trend	
Hudd's Mill Spring	(3/3/93)	(36)	4/6/93	64	(3/3/93)	(3.3)	28/1/93	0.73

Parentheses indicate dubious and/or very low peaks.

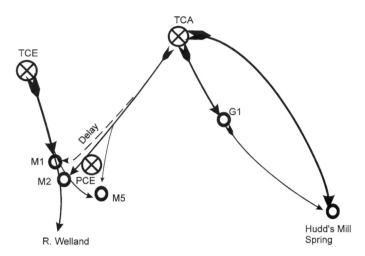

Fig. 11. Schematic model illustrating possible sources and migration pathways of chlorinated solvents (not to scale).

unknown springs. There does not appear to be a rapid flow connection to the Hudd's Mill spring, but the lower concentrations of TCE observed here may be evidence of a more diffuse connection.

The source of PCE appears to be located in the vicinity of M1 and M2, due to the marked differences in concentrations between these boreholes (although this may simply be due to the boreholes intersecting different fissure systems). The PCE concentrations in Hudd's Mill Spring and G1 may indicate another source of PCE up-gradient of these locations.

The borehole at M6 (although only sparsely sampled) gives no indications of solvent levels in excess of $2 \mu g l^{-1}$, suggesting that the former solvent storage area adjacent to the borehole site is unlikely to be a significant source of contamination.

The concentrations of solvents do not exceed relevant EQS levels (10 mg/l for TCE and PCE) or the drinking water limits ($30 \mu g l^{-1}$ for TCE, $10 \mu g l^{-1}$ for PCE) in the River Welland at Uffington, based on data from 1992/93. In the Hudd's Mill Spring, however, concentrations of TCE have occasionally exceeded $10 \mu g l^{-1}$. A biological survey of the Hudds Mill Stream and River Welland (NRA unpublished data) was unable to determine any degradation in aquatic ecosystems that could be attributed to chlorinated solvent contamination.

Summary Examination of the spatial and temporal distributions of TCA, TCE and PCE has suggested that there are at least three apparently independent sources of solvent contamination

in the study area. The sources themselves may be DNAPL pools on the aquifer base, solvent containers or smaller accumulations in pores or fractures. The inferred locations of these sources must, however, be regarded as subject to considerable uncertainty, given the heterogenous, fissured and anisotropic nature of the aquifer. The sources could, in fact, lie in close proximity to each other, e.g. different solvent drums lying in different positions in the landfilled quarry and feeding different fracture systems and, hence, flow pathways.

An alternative explanation for the different solvent behaviours, could be the differing hydrodynamic behaviour, degradation rates and/or retardation of the various solvents. TCE and TCA have similar retardations on organic carbon, whereas PCE is slightly better retarded and less soluble (Table 2). However, given the very rapid groundwater flow rates and the kartsified nature of the aquifer, this is considered an insufficient explanation for the data.

Detections of chlorinated solvents appear to be episodic. The reasons for this may include:

- rusting solvent containers periodically releasing slugs of fresh solvent during progressive deterioration;
- rainfall or rising water levels dissolving solvent contamination from the unsaturated zone;
- rainfall related rises in water-table causing periodically elevated groundwater flow velocities, removing DNAPL phase solvent from fractures, pore spaces or pools within the aquifer.

Conclusions

Hydrocarbon contamination from the filling station

There is no direct evidence of LNAPL hydrocarbon at the water-table. The heterogeneous nature of the aquifer or the delay between the spill and the commencement of monitoring may have led to LNAPL not being observed. Hydrocarbons appear to have been effectively retained in the unsaturated zone (composed of sandy fill material) with some leaching of dissolved components to groundwater for a limited period. Dissolved hydrocarbon contamination from this source is no longer being detected in significant concentrations. The leakage thus presents negligible risk at present and it can be argued that no further action need be taken.

Hydrocarbon contamination from the agricultural depot

There exists no direct evidence of LNAPL hydrocarbon at the water-table, although the placing of the monitoring borehole and/or any delay between the spill and the commencement of monitoring may have led to LNAPL not being observed. Hydrocarbons may have been retained to some extent in the unsaturated zone, with some leaching of dissolved components to groundwater for a limited period. Dissolved hydrocarbon contamination from this source is no longer being detected in significant concentrations at Hudd's Mill Spring and thus presents negligible risk to the River Welland.

It can be demonstrated (Fig. 8) that the first arrival of MTBE contamination at Tallington PS is *not* derived from the depot. Whether this contamination may be derived from the petrol station spillage or whether any of the later (post-1992) contamination is attributable to the depot cannot be adequately resolved on the basis of the existing data.

It remains possible that there is a third (as yet undetected) source of MTBE contamination, resulting in the observed MTBE contamination at Tallington PS. If such a source exists, it may be the same as that responsible for the detected MTBE at Lower Home Farm and Pilsgate Level Crossing. The source may be another fuel station, private fuel storage or runoff from roads, possibly following a spill. However, since concentrations over $c.\,0.4\,\mu g\,l^{-1}$ MTBE have not been detected during 1994 and 1995 at Tallington PS, the contamination probably poses no immediate risk to the source (given the taste/odour thresholds of 2–$3\,\mu g\,l^{-1}$).

Whatever the source of the MTBE, it is possible that other hydrocarbon components could be migrating towards Tallington PS at a more retarded rate than MTBE. As yet, no benzene or toluene have been detected at Tallington PS, possibly due to the higher rates of attenuation and biodegradation of these components relative to MTBE.

Chlorinated solvent contamination

The occurrences of dissolved concentrations of solvents TCE, TCA and PCE appear each to be derived from independent sources in the studied area. These may be geographically separate sources, or sources in relatively close proximity, feeding differing fracture networks. The fact that concentrations are fairly persistent, but with episodes of elevated concentration, suggests that there may be sources of DNAPL solvent within the aquifer, either as pools on the base of the limestone or in containers in the landfill, or as smaller accumulations within fractures or pore spaces. No direct evidence of DNAPL was, however, observed.

The concentrations of solvents do not exceed relevant EQS levels ($10\,\mu g\,l^{-1}$ for TCE and PCE) or the drinking water limits ($30\,\mu g\,l^{-1}$ for TCE, $10\,\mu g\,l^{-1}$ for PCE) in the River Welland at Uffington, based on data from 1992/93. In the Hudd's Mill Spring, however, concentrations of TCE have occasionally exceeded $10\,\mu g\,l^{-1}$. As the Mill Stream is presumed to receive a significant component of its flow, particularly in dry summers, from the spring, it is conceivable that EQS values may occasionally be approached in the Mill Stream itself. A biological survey shows, however, no significant damage to aquatic ecosystems in either the Mill Stream or the Welland. It is concluded that the hazard from chlorinated solvent contamination from Hudd's Mill spring is very low, both measured against regulatory guidelines and on the evidence of field surveys.

Given the possible existence of a flow path towards Tallington, there might be some cause for concern over migration of chlorinated solvents eastward. Given the resistance to degradation of chlorinated solvents, it would be wise to continue sampling for TCE, PCE and TCA on a regular basis along the flow path to Tallington. The remediation of chlorinated solvent contamination is not judged to be feasible given the following:

- the uncertain location and multiple nature of the solvent sources;
- the complex, fractured nature of the aquifer;
- the large volumes of water that would possibly need to be pumped to effect a pump-and-treat solution (with associated disposal problems and possible derogation of springs).

Implications for policy

Some 35% of the population of England and Wales are dependent on groundwater for potable water supply. The majority of this comes from aquifers where the flow mechanism is dominantly via fractures or fissures (solution-enhanced discontinuities), including major aquifers such as the Lincolnshire Limestone, the Chalk, the Jurassic Oolites and the Permian Magnesian Limestone. Due to the historical development of hydrogeology in the UK and the sparse use of true 'hard rocks' (granites, metamorphics) as aquifers, the conventional approach has been to simulate these aquifers as porous continua. Other lands, notably in North America and Scandinavia, have developed fracture network approaches (Odling 1993, Bradbury & Muldoon 1994).

In the UK, the EA implements a groundwater protection policy based (among other strategies) on source protection zones (SPZs) for all major potable groundwater abstractions. These have been analytically and numerically modelled using the porous continuum approach (with models such as FLOWPATH, MODFLOW/MODPATH or WHPA), but the derived SPZs, based on total catchments and 50- and 400-day saturated travel times, have been found to be severely wanting in some lithologies, due to underestimation of transport rates and inadequate modelling of dispersion effects (e.g. Banks et al. 1995). This has obvious and significant consequences in the case of pollution incidents or contaminated land falling outside the modelled, but within the actual, travel time zones.

The Stamford area in the Lincolnshire Limestone demonstrates these points extremely well. Much effort has been expended developing groundwater models, largely for the purposes of groundwater management, of the southern Lincolnshire Limestone. These have proved remarkably successful for that purpose. Guérin & Billaux (1994) have, however, observed that a model which is excellent for modelling groundwater head distributions and fluctuations, may not be satisfactory for modelling contaminant flow and transport. The hydrographs in Fig. 5 demonstrate that the head responses in the aquifer behave as one would expect for a continuous medium: similar amplitudes, timings – in fact, almost identical hydrographs. It is not difficult to imagine that such responses could be effectively modelled using a continuum numerical model. It would be difficult to guess from the hydrographs, that contaminant transport would be so anisotropic and follow such discontinuous paths along discrete fissure systems as has been observed in the course of this study. It is thus concluded that the continuum approach is not adequate for understanding contaminant transport in the Lincolnshire Limestone and similar aquifers.

It is suggested that the time is ripe for research with the following aims:

(i) critical assessment of the existing application of porous continuum codes to British fractured/fissured aquifers;

(ii) development and assessment of a methodology for contaminant transport and SPZ modelling based on stochastic, discrete fracture network techniques;

(iii) application of this approach at a limited number of sites, calibrated against real tracer tests or contamination incidents to test its practical utility;

(iv) production of concise guidelines and recommendations for the estimation of SPZs in fractured/fissured lithologies, which aim to combine ease and rapidity of use with scientifically defensible results.

The study described here was performed for the Anglian Region of the National Rivers Authority (now the Environment Agency). The author wishes to thank the NRA's Lincoln Office for permission to publish this paper and the support of Peter McConvey, Clare Blackledge and John Sweeney during the study. The author thanks Robert Barnes for permission to use material from his MSc thesis and also wishes to acknowledge the contributions of Stephanie Foreman and Nadine Tunstall-Pedoe of Scott Wilson Kirkpatrick to the work.

References

BANKS, D., DAVIES, C. & DAVIES, W. 1995. The Chalk as a karstic aquifer: the evidence from a tracer test at Stanford Dingley, Berkshire. *Quarterly Journal of Engineering Geology*, **21**, S31–S38.

BARKER, J. F., PATRICK, G. C. & MAJOR, D. 1987. Natural attenuation of aromatic hydrocarbons in a shallow sand aquifer. *Groundwater Monitoring Review*, Winter 1987, 64–71.

BARNES, R. C. 1993. *Tracer Study in the Vicinity of Two Chemical Spillages in Stamford, Lincolnshire*. MSc Dissertation, School of Environmental Sciences, University of East Anglia.

BRADBURY, K. R. & MULDOON, M. A. 1994. Effects of fracture density and anisotropy on delineation of wellhead protection areas in fractured rock aquifers. *Applied Hydrogeology*, **2**(3/94), 17–23.

BRADBURY, C. G., RUSHTON, K. R. & TOMLINSON, L. M. 1994. *The South Lincolnshire Limestone catchment – Final report to NRA Anglian Region*. School of Civil Engineering, University of Birmingham.

DRISCOLL, F. G. 1986. *Groundwater and Wells* , 2nd edition. Johnson Filtration Systems, St Paul, Minnesota.

FETTER, C. W. 1994. *Applied Hydrogeology*, 3rd edition. Macmillan.

GEOTECHNICAL ENGINEERING 1987. Geotechnical Engineering (Northern) Limited Report, SA/P/1623N.

GTI 1988. *Preliminary report. Fuel leak at Supermarket, Stamford, Lincolnshire*. Groundwater Technology, Epsom, Surrey, UK, November 1988.

—— 1989. *Petroleum spill, supermarket, Stamford. Review of monitoring and recommendations for further work*. Report for Anglian NRA unit, Lincoln, by Groundwater Technology International Ltd, Epsom, Surrey, UK, September 1989.

GUÉRIN, F. P. M. & BILLAUX, D. M. 1994. On the relationship between connectivity and the continuum approximation in fracture-flow and transport modelling. *Applied Hydrogeology*, **2**(3/94), 24–31.

NYER, E. K. & SKLADANY, G. J. 1989. Relating the physical and chemical properties of petroleum hydrocarbons to soil and aquifer remediation. *Ground Water Monitoring Review*, Winter 1989, 54–60.

ODLING, N. E. 1993. An investigation into the permeability of a 2-D natural fracture pattern. *In*:

Banks, S. B. & Banks, D. (eds) *Hydrogeology of Hard Rocks*, Proceedings of the 24th Congress International Association of Hydrogeologists, Oslo, June/July 1993, 290–300.

RIVETT, M. O., LERNER, D. N. & LLOYD, J. W. 1990. Chlorinated solvents in UK aquifers. *Journal of the Institute of Water and Environmental Management*, **4**(3), 242–250.

RUSHTON, K. R. 1981. Modelling groundwater systems. *In*: Lloyd, J. W. (ed.) *Case Studies in Groundwater Resources Evaluation*. Oxford Science Publications, 150–162.

—— & TOMLINSON, L. M. 1994. *A study of groundwater protection zones in the Southern Lincolnshire Limestone using a regional groundwater model*. Report for NRA Anglian Region, School of Civil Engineering, University of Birmingham, March 1994.

——, SMITH, E. J. & TOMLINSON, L. M. 1982. An improved understanding of flow in a limestone aquifer using field evidence and mathematical models. *Journal of the Institute of Water Engineers and Scientists*, **36**(5), 369–387.

SCOTT WILSON KIRKPATRICK 1996. *Groundwater pollution study, Stamford*. Report for the NRA, Anglian Region, Report No. CAPEP/GEO/333, 18/1/96.

SWEENEY, J., HART, P. A. & McCONVEY, P. J. 1998. Investigation and management of pesticide pollution in the Lincolnshire Limestone aquifer in Eastern England. *This volume*.

SYMINGTON, R., BURGESS, W. G. & DOTTRIDGE, J. 1994. Methyl tertiary butyl ether (MTBE): a groundwater contaminant of growing concern. *Proceedings of the 3rd Annual Conference on Groundwater Pollution*. IBC Technical Services/ Royal Holloway, University of London, 16/17 March 1994.

Investigation and clean-up of jet-fuel contaminated groundwater at Heathrow International Airport, UK

L. CLARK & P. A. SIMS

WRc alert, Henley Road, Medmenham, Marlow, Buckinghamshire SL7 2HD, UK

Abstract: British Airways in 1987 discovered a leak of jet fuel (kerosene) at Heathrow Airport, west of London. The leak had led to free-phase kerosene floating on top of the shallow water-table beneath the site, with thicknesses of product of almost 1 m in places. WRc alert was employed to undertake a programme of investigations to delimit the free product. Work was then initiated to remove the free product from the water-table and restore the aquifer. This has been successful, with a total of 39 400 l of kerosene being recovered from the site. The recovery of kerosene took about four years of continuous work, illustrating the difficulty of aquifer remediation and the long time involved in remediation programmes.

British Airways discovered in 1987 that a leak of jet fuel (kerosene) adjacent to Technical Block L at Heathrow Airport, 15 miles west of Central London, had led to the fuel reaching the water-table and floating on the groundwater. The leak was from a cracked fuel pipe leading to an engine maintenance facility, and was believed to have taken place over a number of years. The leak was discovered when fuel was observed floating on drainage water in a manhole north of Technical Block M (Fig. 1). British Airways installed a large concrete-lined well (Well 1), about 1.5 m in diameter, close to the manhole and this showed about 10 cm of kerosene floating on the groundwater. The leak was also traced to the cracked pipe and the fracture repaired.

WRc alert was engaged in 1987 to make an assessment of the size of the hydrocarbon leak and advise on necessary remedial action. This paper describes the assessment of the leak and the subsequent groundwater remediation programme. The remediation has met problems from having to take place in a built-up area of a large working airport but has been successful through the active co-operation of British Airways.

Hydrogeology

Heathrow Airport is built on the Taplow Terrace, a fluvio-glacial terrace of the River Thames, adjacent to the floodplain of the river. Beneath the airport approximately 4.5 m of coarse clean gravels overlie low-permeability London Clay. The London Clay is an aquiclude up to about 50 m in thickness, and overlies the regional Chalk aquifer in the central London Basin. The water-table in the gravels is shallow (about 2.5 m from the surface), and groundwater flow tends to be from beneath the airport

towards the Thames at Shepperton some 9 km to the south (Fig. 2).

Site assessment

The site assessment was phased to determine the geology and groundwater flow beneath the site and then delimit the extent of the plume from the hydrocarbon leak. The behaviour of hydrocarbons following a leak has been described in numerous publications (Fussel *et al.* 1981). Briefly, provided a sufficient volume of hydrocarbon leaks, it will percolate vertically downwards to the water-table and then spread outwards from the leak location as a thin layer ('pancake') floating on the water-table. Hydrocarbon will dissolve from the pancake into the underlying groundwater to pollute the water. The floating product will tend to move very slowly down the groundwater gradient, but on a very flat water-table, if the pancake is thick enough, it may move up the groundwater gradient under the head generated by the leak. The thickness of the hydrocarbon is commonly monitored by boreholes or wells, but it is important to recognize that the thickness observed in a borehole is *not* the same as the thickness of hydrocarbon in the adjacent aquifer; the thickness measured in a borehole may be *four times* the thickness in the adjacent aquifer.

Four observation boreholes, Boreholes 1–4 (Fig. 1), were drilled to assess the spread of kerosene east, west, south and north of the leak. The presence of the floating free-phase kerosene 'pancake' in Borehole 1 and Well 1, against the groundwater gradient, suggested that the kerosene may have been coming from a totally different source somewhere to the northwest. A further eight boreholes, Boreholes 5–12, were drilled to delimit the extent of the

CLARK, L. & SIMS, P. A. 1998. Investigation and clean-up of jet-fuel contaminated groundwater at Heathrow International Airport, UK. *In*: MATHER, J., BANKS, D., DUMPLETON, S. & FERMOR, M. (eds) *Groundwater Contaminants and their Migration*. Geological Society, London, Special Publications, **128**, 147–157.

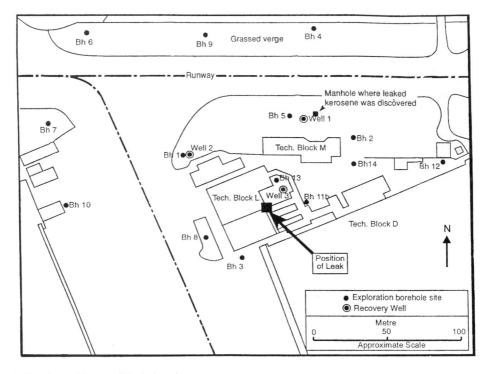

Fig. 1. Heathrow Airport oil leak: location map.

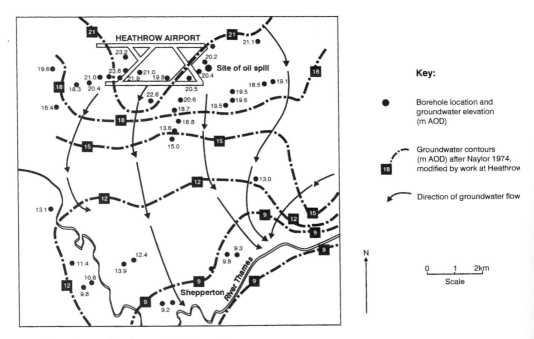

Fig. 2. Groundwater flow beneath Heathrow Airport.

floating hydrocarbon more closely. Boreholes 11 and 12 were drilled along the edge of Technical Block D to give more information on the extent of the floating hydrocarbon along the foundations of the Block. A final two observation boreholes, Boreholes 13 and 14, were drilled to give more information on the 'pancake' of free product and to monitor the effects of the remediation programme.

Observation Boreholes 1–10 were drilled to different depths but to a similar design. Initially, due to a subcontractor's error, Boreholes 1–4 were lined with solid casing to 1 m below the water-table, which effectively cut them off from any floating product. These holes were redrilled. In their final design, all boreholes penetrated the London Clay and were lined with perforated screen extending to about 1 m above the water-table, allowing free product floating on the water-table to enter the borehole. The boreholes with faulty designs were numbered 1A to 4A; their replacements were labelled 1B to 4B. The completion of observation Borehole 10 is shown in Fig. 3. Observation Boreholes 11–14 were completed with 50-mm ID piezometer pipe instead of 150-mm ID casing.

The observation boreholes showed that the undisturbed groundwater flow beneath the spill site is easterly or southeasterly (Fig. 4). The 'pancake' of floating kerosene was about 100 m in diameter and at its thickest point measured 0.95 m in thickness (Borehole 5 on 16 September 1988). Free product has been measured in Boreholes 1B, 5, 11 and 13 and in Well 1. Although no free product was detected during the construction of Boreholes 3B, 9 and 10, kerosene odour was reported during the drilling operation. No kerosene has been detected in the outlying observation boreholes since monitoring began. The absence of free kerosene in Boreholes 3B and 8 when they were drilled was attributed to the easterly groundwater gradient and strong southeasterly groundwater flow at the western edge of the site preventing pollution migration in that direction. No kerosene had been detected in Borehole 2B but it was suspected that some migration may have been directed along the north face of the foundations of Technical Block D, so Boreholes 11 and 12 were drilled. No kerosene was found in Borehole 12. It should be noted that free kerosene has been detected in Borehole 2 in the later stages of the clean-up, most likely due to changes in the local groundwater flow induced by groundwater abstraction.

The site assessment has confirmed the essential simplicity of the site geology but has shown up some variation in lithology. The alluvial gravels thicken and coarsen eastwards and a thicker lens of coarser gravels extends southeastwards from Borehole 6 to Borehole 1B. This lens could act as a preferential flowpath for groundwater, giving rise to the strong flow mentioned above. Observation Boreholes 1B, 2B and 5

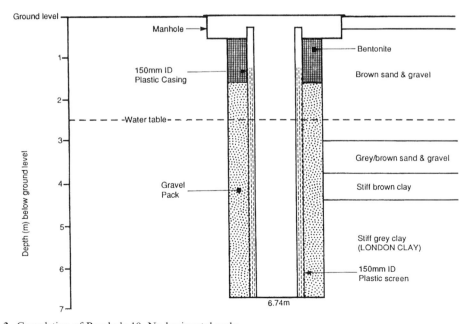

Fig. 3. Completion of Borehole 10. No horizontal scale.

were pump tested by short constant discharge tests to obtain an estimate of the aquifer characteristics and to help to establish design criteria for the remediation works. The aquifer transmissivity ranged from $100\,m^2\,day^{-1}$ (Borehole 5) to $1000\,m^2\,day^{-1}$ (Borehole 2B), both of which are high values for such a thin aquifer. The specific capacities, in $m^3\,day^{-1}\,m^{-1}$, were 210 (Borehole 1B), 330 (Borehole 5) and 1900 (Borehole 2B).

The sampling and analysis of the groundwater and free product were aimed at identifying the product beyond reasonable doubt and obtaining the water quality information needed to implement effective clean-up. Product identity was necessary to confirm that the source of the leak had indeed been identified and stopped. Water quality information was needed to ensure that effluent discharge consent conditions imposed by Heathrow Airport Limited for the waste water could be met. The water from the discharge boreholes was disposed to the surface drainage system which empties to balancing lakes before flowing to the Thames under an NRA (now Environment Agency) discharge consent.

Samples of free product floating on the water in Borehole 1B and Well 1 (Fig. 4) were compared with a sample of fresh jet fuel supplied by British Airways. The analyses were undertaken at WRc's Medmenham laboratory using capillary gas chromatography with flame ionization detection (CGC-FID) and confirmed that the free product was jet fuel. A sample of water from Borehole 3B, in which a kerosene odour could be detected, was analysed by the same method and compared with groundwater spiked to saturation with jet fuel. The dissolved hydrocarbon in Borehole 3B was shown by this means to be derived from jet fuel.

Samples of groundwater from Boreholes 1B, 2A and 2B (two boreholes drilled to different designs at the same place; see above), and Borehole 5 were analysed for dissolved hydrocarbons. Boreholes 2A and 2B both showed no detectable dissolved hydrocarbons; Borehole 1B contained about $1\,mg\,l^{-1}$ dissolved hydrocarbon; and the sample from Borehole 5 contained suspended free product. The concentration in the water from Borehole 1B, in which free product occurs on the water-table, may be representative of that to be expected close to a kerosene spill. The solubilities of kerosene and JP-4 jet fuel are both $<1\,mg\,l^{-1}$ (Cole 1994).

The groundwater samples from Boreholes 1B and 2B were analysed for major ions to determine the difference between water contaminated by kerosene, and water with no detectable

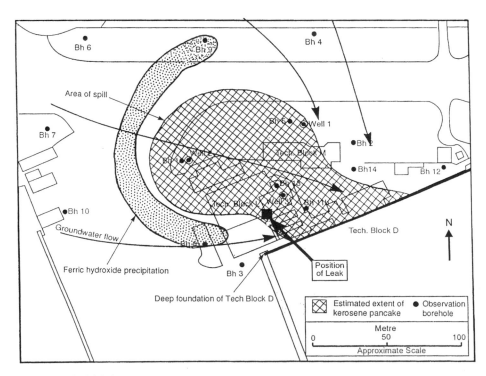

Fig. 4. Summary of initial site assessment.

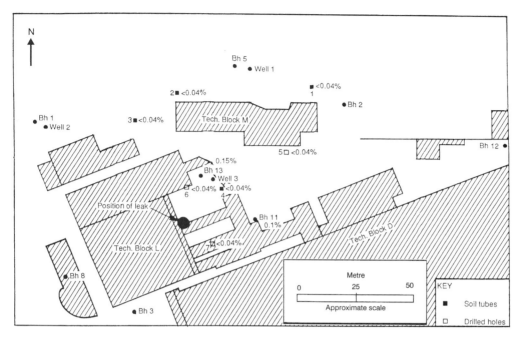

Fig. 5. Kerosene concentration in soil gas in the vicinity of Technical Block D.

pollution. The main differences between the two samples are shown in the iron and nitrate concentrations. In kerosene-contaminated water (Borehole 1B) the iron was high ($0.54\,\mathrm{mg\,l^{-1}}$) and the nitrate low ($0.66\,\mathrm{mg\,l^{-1}}$ N), while in the uncontaminated water the iron concentration was comparatively low ($0.04\,\mathrm{mg\,l^{-1}}$) and the nitrate high ($4.99\,\mathrm{mg\,l^{-1}}$ N). The low nitrate and high iron concentrations in the samples from Borehole 1B suggest that the groundwater is deoxygenated, with a potential for denitrification and the formation of soluble ferrous iron from the aquifer material. The high concentration of dissolved iron was later to present problems during clean-up operations.

A survey of dissolved oxygen in groundwater in all the observation boreholes did confirm water with low levels of dissolved oxygen to be present across an area corresponding generally with the kerosene 'pancake'. During this survey, iron hydroxide precipitate was noted in water samples from Boreholes 8 and 9. This could have been produced by the mixing of oxygenated and deoxygenated water along the margins of the dissolved contaminant plume (Fig. 4).

To complement the survey of groundwater quality and the distillation of free product, a soil gas survey was undertaken. Four small-diameter tubes were installed in the open ground between existing observation boreholes

(Fig. 5) and three holes were drilled through the concrete floors of Technical Blocks M and L. The kerosene concentrations in the soil atmosphere in these tubes and holes, together with those concentrations in observation Boreholes 11 and 13, were measured using Draeger tubes. Kerosene was detectable (detection limit 0.04%) only in the two observation boreholes. Samples of the soil atmosphere taken for laboratory analysis by gas chromatography (GC) showed a significant depletion of oxygen, and enrichment in methane and carbon dioxide in the samples from the two observation boreholes and that from tube 2 (between Boreholes 1 and 5). The source of the methane and carbon dioxide was believed to be the biodegradation of the underlying kerosene. Tubes 1 and 3 and drilled holes 5 and 7 showed only elevated concentrations of carbon dioxide relative to the other sampling points, possibly also derived from biodegradation.

The assessment phase illustrates a problem commonly met in remediation work in a built-up area: accessibility. The leak took place adjacent to Technical Block L but drilling of observation boreholes close to the point of the leak was not possible because the area is in constant use by road traffic. The extent of the spill beneath this Block, shown in Fig. 4, must therefore be conjectural.

Spill remediation

The decision to clean up the kerosene spill at Heathrow was generated internally within British Airways in response to discussions with the British Airports Authority. There was no formal risk assessment nor the identification of potential impacts on receptors. However, the unacceptability of allowing free kerosene floating in the drainage system, as well as the pollution potential and fire hazard, were clearly recognized. The aim of the clean-up was to remove the free product and thus to remove the pollution potential and fire hazard represented by the floating kerosene.

Remediation structures

The basic remediation structures used at Heathrow are large-diameter wells lined with perforated concrete rings about 1.5 m in diameter. The wells were installed by excavating pits by back-hoe as deeply as possible, about 2 m below the water-table, lowering the rings into position and then backfilling around these using the gravel excavated from the pit (Fig. 6).

The first well, Well 1, was installed by British Airways shortly after the spill was detected. Well 2 was installed (Fig. 6) close to Borehole 1B where a considerable thickness of fuel was shown to be floating on the water-table. These

two wells were then used to begin the recovery of the floating kerosene.

The shallow water-table enabled much of the remediation equipment to be surface-mounted. The kerosene was removed by floating oil-skimmer pumps manufactured by Oil Recovery Systems. Surface-mounted centrifugal pumps were used to lower the water-table and produce as large a cone of depression as possible.

The oil skimmer pumps operate on a system whereby the installation floats on the water-table, so that the kerosene layer is in contact with a hydrophobic mesh. The kerosene filters through the mesh into a chamber which, when full, is automatically emptied to a waste oil tank by a small oil pump. The systems operate automatically and have worked with very little trouble throughout the period of the clean-up operation.

The water pumps were installed initially in Boreholes 1B and 2B with their intakes at the bottom of the boreholes. The high discharge rates needed to produce a cone of depression in the adjacent wells resulted in kerosene mixing with groundwater so that the waste water held unacceptable (commonly over $1\,g\,l^{-1}$) concentrations of dissolved and entrained hydrocarbon.

As a result, the water pumps were re-installed with their intakes in the wells in order to reduce the intake turbulence because of the larger diameter of the wells. This resulted in an instant improvement in water quality, although levels of

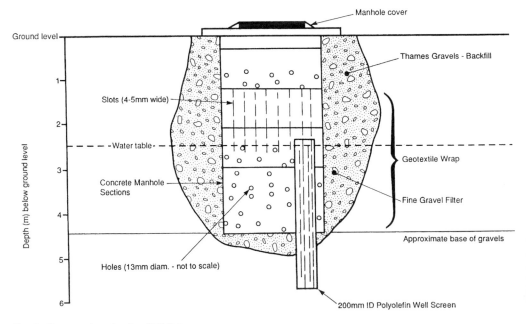

Fig. 6. Construction details of Well 2.

$1\,\mathrm{mg\,l^{-1}}$ of dissolved hydrocarbon were still found and were seen as unacceptably high. In addition to the high levels of hydrocarbons, ferric iron precipitated in thick deposits on the pump intake filter through the mixing and oxygenation of an anoxic groundwater.

The design of the wells was then altered by drilling a borehole in the base of each well into the London Clay. These boreholes were lined with 200 mm ID screen and served as pumping sumps. Pumping from this deeper level resulted in further improvements in water quality and reduced the problem of iron precipitation.

The groundwater levels were monitored regularly and the cone of depression produced by the pumping of Wells 1 and 2 encompassed the estimated area of the kerosene 'pancake' within two months of the pumping starting. It was anticipated, therefore, that these two installations would be sufficient to remove all the floating kerosene.

The initial rate of recovery was extremely successful and 19 200 litres of free phase kerosene were removed. The recovery rate then dropped substantially, yet the kerosene layer in Borehole 11 still remained unaltered, suggesting that Wells 1 and 2 were not affecting the southern part of the kerosene 'pancake'. In January 1990, a third well, Well 3, and an adjacent observation borehole, Borehole 13, were installed. The oil skimming pump from Well 2, where oil recovery was complete, was then installed in Well 3. Initially, kerosene recovery was poor but with the installation of a larger water pump used to increase the drawdown in the well, recovery improved. It is believed that Well 3 tapped a 'pool' of kerosene isolated from the effects of Wells 1 and 2 by foundations of the Technical Blocks. This illustrates a further difficulty in cleaning up a spill in a built-up area where isolated pockets of pollution can be held in building substructures.

The enhancement of recovery using the waste water to flush the kerosene towards the wells by recharging the water to the aquifer upgradient from the well was considered. A recharge system was installed near Well 3 on a trial basis but the waste water precipitated excessive amounts of iron hydroxide, which blocked the pipework and gravel surface and created an eyesore. The trial was abandoned when no improvement in recovery could be measured.

An alternative to oil-skimming pumps is a submersible dual-pump recovery system which is commonly used in situations where floating product is not close to the surface, i.e. deeper than 6–7 m. Such a system was used at Heathrow on a trial basis but was found to be unsuitable in this situation. The float system that controlled the switching of the oil recovery pump was not sufficiently precise to ensure the kerosene layer was removed and the system required a high maintenance staffing. In view of the reliability and low-maintenance requirements of the skimmer system, the trial was abandoned.

Remediation results

The recovery system was active for four years up to early 1993, during which 29 300 l of kerosene were recovered, of which 10 100 l came from Well 3 (Fig. 7). The kerosene was sold to be blended into commercial heating oil.

After operations began, the recovery from Wells 1 and 2 continued at a fairly constant rate of about 40 l day^{-1} for some 480 days but then dropped markedly. At this stage the fuel entering Well 2 was minimal, although a thin film of kerosene was visible on the water in the well. The recovery from Well 1 continued but at a low rate of 0.3–0.5 l day^{-1}.

The removal of fuel within the influence of Well 2 was considered complete and the equipment was moved from this well to Well 3 in January 1990. The specific capacity of Well 3 was much higher than the other two wells and a larger pump unit had to be installed to lower the water-table sufficiently to allow kerosene to flow into the well. The recovery in Well 3 started slowly but, after the large water pump had been installed, followed a similar pattern to that of Wells 1 and 2, with a good recovery rate continuing for about 200 days before then falling off.

The recovery system in Well 1 was removed in May 1991 and slow recovery by the oil skimmer continued from Well 3 up to April 1993. The recovery from all wells and any borehole containing measurable kerosene then continued for a short time by passive systems: either absorbent mops or a floating bucket with hydrophobic mesh sides. The removal of the original kerosene 'pancake', as far as could be ascertained, was then considered complete.

The effects of the long-term water pumping from the wells can be seen in Fig. 8. This cone of depression was virtually the same as that achieved after only two months of pumping.

Until July 1990, the quality of the groundwater abstracted from the wells was monitored weekly for BOD, suspended solids and total dissolved hydrocarbons – a requirement of Heathrow Airport Limited. Initially the hydrocarbon concentration was high, commonly exceeding $1\,\mathrm{mg\,l^{-1}}$, due to entrapped free kerosene. Following changes to the well design, the concentrations

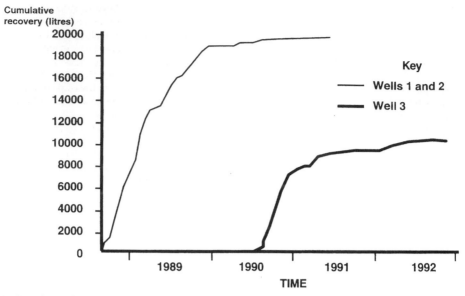

Fig. 7. Cumulative free phase oil recovery from Wells 1, 2 and 3 at Heathrow Airport.

dropped and rarely exceeded the detection limit of $0.06 \, \text{mg} \, \text{l}^{-1}$. The waste water from the remediation has been discharged to the drainage system of the airport, which leads to a balancing reservoir before flowing into the Thames. The airport has a consent imposed by the National Rivers Authority (now the Environment Agency) on the discharge from the balancing reservoir to the Thames. The balancing reservoir provides settlement and dilution for the remediation

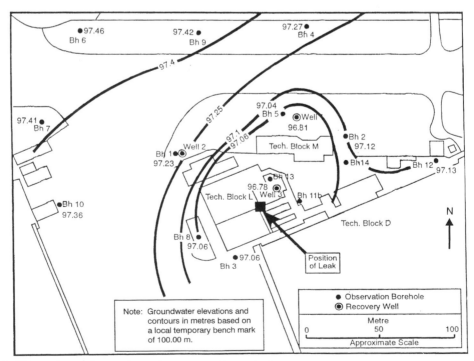

Fig. 8. Effects of groundwater abstraction on groundwater levels.

waste water and the remediation operations have caused no problems with the quality of the discharge to the Thames.

Because of the presence of a large degreasing facility in Technical Block D, it was suspected that there may be local solvent contamination of the groundwater. At the request of British Airways, the water from the wells was monitored monthly for a suite of solvents including 1,1,1-trichloroethane (TCA) and trichloroethene (TCE). The analyses showed that the groundwater at the beginning of the clean-up was significantly contaminated with TCA (up to $194 \mu g l^{-1}$) and TCE (up to $40 \mu g l^{-1}$). There

was no reason to believe that the solvents were connected to the kerosene leak but the monitoring results for the solvents are shown in Fig. 9 because of their significant trend throughout the clean-up period. The concentrations of TCA and TCE from both Wells 1 and 2 declined substantially during the clean-up. It appears that the kerosene clean-up was also removing a local solvent problem at the same time.

The active remediation by pump and treat ended early in 1993 though kerosene recovery through passive collectors in the three wells continued for about another year. During this period, after pumping stopped, the monitoring

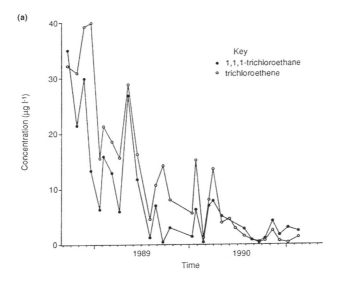

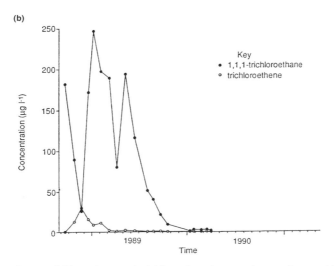

Fig. 9. Concentration of 1,1,1-trichloroethane and trichloroethene in groundwater abstracted from (**a**) Well 1 and (**b**) Well 2, Heathrow Airport.

of groundwater also ended. The clean-up project officially ended in 1994.

Discussion

The experience of WRc alert in working with the Heathrow kerosene leak, together with similar incidents in the UK and abroad, has taught us lessons that are applicable to most hydrocarbon spills or leaks.

In the case of a large long-term leak, such as Heathrow, where the hydrocarbon has already reached the water-table, there is no need for emergency action other than to find the source of the leak and stop it. The hydrocarbon 'pancake' in *granular aquifers with low ground-water gradients* will not move any significant distance and the main pollution front, that of the dissolved hydrocarbons in groundwater, will move only slowly. At Heathrow, the edge of the 'pancake' after several years was still less than 100 m from the source and the hydro-carbon-polluted water was present only in the same general area. The groundwater in Borehole 2B (Fig. 4), which is only metres from the edge of the floating fuel, was uncontaminated.

The fact that emergency action is not required does not mean that nothing needs to be done. However, it provides sufficient time for the necessary remediation action to be carefully planned. At Heathrow, the driving force behind the remediation was internally generated in British Airways; elsewhere, however, the driving force could be from regulatory authorities or demands for clean-up from agencies responsible for funding redevelopment or land sales.

In almost all cases the first phase of the remediation programme will be a site investiga-tion to establish the site land-use, geology, hydrogeology and extent of the effects of the spill or leak. The importance of conducting this investigation phase thoroughly cannot be over-emphasized, because the data collected will provide the basis of the remediation system design. The success of the subsequent remedia-tion will depend on the quality of data obtained from the site investigation.

On a green field site, the site investigation can be aided by geophysical surveys but most spills take place in built-up industrial areas where such techniques are inappropriate. Our Heathrow experience show that in such sites, where the water-table is shallow enough for foundations to extend below it, the foundations can act as cells, isolating pockets of an oil 'pancake' from remediation pumping effects. Similarly, buried services may act as pollution conduits. Access to detailed site plans, and much more detailed site investigations, will be needed in such cases. The Heathrow study has also shown the problems of working on an active industrial site: however desirable, it is not possible to drill in an active airport runway, nor is it feasible to drill through the floor of a precision engineering shop. The detailed 'necessary' investigations then have to be designed through discussions with the client and owners of affected buildings.

The actual remediation programme used will be site-specific. The programme at Heathrow in 1987 focused sharply on the removal of floating product from the water-table. The installations and techniques used at Heathrow were selected because they were eminently suitable for areas where the water-table is shallow and to meet the aims of the project. In all cases, however, the design and distribution of recovery wells should be based on the extent of the spill and the hydrogeology of the site. In built-up areas, extra recovery systems should be installed for any isolated pools of hydrocarbon which, in an ideal world, will have been identified during the investigation phase.

The exact time needed for remediation will be the result of a balance between a number of factors, including the number of recovery wells installed, the acceptable intrusion of site works, and the time available for the clean-up. It is certain, however, that the clean-up of a large leak will take some time, probably several years. This means that the equipment used for the clean-up should not be treated as temporary; it should be bought and installed to fit a permanent role.

All hydrocarbon remediation programmes will produce waste streams, mainly oil and water, and plans must be made for the disposal of such wastes. As in this case, waste oil can be sold to a disposal contractor for use as a low-grade fuel. The disposal of waste water can be more problematical and the relevant regulatory authority should be consulted in the planning stage of remediation. Some quality controls will almost certainly be set for the waste water. If the waste water is to go to sewer then the relevant operating company for the area should be consulted to establish if it will accept the water and at what price. If the discharge consent cannot be met, then on-site treatment tech-niques for the waste water will need to be considered.

The authors would like to thank British Airways for their co-operation throughout this project, and for their permission to publish this paper.

References

COLE, G. M. 1994. *Assessment and Remediation of Petroleum Contaminated Sites.* CRC Press (Lewis), Boca Raton, USA.

FUSSEL, D. R., GODJEN, H., HAYWARD, P., LILIE, R. H., MARCO, A. & PANISI, C. 1981. *Revised Inland Oil Spill Clean-up Manual.* Concawe, Den Haag, Holland.

NAYLOR, J. A. 1974. *The Groundwater Resources of the River Gravels of the Middle Thames Valley.* Water Resources Board, Reading, UK.

Remediation of hydrocarbon leakage from a service station at Wansford, Cambridgeshire, UK

R. G. CLARK

CL Associates, Prospect House, Prospect Road, Halesowen,
Birmingham B62 8DU, UK

Abstract: Two service stations have been built at the location of the present Nene Service Station on the A47 Trunk Road at Wansford, Cambridgeshire, UK; one replacing the other. Both have been affected by ground movements resulting from the reactivation of a periglacial landslip. During the investigation of these ground movements, hydrocarbon contamination was found in the flood plain gravels adjacent to the River Nene, which is downslope of the present service station. The paper describes the investigation work undertaken, with emphasis on soil vapour monitoring to map the contaminant plume. Remedial measures carried out included pump-and-treat of contaminated groundwater from a recovery pit, together with removal of LNAPL phase petroleum by means of absorbent pads.

It is recognized that a large proportion of petrol and diesel filling stations have at some time in their history experienced some leakage of fuels into the ground. This results from spillages on the forecourt, leaking pipe joints and, in extreme cases, the rupturing of fuel lines or tanks. At the Nene Service Station at Wansford (the present station was built in 1988) there was an additional problem whereby the promontory of fill on which the service station stands was experiencing ground movements. Displacements of up to 50 mm horizontally and 40 mm vertically had caused both the forecourt slab and the brick sales building to crack. These movements were first noted shortly after construction and continued in the following years. A previous service station at the same site (date of construction unknown) had also been subject to similar damage in 1983 and the fuel tanks had rotated in the ground.

During the investigations in 1992 and 1993 to determine the cause of the ground movements, hydrocarbon contamination was found in soil and groundwater. This had to be remediated before any ground stabilization measures could be implemented.

The site and geology

The site is located immediately to the south of the A47 Trunk Road about 1 km to the east of Wansford in Cambridgeshire (Grid Reference TL 083 997). Here, the A47 runs on a low embankment near to the crest of a slope, and the service station is constructed on a promontory of fill extending out from the A47 over the slope.

The natural slope is about 9 m high and falls to the south at an inclination of about 15° to the horizontal, to the flood plain of the River Nene. At this point, the strip of land forming the flood plain is about 45 m wide and is at an elevation of approximately 10 m above sea level. The River Nene flows from west to east.

At the crest of the slope on which the promontory of fill stands there is Lincolnshire Limestone, which is either *in situ* or occurs as broken fragments (Head) derived from the *in situ* limestone which is further upslope. The slope comprises alluvium over Terrace Gravels and Northampton Sand directly overlying solifluced (reworked and destructured) Lias Clay. There is some alluvium on the flood plain, which also lies directly on the Upper Lias Clay.

A plan and section through the site are shown in Figs 1 and 2 respectively.

Mechanism of failure

Following the incidence of movement and cracking at the present service station, a ground investigation was carried out comprising shell and auger boring, rotary drilling, trial pits and piezometer installation. The investigation positions are shown in Fig. 1.

The results of the investigation indicated that the promontory of fill had moved as a result of slippage along pre-existing shear surfaces within a solifluction layer. There was direct evidence of these shear surfaces from examination of borehole cores and the sides of trial pits. A toe bulge had also appeared at the toe of the promontory of fill.

The geological map for this location (British Geological Survey 1978) confirms this to be an

Clark, R. G. 1998. Remediation of hydrocarbon leakage from a service station at Wansford, Cambridgeshire, UK. *In*: Mather, J., Banks, D., Dumpleton, S. & Fermor, M. (eds) *Groundwater Contaminants and their Migration*. Geological Society, London, Special Publications, **128**, 159–163.

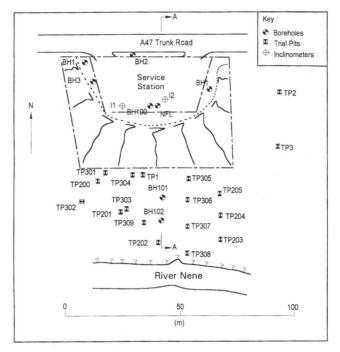

Fig. 1. Site plan showing exploratory hole positions.

area of landslipping and ground instability. Also, during excavations for an adjacent Anglian Water pumping station, which extracts water from the river, severe instability was experienced which affected the A47 and the service station site. The work associated with the pumping station has been reported by Chandler (1979).

Hydrocarbon contamination

Whilst investigating the cause of instability at the service station site, hydrocarbon contamination

was found in the gravels on the flood plain. Consequently, a specific investigation comprising boreholes, trial pits and probeholes was carried out to investigate the extent and degree of this contamination. The probes passed through the flood plain gravels into the clay beneath.

Thirty-four probes were put down on an approximate 10 m grid to a maximum depth of 2.6 m using a Marlow Probe with 30 mm diameter rods and attached flow-through sampler. The small 'undisturbed' samples of clay thus obtained were then examined for visual and

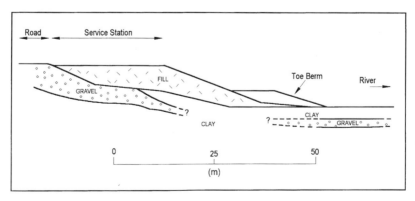

Fig. 2. Idealized cross-section A–A.

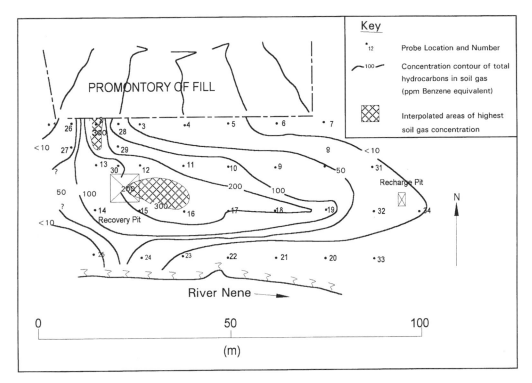

Fig. 3. Hydrocarbon plume.

olfactory evidence of hydrocarbon contamination prior to being sealed and taken to the laboratory for testing. Sampling of the gravels was not possible by this method. Soil-gas concentrations were obtained in the probeholes using a portable photo-ionization analyser immediately after the completion of every probe. The soil atmosphere was sampled for volatile aromatic hydrocarbon fractions at a depth of 0.25 m below ground level after sealing the top of the hole to prevent dilution of the soil atmosphere. The probehole locations and concentration contours of total hydrocarbons in soil gas (expressed as ppm benzene equivalent) are shown in Fig. 3.

Trial pits were excavated in the flood plain to enable a visual examination to be made of the nature of the contamination and so that water samples could be obtained from any issues of groundwater into the pits, taking care to avoid any cross contamination resulting from the excavation of the pits. The water samples were subjected to quantitative gas chromatography analyses to determine total hydrocarbon concentrations.

The subsurface behaviour of petroleum is difficult to predict because of its low relative density and its tendency to float on the water-table.

Floating petroleum can often move in directions other than the regional hydraulic gradient. For this reason, a large number of dynamic probeholes were put down to enable a large area to be tested for volatile aromatic hydrocarbons in the soil gas. It can be seen from Fig. 3 that there was a well-defined plume of petroleum contamination generally following the direction of flow in the river. This figure also indicates where the petroleum contamination was exiting as a below-ground seepage from the toe of the fill slope. Although the figure suggests that contamination was migrating towards the river, none was observed in the river.

Hydrocarbon remediation

The method of hydrocarbon remediation was essentially a pump-and-treat system using a combination of absorbent pads and circulation of contaminated groundwater through a granular activated carbon (GAC) filter. This combination of methods was chosen because the contamination was substantially constrained within a relatively shallow granular layer in the flood plain.

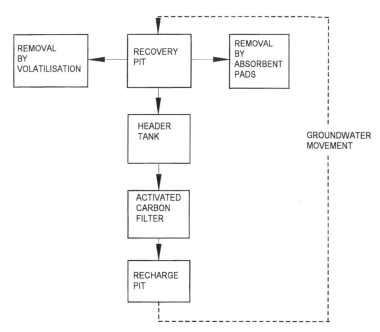

Fig. 4. Flow diagram of the remediation process.

A recovery pit (see Figs 3 and 4) approximately 2.5 m deep and with plan dimensions of 7.5 × 7.5 m at the surface was excavated at the location where the highest concentrations of petroleum product had been found in the soil vapour. A recharge pit approximately 2 × 2 m in plan and deep enough to penetrate the gravel was excavated downstream of the recovery pit at the eastern edge of the contamination plume.

As expected, the excavation of the recovery pit caused a considerable inflow of free-phase hydrocarbon contamination into the pit. This was effectively removed by the use of floating absorbent pads (type OPEC RP18) which were placed on the surface of the water within the pit. The pads were supplied in 450 mm wide rolls perforated every 450 mm and the rolls had a total length of 45 m. The pads were replaced as necessary during the subsequent circulation of groundwater. The used pads were sent to a licensed disposal facility.

A submersible pump (capacity approximately 5 m³ h⁻¹, 230 V, 0.5 kW) was installed in the recovery pit just below the surface water level. The pump was connected to a float switch to ensure that free hydrocarbon product floating on the surface of the water was not drawn into the pump. This free product was removed by the pads.

The pump discharge of groundwater containing dissolved hydrocarbon was directed to a GAC Disposorb F300 filter(diameter 1100 mm and overall height 1700 mm) via a header tank. The pump discharge from the recovery pit entered the drum filter through a nozzle, percolated downward through the carbon bed, and was then pumped by an internal pump back to the exit nozzle at the top of the absorber. The treated water was then discharged into the recharge pit as relatively clean, clear water.

The capacity of the pump was sufficient to cause a groundwater flow in the gravels from the recharge pit to the recovery pit in opposition to the direction of natural groundwater flow. In theory, this provided a back-flushing of hydrocarbons towards the recovery pit. Flow also took place from the promontory of fill towards the recovery pit.

During the decontamination, extra trial pits were excavated in the flood plain to ensure that the recovery pit was located at an appropriate place within the contamination plume and that at the end of the decontamination process the water in the recharge pit was representative of the surrounding groundwater.

Total dissolved BTEX (benzene, toluene, ethylbenzene, xylene) concentrations in groundwater were reduced from a pre-clean-up value of about 1850μg l⁻¹ to a post clean up value of about 5μg l⁻¹ in the recovery pit.

A further cause for concern in respect of groundwater quality is the presence of methyl tertiary butyl ether (MTBE). This is an oxygenate additive (i.e. an octane enhancer) to unleaded fuels, introduced into the UK in the mid 1980s. It is permitted to comprise up to 10% by volume of unleaded petrol. It is not especially toxic in water but imparts a taste or odour above a concentration of $2–3\mu g\,l^{-1}$ (Symington *et al.* 1994). It is highly soluble in water, has a low vapour pressure and is not readily retarded by sorption. Thus it migrates very rapidly in groundwater, can reach substantial concentrations and is not readily biodegraded. Garrett *et al.* (1986) report that when petrol and MTBE leak into groundwater, the MTBE spreads both further and faster than the petrol, and the concentration of petrol dissolved in groundwater increases. Thus the MTBE plume will appear as a 'halo' around the dissolved petrol plume, which in turn appears as a 'halo' around the free product plume.

The removal of MTBE was not a specific requirement of this project. In any case, due to its relatively high solubility, it is poorly absorbed by GAC (Garrett *et al.* 1986). The concentration in the groundwater samples varied considerably during the remedial works, from $130\mu g\,l^{-1}$ to below the detection limit which would be consistent with the rapid mobility of this contaminant. There is no drinking water limit for this parameter in the UK.

The decontamination works were carried out in late 1994 and lasted for a period of six weeks. At the end of the period both the recovery pit and the recharge pit contained relatively clean, clear water and no LNAPL phase free petroleum was entering the recovery pit. The ground stabilization works, in the form of a toe-weighting berm, then proceeded.

Conclusions

A simple and effective means of remediation has been used to remove hydrocarbon contamination in the flood plain gravels adjacent to a service station. This was possible because the contamination was confined to the gravels by low-permeability clays beneath the gravel thus avoiding the need for any deep remediation.

Subsequent testing of groundwater samples showed that the combination of absorbent pads and a pump-and-treat system based on a granular-activated carbon filter was able to achieve of the order of 99% reduction in dissolved BTEX concentrations in groundwater.

MTBE concentrations are also of concern in groundwater. The remediation did not specifically target this constituent and activated carbon filters are not particularly effective in this respect.

The author wishes to thank Heron Garage Properties and Snax 24, together with the Directors of CL Associates, for permission to publish this paper. The consulting structural engineers for the remediation were Cameron Taylor Bedford and the civil works were carried out by Weldon Construction. The absorbent pads were supplied by Oil Pollution Environmental Control Limited and the activated carbon filter was manufactured by Chemviron Carbon Limited.

References

BRITISH GEOLOGICAL SURVEY 1978. *Sheet 157: Stamford 1 : 50 000*. Geological Map (Solid and Drift). British Geological Survey.

CHANDLER, R. J. 1979. Stability of a structure constructed in a landslide: selection of soil strength parameters. *In: Proceedings of the 7th European Conference on Soil Mechanics and Foundation Engineering*, Vol. 3. Brighton, 175–182.

GARRETT, P., MOREAU, M. & LOWRY, J. D. 1986. MTBE as a groundwater contaminant. *In: Proceedings of the National Water Well Association Conference on Petroleum Hydrocarbons and Organic Chemicals in Groundwater: Prevention, Detection and Restoration*. Houston, Texas.

SYMINGTON, R., BURGESS, W. G. & DOTTRIDGE, J. 1994. Methyl tertiary butyl ether (MTBE): a groundwater contaminant of growing concern. *In: Proceedings of the 3rd Annual Conference 'Groundwater Pollution'*. IBC Technical Services/ Royal Holloway, University of London, 16/17 March 1994.

Remediation of a petroleum spill to groundwater at a fuel distribution terminal (Long Island, USA) using pump-and-treat and complementary technologies

JOHN M. W. HOLDEN[1] & NADINE TUNSTALL-PEDOE[2]

[1]*Scott Wilson CDM, Bayheath House, Rose Hill West, Chesterfield, Derbyshire S40 1JF,UK*
[2]*Celtic Technologies, CBT Centre, Senghenydd Road, Cardiff CF2 4AY, UK*

Abstract: A petroleum leakage to groundwater of up to 4000 m^3 was discovered in November 1987 at a fuel terminal site on Long Island, USA. A pure petroleum product pool of up to 1 m in thickness was floating on the water-table, and associated dissolved hydrocarbon and halogenated hydrocarbon plumes appeared to extend in three directions from source areas within the site. Risk assessments indicated that some local private wells were threatened by the migration of a dissolved petroleum plume. The petroleum spill is being treated using pump-and-treat technology with associated free product recovery, soil vapour extraction, air sparging wells and intrinsic bioremediation techniques. The treatment is proving to be successful, with half the petroleum free product being recovered and protection of the local residents and groundwater achieved.

The site is occupied by a terminal for the storage and distribution of liquid petroleum products, covering an area of 14 ha on Long Island, USA. (Fig. 1). The storage volume available at the site is estimated to be 400,000 m^3. The surrounding land-use consists primarily of undeveloped woodlands and residential developments. A public water supply well-field borders the woodland some 600 m to the east. The terrain at the terminal is relatively flat, with some moderately sloping grades.

During November 1987, a petroleum spill was detected whilst wells for groundwater monitoring were being installed to comply with state regulations. The groundwater beneath the terminal was found to be contaminated by a layer up to 1 m thick of floating petroleum (referred to as the light non-aqueous phase liquid (LNAPL) pool), which had leaked from a single underground pipe in the vicinity of the truck loading area. The pool was found to be lying on the water-table within an area of approximately 500 × 300 m, part of which was located outside the terminal boundary.

A separate leak of chlorinated hydrocarbons was found from the vapour recovery system in the truck loading bay. The main compounds involved here were methylene chloride (MC – CH$_2$Cl$_2$) and trichloroethene (TCE), used as a refrigerant and deicer, respectively. In their pure form, these are dense, non-aqueous phase liquids (DNAPLs). It is possible that DNAPL phase contamination arose from these leaks which, due to their density, sank through the water-table, creating complex source areas and associated plumes of dissolved chlorinated hydrocarbons. Chlorinated solvents were also likely to have dissolved in the petroleum where present.

A complex picture of contamination gradually emerged. In addition to the main hydrocarbon plume, several small isolated zones of dissolved TCE and dichloroethene (EDC) were discovered to the north and east of the terminal, respectively. The source of these is unclear, but EDC, together with EDB (dibromoethene), are used as petrol additives to prevent lead scaling of engine cylinders (Thomas & Farago 1973).

There are residential areas to the west and north of the site, in which some houses abstract water from shallow private wells. These wells were considered to be under threat from the contaminated groundwater. In addition, the closest properties were believed to be at risk from petroleum vapour migration.

Geology and hydrogeology

The site lies on a native sandy soil, which allows rapid percolation of precipitation and contaminants to the groundwater. The aquifer system comprises predominantly sandy strata of the Upper Glacial sequence and the underlying unconsolidated sands and gravels of the Magothy Beds, of Cretaceous age, as illustrated in Fig. 2. The Magothy Beds are, in turn, underlain by the Raritan formation. The Upper Glacial formation is up to 90 m in thickness, essentially being a thick sand unit of high porosity and permeability, with some silty-clay lenses. The most prominent silty-clay layer within the

HOLDEN, J. M. W. & TUNSTALL-PEDOE, N. 1998. Remediation of a petroleum spill to groundwater at a fuel distribution terminal (Long Island, USA) using pump-and-treat and complementary technologies. *In*: MATHER, J., BANKS, D., DUMPLETON, S. & FERMOR, M. (eds) *Groundwater Contaminants and their Migration*. Geological Society, London, Special Publications, **128**, 165–180.

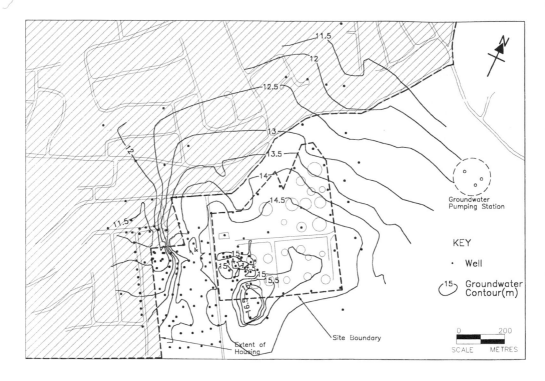

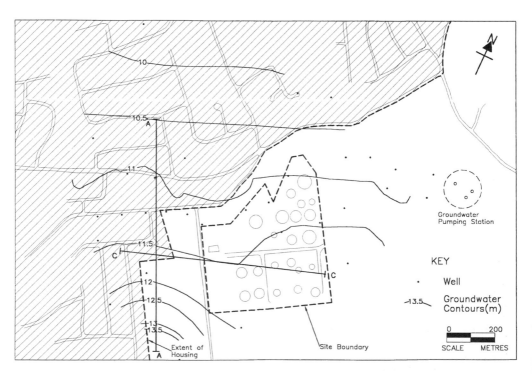

Fig. 1. Maps of terminal site showing the monitoring well network and (**a**) groundwater head in the C horizon, (**b**) groundwater head in the A horizon.

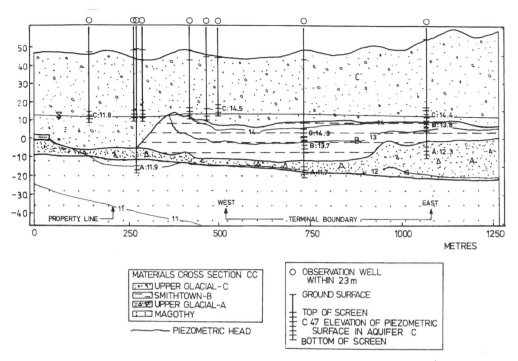

Fig. 2. Schematic west–east section through the study area, showing the protrusion of the Smithtown Clay above the water-table which modifies the migration of the western plume. The figure also shows actual, observed groundwater heads (March 1988) posted at well screens, together with simulated groundwater head contours.

Upper Glacial Sands occurs impersistently at depths of 30–46-m. It is locally known as the Smithtown Clay, is on average 10–12 m thick and behaves as an aquitard. Local groundwater flow is greatly influenced by this formation as the Smithtown Clay becomes laterally discontinuous to the north, east and west of the terminal area. Beneath the terminal, the Clay is at its greatest thickness of approximately 25 m, protruding above the water-table at several points, most importantly in an area to the southwest of the terminal boundary, thus impeding groundwater flow. There are occasional apertures in the Smithtown Clay which are thought to allow downward migration of groundwater and contaminants under the prevailing relative density and/or vertical head gradient conditions. The hydraulically distinct strata have been classified alphabetically as follows:

- the Upper Glacial Sands above the Smithtown Clay and containing the water-table and unsaturated zone are termed the 'C horizon';
- the Smithtown Clay is the 'B horizon';
- the Upper Glacial Sands below the clay are the 'A horizon';

- the 'D horizon' contains the Magothy aquifer.

The water-table lies some 30–40 m below the ground surface at the site and the local groundwater flow is generally in a northerly direction towards Long Island Sound, from a groundwater divide to the south of the site. It will be seen from Fig. 1 that the groundwater heads in the A and C horizons exhibit somewhat differing distributions, the water-table in the C horizon being influenced, among other factors, by the topography of the upper surface of the Smithtown Clay, the absence of the Smithtown Clay in places, and by pumping from the public abstraction wells to the east of the site. It will be seen from the head contours on the section in Fig. 2 that there is a general downwards vertical component of the hydraulic gradient, in addition to the northward lateral component. This has significant implications for the migration of contaminants.

The aquifer system is an important source of water. Several public water supply wells draw water from the basal Upper Glacial Sands and Magothy Beds. There are also many private wells for both domestic and industrial usage

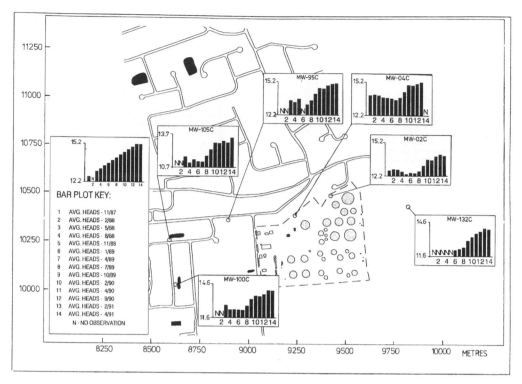

Fig. 3. Time history of groundwater heads in selected monitoring wells in the C horizon (1987–1991).

which tend to be shallower, abstracting their water from the Upper Glacial Beds.

Groundwater heads gradually increased (Fig. 3) throughout the study area during the study period 1987–1991. This appeared to be part of a regional pattern. The rise was in excess of 2 m in several wells.

Site investigation

Following detection of the spill, extensive investigations revealed the extent of the problem. These investigations included the following elements:

- Groundwater monitoring wells, which are screened in the relevant horizons. These were used to monitor the groundwater head distribution (Fig. 1) and to estimate the floating product thickness. Selected wells were originally sampled quarterly, and are now sampled on a six-monthly basis.
- A total of 113 vapour monitoring wells, 56 of which were sampled bi-weekly from 1989 to 1994. The samples were tested for VOCs (volatile organic compounds), including BTEX products (benzene, toluene, ethylbenzene, and

xylenes). Since June 1994, the frequency of sampling has dropped to quarterly in 13 wells.
- Long-term pumping tests.
- A data management system.

The main LNAPL petroleum pool was found to have spread within a source area of approximately 500 × 300 m and, when first investigated, lay between 1 mm and 1 m thick on the surface of the groundwater. These estimates of the product thickness in the aquifer, and the estimate of the leaked volume of petroleum, were based on the actual thickness of product measured in the wells, multiplied by 0.25, which is a conversion factor derived from the CONCAWE report of 1979, taking the aquifer porosity into account. This pool was found to have migrated westwards around the protrusion of the Smithtown Clay layer above the water-table (Fig. 2), and then travelled with the groundwater flow in a northerly direction. A plume of dissolved petroleum hydrocarbons extended northwards by up to 400 m from the northern extremity of the product pool. This dissolved plume, which is referred to as the western plume, continued to migrate northwards in a layer of high conductivity within the A horizon, having migrated downwards with the hydraulic gradient from

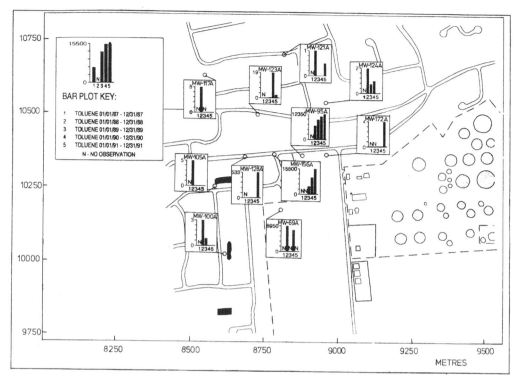

Fig. 4. Time history of dissolved toluene concentrations (μg l^{-1} in the western plume (1987–1991) in the A horizon.

the pool in the C horizon through an aperture in the Smithtown Clay.

The distribution of organic contaminants was highly complex, both in the three spatial dimensions and in time.

The main western plume consisted of petroleum hydrocarbons including benzene, toluene, ethylbenzene and xylenes (BTEX). This petroleum plume was associated with related parameters such as dichloroethene (EDC) and dibromoethene (EDB), methyl tertiary butyl ether (MTBE – an additive to unleaded fuel; Symington *et al.* 1994) and organic lead complexes (used as anti-knock agents in leaded fuels). As an example of the evolution of contaminant concentrations with time, Fig. 4 shows the variation with time of dissolved toluene concentrations in the western plume, which exceed 10 000 mg l^{-1} in some wells.

Other contaminants were also discovered in other parts of the site. Properties of selected organic contaminants at the site, which influence their potential to migrate in groundwater, are summarized in Table 1. Initial computer simulations were based on the assumption of two plumes emanating from sources of TCE and EDC within the site:

- a northern plume, dominated by dissolved trichloroethene (TCE); and
- an eastern plume, dominated by dissolved dichloroethene (EDC) at levels of several hundred milligrams per litre beyond the site boundary, although aromatic hydrocarbons (e.g. benzene) were also found to be present at levels of up to several thousand milligrams per litre in 1990, at the eastern boundary of the site (Fig. 5).

The assumed sources were included in initial contaminant transport simulations (Fig. 6) although subsequent data failed to support the discrete EDC and TCE plumes, the actual distributions being far more complex.

As further examples of the complex three-dimensional distribution of dissolved contaminants, Figs 7 and 8 show the distributions of total volatile organic pollutants and of MTBE, respectively, in the C and A horizons.

An initial risk assessment identified the major threat to the residential area as being the migration of the western dissolved petrol plume towards the private wells, and the remainder of this paper will focus on the modelling and remediation of this plume.

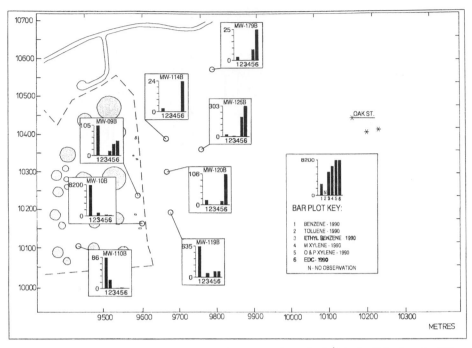

Fig. 5. Fingerprint plots of selected contaminants in the eastern plume ($\mu g\,l^{-1}$) in the B horizon in 1990.

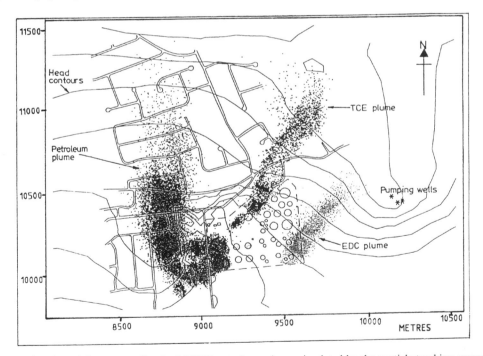

Fig. 6. Migration of the western dissolved BTEX petroleum plume, simulated by the particle tracking computer program for contaminant transport modelling, DYNTRACK. This plume was thought to have migrated in a westerly direction initially within the C horizon, then dropping down through a hole in the Smithtown Clay into the A horizon where the groundwater flows in a northerly direction. The simulated TCE and EDC plumes were not fully supported by subsequent data.

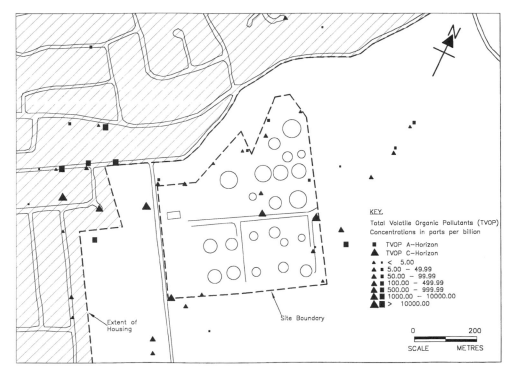

Fig. 7. TVOP (total volatile organic products) concentrations in the C and A horizons, October 1994.

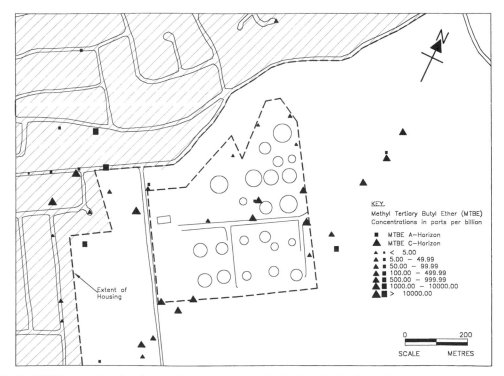

Fig. 8. MTBE (methyl-tertiary-butyl-ether) concentrations in the C and A horizons, October 1994.

Modelling studies

The data from sampling and analysis of the groundwater monitoring wells, hydraulic testing, well logs and data from two long-term pumping tests were used to develop a conceptual model of the subsurface environment at the site together with a groundwater flow and contaminant transport computer model. The modelling package employed was Camp Dresser & McKee's own in-house modelling suite DYNSYSTEM, which contains the following integrated modules:

- DYNFLOW, a three-dimensional finite-element model which simulates the head distribution and bulk groundwater flow in aquifers;
- DYNTRACK, a particle tracking model which simulates dispersion, diffusion and simple adsorption and biodegradation reactions;
- DYNAPL, which simulates the non-aqueous phase liquid (NAPL) flow.

These modules were used in conjunction with the GIS (geographical information system)-based data management package, DYN-EDM. The model was used to predict:

- the extent and rate of migration of the various plumes, both vertically and horizontally;
- the persistence or retardation of the contaminants;
- the likely impacts of potential remediation techniques.

One particularly interesting feature of the modelling exercise was that it confirmed field evidence that a significant proportion of the dissolved phase hydrocarbon in the western plume was migrating downwards from the 'C' to the 'A' horizon via an aperture in the laterally discontinuous Smithtown Clay (Fig. 9), under the prevailing downward hydraulic gradient. This was a particularly important result as it indicated that the majority of the migration would not be intercepted by most of the groundwater abstractions, the public supply wells predominantly abstracting from the deeper (lower 'A' and 'D') horizons and the private wells predominantly abstracting from the shallow 'C' horizon.

The modelling also demonstrated that if the petroleum pool, and the adjacent zone of high concentrations of dissolved petroleum within the plume, were controlled and treated, then the plume would eventually start diminishing in size and the concentrations of contaminants within it would decrease due to natural biodegradation (or intrinsic bioremediation).

These findings underline the importance of three-dimensional groundwater modelling in

Table 1. *Properties of selected organic contaminants at the terminal site in order of decreasing mobility, aqueous solubility and soil-water partition coefficient (after Symington* et al. *1994, Fetter 1994)*

Compound	Solubility (ppm)	K_{oc} (ml g^1)
TBME	48 000	
Methylene chloride	13 200	25
1,2-Trans-dichloroethene	6 300	39
Benzene	1 780	97
Trichloroethene	1 100	152
1,1-Dichloroethene	400	217
Toluene	500	242
o-Xylene	170	363
p-Xylene	156	552
m-Xylene	146	588
Ethyl benzene	150	622

developing conceptual models and remediation strategies in complex multi-layered aquifers.

Remedial actions

Emergency response

The immediate actions taken for emergency remediation included sealing the basements of the houses in the area found to be threatened by the migration of the petrol vapour, and installing under-floor ventilation systems.

Phase 1

The recovery of the free product petroleum hydrocarbon was quickly initiated using 20 pneumatic ejector pumps in up to 40 of the groundwater monitoring wells. The water-table was depressed by pumping from a single well located near to the northern extent of the product pool, which was also equipped with a second pump to remove the floating free product and prevent further migration. This immediate Phase 1 remediation successfully recovered approximately 1800 m^3 of petrol (almost half of the estimated volume of the initial spill) during the Phase 1 remediation, mostly within the first two years. The rate of product recovery decreased exponentially, partly because the water-table rose by up to 2 m regionally during the period. This rise 'smeared' the petrol product within the zone of fluctuation (i.e. the lower reaches of the former unsaturated zone). This Phase 1 system of pure product recovery was decommissioned in 1991. A Phase 2 SVE (soil vapour extraction) system, coupled with a pump-and-treat system, was designed in its stead (Fig. 10).

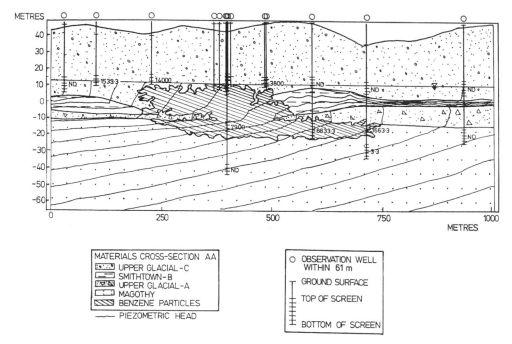

Fig. 9. Cross-section showing simulated benzene particle transport and observed concentrations in the western plume in 1990. Note the downward migration of benzene through an aperture in the Smithtown Clay.

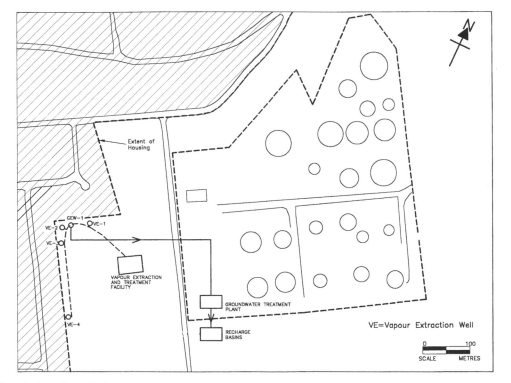

Fig. 10. Locations of Phase 2 interim remedial measures on the site in 1991.

Treatability studies undertaken to assist with the design of Phases 2 and 3 of the remediation scheme

The design of Phases 2 and 3 of the remediation scheme involved several pilot studies to address two problems: (i) the remediation of the groundwater contamination, and (ii) the removal of petroleum product and other VOCs typically remaining at residual saturation levels in the soils above and immediately below the free water-table.

Pilot studies for the remediation of the groundwater contamination   Several treatment options are available for treatment of hydrocarbon-contaminated groundwater (Stover 1989) and the following technologies were assessed for treating groundwater discharged from a long-term pump-and-treat system at this site:

- air stripping;
- advanced oxidation processes (AOP) using ozone and hydrogen peroxide;
- metal precipitation by pH adjustment;
- carbon adsorption;
- ion exchange;
- submerged fixed film biological treatment;
- powdered activated carbon treatment (PACT);
- granular activated carbon (GAC) in a fluidized bed.

A pilot test showed that air stripping was effective at removing most of the organic contaminants, but treatment of EDB and MTBE proved expensive (because they required a high air:water ratio) and the emissions from the stripper required additional treatment for vapour phase carbon.

The AOP system was also pilot tested, but was found to be unsuitable. Even with high peroxide and ozone loadings, it could only achieve 15–50% BTEX removal, thus falling short of the desired discharge limits.

The groundwater abstracted by the pump-and-treat system contained high levels of iron and manganese, presumably partially mobilized under reducing conditions promoted by the high levels of hydrocarbon present in the aquifer (Baedecker *et al.* 1993). By adjusting pH to 10.5 and filtering, the iron and manganese were removed from the groundwater, but lead (related to petrol additives) could only be removed to below the acceptable limit of $0.05\,\mathrm{mg\,l^{-1}}$ using ion exchange.

Water phase carbon adsorption was found to be a successful but costly treatment.

Biological treatment using a submerged fixed film reactor was tested in the field, with groundwater fed into the system at a rate of 0.25–$0.321\,\mathrm{s^{-1}}$ and a nutrient solution fed into the influent. Except when hindered by excessive biomass growth, the system worked successfully during normal weather conditions. However, the BTEX removal was only of the order of 40–70%, so additional treatments were considered necessary.

Alternative biological treatments proved to be more promising. The PACT process removed BTEX with an efficiency rate of 99.8%, EDB and EDC to 96%, manganese to 93%, iron to 64% and lead to 50%. The process combines suspended growth activated sludge biological treatment with powdered activated carbon adsorption.

The process involving GAC in a fluidized bed was found, however, to be more cost-effective. It achieved removal levels of 98% for BTEX, with significant removal of EDB, EDC and MTBE. It operates by combining attached growth biological treatment with carbon adsorption. Both PACT and GAC also showed a greater level of stability at low levels of BTEX than the submerged fixed film reactor.

The fluidized bed GAC unit was regarded as being the most promising technique and was scaled up from laboratory testing into pilot testing in the field. A $1.21\,\mathrm{s^{-1}}$ GAC fluidized bed unit was operated for 8 months to demonstrate its effectiveness. It removed BTEX to barely detectable traces and EDB, EDC and MTBE were successfully biodegraded. Lead was removed with an efficiency rate of approximately 50%.

However, eventually, air stripping was chosen for water treatment in Phases 2 and 3 because there was far more previous experience of its use compared with GAC in a fluidized bed.

Pilot treatability study of extracted soil vapours
The technologies tested for the final soil vapour extraction design included:

- thermal oxidation;
- catalytic oxidation;
- activated carbon treatment.

The activated carbon treatment was eliminated in the initial stages of the study due to the high cost of treating such high contaminant levels in the petroleum vapour, as measured in the groundwater monitoring wells.

The parameters measured in the pilot testing of the two other treatment techniques were temperature, gas flow rates, VOC concentrations and vacuum/pressure. The catalytic oxidation unit produced removal efficiencies of 99.5% or better, at influent concentrations in the range 1.7–2.1% hydrocarbons (as methane) but this rate dropped as concentrations decreased. The thermal oxidation unit, however, operated at the same efficient removal rate consistently over a wider range of influent concentrations. The combustion temperature of 850–900°C was easily maintained, although at lower influent concentrations, a higher rate of propane assist gas was required. Thermal oxidation was therefore the treatment process selected for use in Phases 2 and 3.

Phase 2 interim remedial measures

The remedial objectives identified for Phase 2 of the remediation strategy were

- to hydraulically control further spreading of the dissolved petrol plume;
- to remove free product petrol from the groundwater surface.

As part of the Phase 2 interim remedial measures (IRM), a pump-and-treat system was installed to achieve hydraulic control of the free product pool and dissolved plume at the northern tip of the petrol pool. This had a 4300 m^3 day^{-1} capacity and consisted of a groundwater recovery well (GEW-1), from which water was pumped and transmitted to a water treatment plant (Fig. 10). The first stage in the plant comprised an oil/water separator. The water then passed to an air stripping tower, in which VOCs were volatilized and transferred to the air flow. The resultant off-gases were treated using a vapour phase GAC (granular activated carbon) unit. The treated water was directed into one of three recharge basins which returned the water to the aquifer system. This pump-and-treat system was operated between 1991 and April 1995 when it was decommissioned. Over this period it successfully treated more than 267 000 m^3 of water, removed 27 220 kg of organic contaminants, and successfully achieved hydraulic control of the petroleum pool and slowed the growth of the dissolved contaminant plume.

The Phase 2 SVE scheme was constructed in 1991 as a system of four 300 mm diameter vapour extraction wells and they have been subsequently incorporated into the Phase 3 scheme. They have a radius of influence of 60–120 m and are fitted with a slotted screen in the bottom 7.6 m of each well. The locations of these wells (VE-1 to VE-4) are shown in Fig. 10. The slotted screen penetrates down into the water-table to allow for fluctuations in the piezometric surface.

Pipework connects these wells to three vacuum extraction blowers with a maximum total air flow capacity of 0.9 m^3 s^{-1}. The wells operate at air flow rates of up to 0.5 m^3 s^{-1}. Extracted hydrocarbon vapours pass from the blowers to two bottom burning thermal oxidation units through water seal units (which prevent blow back). During Phase 2, these achieved a destruction efficiency of 99.8% of petroleum vapours at a flow rate of 0.5 m^3 s^{-1} and removed approximately 3400 kg week^{-1} of hydrocarbons. Propane was used to assist combustion when hydrocarbon concentrations dropped to low levels, and this procedure is still being used. The progress of the SVE system in removing floating product and product trapped in the unsaturated zone was checked by a series of 50 vapour monitoring wells.

Remediation modelling, remedial measures in Phase 3, and the role of intrinsic bioremediation

Modelling of the development of groundwater contamination and of the effects of remediation technologies, suggested that intrinsic bioremediation (or natural biodegradation) would play an important role in reducing the plume concentrations (Barker et al. 1987). The natural decay rates are thought to be sufficient to ensure that the majority of the aqueous plume does not reach public or private water abstractions. Due to vertical migration down through the aperture in the Smithtown Clay (Fig. 9), most of the plume should pass under the base of the shallow private wells and above the public supply wells which tend to abstract water from greater depths. It is conceivable, however, that the fringes of the plume may reach some of the private wells. These potentially affected residences have been connected to the public water supply.

Figure 11 shows the predicted reduction in the concentration of contaminants (in this case, dissolved benzene) in the plume due to natural biodegradation, employing estimated biodegradation half-lives of 4.5 months and 2 years, respectively, representing values falling in half-life ranges cited in the literature.

Using the module DYNAPL, it was also possible to simulate the development of the petroleum free product pool under the influence of the remedial actions. The simulation results are indicated in Fig. 12(a) with the measured free product thickness in 1995 in Fig. 12(b). The degree of correspondence is found to be rather

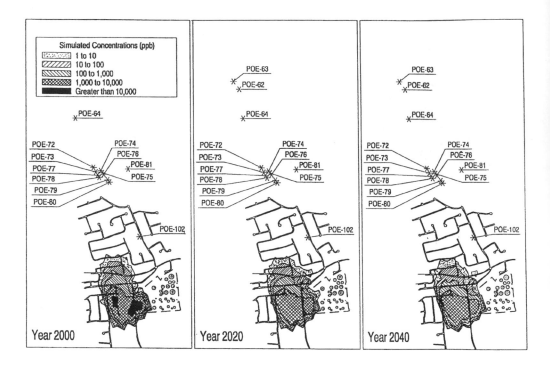

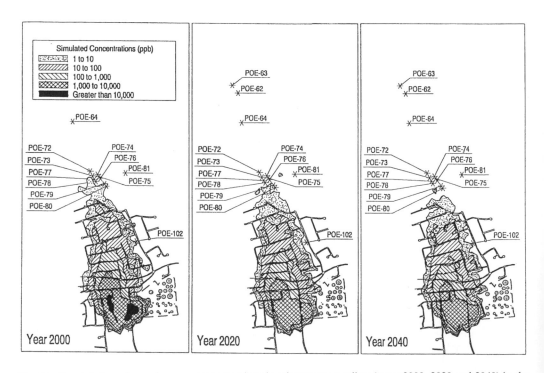

Fig. 11. Simulated maximum benzene concentrations in micrograms per litre (years 2000, 2020 and 2040) in the western plume using a biodecay rate of (**a**) 0.005/day (half-life 4.5 months) and (**b**) 0.001/day (half-life 2 years). Concentrations shown are maximum model layer concentrations for all aquifer horizons.

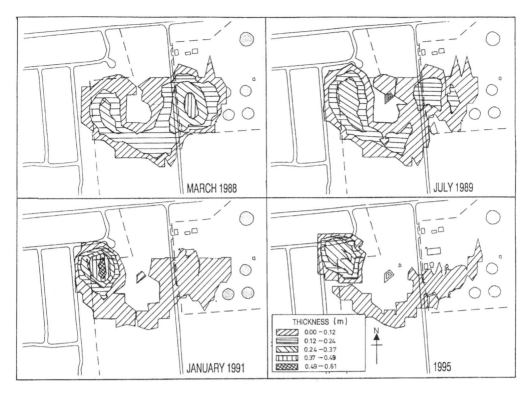

Fig. 12. (a) Simulated LNAPL thickness in the western plume (1988–1995). (b) Observed LNAPL thickness in the western plume in 1995.

good, although observed maximum thicknesses in wells are locally higher than modelled thicknesses, due to exaggeration of product thicknesses in wells compared with aquifers.

It was decided, in conjunction with the regulatory authorities, that significant reliance should be placed on intrinsic bioremediation in Phase 3 of the remediation strategy. A relatively small capacity pump-and-treat system was adopted with the following objectives:

- controlling the spread of the free product pool and reducing the mass of contaminants entering the plume;
- reducing contaminant concentrations in the part of the plume with the highest concentrations immediately to the north of the product pool.

An upgraded SVE system with air sparging below the water-table was adopted to treat the source of contamination.

As a further precaution within the Phase 3 treatment strategy, a wellhead protection programme was instigated at the adjacent public supply wells.

It was agreed with the regulators that Phase 3 of the treatment strategy should continue in operation until:

- the SVE system eliminates completely the film of product on the water-table at the monitoring installations; and
- the pump-and-treat system reduces BTEX concentrations to less than $20\,\mathrm{mg\,l^{-1}}$ (or ppm); for comparison drinking water quality limits are $5\,\mathrm{\mu g\,l^{-1}}$ (or ppb) benzene, $100\,\mathrm{\mu g\,l^{-1}}$ toluene and $10\,\mathrm{mg\,l^{-1}}$ xylene.

Final Phase 3 remediation scheme

The Phase 3 pump-and-treat system with a capacity of $2000\,\mathrm{m^3\,day^{-1}}$ began operation in June 1995. The system included the following:

- wells (GEW 1–4) in the western plume area of 50–60 m depth (Fig. 13);
- a surface treatment system in the south-east corner of the terminal comprising:
  – an oil/water separator,
  – an air stripping tower;

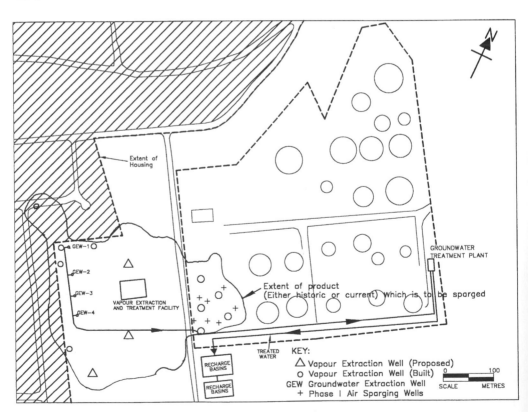

Fig. 13: Locations of final (Phase 3) remedial installations at the site.

– a 30 m^3 air phase GAC unit, which will be used if odours present a problem; this includes a heater to reduce the moisture content of the influent gas to the GAC unit.
• two large recharge basins.

This system was initially operated at 400 m^3 day^{-1}, with only one well pumped at a time, because of concerns of the water level rising further and increasing the level to which 'smearing' of the petroleum product occurs. This latest pump-and-treat system achieves hydraulic control and improves the water quality in the source zone and in that part of the plume near the source. It is expected to run until the year 2005. To achieve the total treatment of the source and associated plume would have required a larger plant, with an estimated capacity of 13 000 m^3 day^{-1}. The disadvantage of such a large plant, apart from its higher cost, would have been the effect of such a high pumping rate on the groundwater abstractions in the area.

In 1995, the SVE system was upgraded by the installation of six new vapour extraction wells (Fig. 13). A minimum of three additional SVE wells are proposed to complement the air sparging system.

The problem of additional removal of petroleum from the phreatic zone, where the fluctuating water-table traps the petroleum within the pore spaces, was addressed with air sparging technology (Marley *et al.* 1992), and seven (Phase I) air sparging wells were also installed at this time in the southwest corner of the terminal. Air sparging involves the injection of air below the water-table to entrain hydrocarbon vapour in the sparging gas, forcing the contaminants into the unsaturated zone, where they can then be extracted via the SVE unit. Eventually, up to 8 ha of the source area will be sparged, by 47 wells with a radius of influence ranging from 25 to 75 m and located at approximately 50 m centres.

Costs

The total costs of treatment are outlined in Table 2. In addition to the costs of remediation, the site owners have had to pay a large 'sum of compensation' to the State regulatory authority. This in part recognizes that the use of intrinsic bioremediation as a method of treating the part of the plume with lower concentrations of dissolved petroleum will result in a longer period of derogation of the groundwater resource compared with an active treatment system (such as pump-and-treat).

Conclusions

From the success of this project, several conclusions about the use of pump-and-treat technologies for groundwater remediation can be drawn. Free product recovery in Phase 1 was demonstrably successful in removing, within a relatively short period, a significant proportion of the free product pool that was floating on the water-table. Pump-and-treat was used successfully to establish hydraulic control of the pool. The removal rate did decrease as less pure product became available.

In Phase 2 an interim pump-and-treat system of fairly small capacity ensured hydraulic control of the product pool and dissolved petroleum plume, thereby preventing them from spreading further. In addition, the system removed a significant amount of petroleum from the subsurface. The design of the final Phase 3 treatment system acknowledged that pump-and-treat is a relatively inefficient process particularly when a problem involves some or all of the following:

• contamination by NAPLs;
• strongly adsorbed contaminants;
• heterogeneities within the aquifer;

Table 2. *Costs of the remediation treatments used at the terminal (millions of US dollars)*

Engineering and project management	$1–3 M
Operation and maintenance of the SVE system for five years	$1.175–1.925 M
Construction, operation and maintenance of 2000 m^4 day^{-1} pump-and-treat facility with GAC system	$3.9–4.2 M
Design, construction, operation and maintenance of a full-scale air sparging programme for five years	$1.825–2.475 M
Environmental monitoring of groundwater, product and vapours for five years	$0.7–1.1. M
Post-remediation demolition, and restoration (wells, pump-and-treat system and SVE system)	$1.3–1.8 M
Connection of private well owners to the public water supply system	$0.9 M
Wellhead protection programme for ten years	$10 M
Groundwater preservation programme with land acquisition	$9 M
TOTAL	$27.99–32.59 M

- a requirement to clean up to very low concentrations of contaminants.

Instead of trying to remediate the whole of the dissolved petroleum plume to low concentrations (established by risk assessment to be acceptable) using a pump-and-treat system, it was decided to only use pump-and-treat in the most highly contaminated part of the plume nearest the former position of the pool of free product. In the larger part of the plume lying further to the north (i.e. down the hydraulic gradient), risk assessment demonstrated that intrinsic bioremediation (or natural biodegradation) would achieve these treatment targets in due course and the potential cost savings are large. However, with this approach monitoring becomes very important to ensure the plume degrades as expected and that unacceptable impacts are avoided.

The use of complementary remedial techniques, such as soil vapour extraction and air sparging to treat petroleum LNAPLs within the source area, has considerably improved the effectiveness of the overall treatment strategy, and has shortened the period required to achieve reasonable treatment targets. Soil vapour extraction, in particular, has proved to be an economically viable and rapid method of continuing the removal of the pure petroleum product, after free product recovery by pumping becomes less effective. Soil vapour extraction is also effective at removing petroleum LNAPLs from the unsaturated zone.

Finally, the modelling systems used have proved themselves useful in simulating conceptual models for the migration of the pure product and associated dissolved product plumes and in predicting the success of the various remedial technologies. In particular, the use of three-dimensional contaminant transport modelling was regarded as being essential in predicting contaminant behaviour in a multi-layered aquifer system.

The authors are indebted to all their friends at Camp Dresser & McKee who have helped in the production of this paper and, particularly, to Mr David Keil.

References

BAEDECKER, M. J., COZZARELLI, I. M., EGANHOUSE, R. P., SIEGEL, D. I. & BENNETT, P. C. 1993. Crude oil in a shallow sand and gravel aquifer – III. Biogeochemical reactions and mass balance modelling in anoxic groundwater. *Applied Geochemistry*, **8**, 569–586.

BARKER, J. F., PATRICK, G. C. & MAJOR, D. 1987. Natural attenuation of aromatic hydrocarbons in a shallow sand aquifer. *Groundwater Monitoring Review*, Winter 1987, 64–71.

CONCAWE 1979. *Protection of groundwater from oil pollution*. CONCAWE Report No. 3, The Hague, Netherlands.

FETTER, C. W. 1994. *Applied Hydrogeology*, 3rd edition. Macmillan, New York.

MARLEY, M. C., HAZEBROUCK, D. J. & WALSH, M. T. 1992, The application of in-situ air sparging as an innovative soils and ground water remediation technology. *Ground Water Monitoring Review*, Spring 1992, 137–145.

STOVER, E. L. 1989. Coproduced groundwater treatment and disposal options during hydrocarbon recovery operations. *Ground Water Monitoring Review*, Winter 1989, 75–82.

SYMINGTON, R., BURGESS, W. G. & DOTTRIDGE, J. 1994. Methyl tertiary butyl ether (MTBE): a groundwater contaminant of growing concern. *In*: *Proceedings of the 3rd Annual Conference on Groundwater Pollution*, 16–17 March 1994, London. Royal Holloway College/IBC Technical Services Ltd.

THOMAS, R. W. & FARAGO, P. 1973. *Industrial Chemistry*. Heinemann, London.

Section 5: Groundwater pollution by chlorinated solvents

Chlorinated hydrocarbons are considered to be amongst the most problematic and pervasive groundwater contaminants in the industrialised world. They are widely used in a variety of contexts including the metals industry, vehicle and components manufacturing, plastics, electronics, textiles, vehicle maintenance and dry-cleaning.

Chlorinated solvents are typically DNAPLs (dense, non-aqueous phase liquids). These are denser than water, sinking through the unsaturated and saturated zones and accumulating at the base of aquifer systems and in blind fractures. In their free phase, their movement is controlled in the saturated zone by gravitational potential gradient rather than hydraulic gradient (i.e. groundwater flow). Chlorinated solvents are also able to dissolve in groundwater to a significant extent, are rather poorly biodegradable under many conditions and are tolerated in drinking water at concentrations of only a few tens of µg/l. These features, together with the fact that small unregistered frequent spills are often as problematic for groundwater as a single major leakage, make chlorinated solvent pollution incidents very difficult to investigate adequately, to model and (especially) to remediate.

Some of the most widely publicised groundwater pollution incidents in the UK, USA and elsewhere have been related to chlorinated solvents (Folkard 1986). Researchers at Birmingham University caused considerable concern when they revealed that the unconfined Permo-Triassic sandstone aquifer beneath Birmingham was subject to widespread solvent contamination, largely caused by engineering industries and dry cleaning (Rivett *et al.* 1990*a,b*). The notorious 'Eastern Countries Leather vs. Cambridge Water Company' case also involved such solvents (see **Misstear** *et al.* and **Ashley**, section 1). Extensive chlorinated solvent contamination has also occurred in the Chalk Downs of Southern England between Harwell and Blewbury in Oxfordshire. Evidence is pointing towards the source being a research establishment at Harwell which operated a waste chemical disposal involving unlined lagoons in the Chalk (Fellingham *et al.* 1993).

We have already met chlorinated solvents in conjunction with hydrocarbon spills in papers in the previous section by **Banks** and **Holden & Tunstall-Pedoe**. Here we meet them again: **Misstear** *et al.* describe yet another solvent pollution incident in the Chalk of Eastern England while **Bishop** *et al.* also describe a site in a fractured sandstone sequence in the industrialised English Midlands. Finally, **Muldoon** *et al.* give some insight into how chlorinated solvent contamination may be remediated using pump-and-treat techniques at an electronics factory in the USA.

FELLINGHAM, L. R., ATYEO, P. Y. & JEFFERIES, N. L. 1993 Investigation and remediation of the groundwater pollution at Harwell laboratory. *Proc. Conf. 'Groundwater Pollution', 16th–17th March 1993, Royal Lancaster Hotel, London. IBC Technical Services / Royal Holloway University of London. IBC conference documentation* **E0136**.

FOLKARD, G. K. 1986. The significance, occurrence and removal of volatile chlorinated hydrocarbon solvents in groundwaters. *Water Pollution Control*, **85**, 63–70.

RIVETT, M. O., LERNER, D. N., LLOYD, J. W. & CLARK, L. 1990*a*. Organic contamination of the Birmingham aquifer, U.K.. *Journal of Hydrology*, **113**, 307–323.

RIVETT, M. O., LERNER, D. N. & LLOYD, J. W. 1990*b*. Chlorinated solvents in U.K. aquifers. *Journal of the Institution of Water and Environmental Management*, **4**, 242–250.

Groundwater remediation of chlorinated hydrocarbons at an electronics manufacturing facility in northeastern USA

D. G. MULDOON[1], P. J. CONNOLLY[2], A. W. MAKOVITCH[3],
J. M. W. HOLDEN[4] & N. TUNSTALL-PEDOE[4]

[1] *Environmental Consultant, 22 Fern Drive, Walpole, Massachusetts 02081, USA*
[2] *Camp Dresser & McKee, Ten Cambridge Center, Cambridge, Massachusetts 02142, USA*
[3] *AT&T*
[4] *Scott Wilson Kirkpatrick, Bayheath House, Rose Hill West, Chesterfield,
Derbyshire S40 1JF, UK*

Abstract: An extensive investigation was undertaken to determine the degree and extent of contamination of three groundwater production wells by chlorinated solvents, and to design a full-scale remediation programme. Four major potential contaminant sources were identified. Hydrogeological investigations included drilling and multi-level groundwater sampling of 21 wells at 11 locations, monitoring of road-salt chloride as a groundwater flow tracer, and a seven-day pumping test to confirm the conceptual hydrogeological model. Immediate remedial measures were recommended to minimize further groundwater contamination. Camp Dresser & McKee's three-dimensional models DYNFLOW and DYNTRACK were used to develop a contaminant transport model which was used extensively in a feasibility study for long-term remediation. The implemented remedial programme included installation of a pump-and-treat system involving two extraction wells and surface treatment of the water in an air stripper. Operational and performance data for the remedial system are given for the first three years of operation. In this time progress towards treatment standards has been slow. However, treated water has been used in the manufacturing plant and the system has acted as a hydraulic control, preventing the dissolved contaminant plume reaching one of the original abstraction wells which supplies water to the plant.

The AT&T Merrimack Valley Works began manufacturing transmission equipment at its present site in 1956. This site comprises 68 ha in an industrial area of North Andover, Massachusetts. The Merrimack Valley Works is separated from the Merrimack River by a Boston and Maine Railroad (B&M) easement; the City of Methuen is directly across the river. The Merrimack Valley Works currently employs approximately 7500 people. An additional 1200 employees work at Bell Laboratories, an AT&T subsidiary, at the same location.

Process water for the Merrimack Valley Works is obtained from an on-site de-ionized water plant (DI plant). The source of water for this plant is three production wells: Production Well 1 is on AT&T property, and Production Wells 2 and 3 are located on the Boston and Maine Railroad easement between the Merrimack Valley Works and the Merrimack River. The three wells discharge through a common line to the DI plant. After the discovery of low levels of volatile organic compounds (VOCs) in the water coming from the production wells, State regulatory authorities were notified immediately, and Camp Dresser & McKee (CDM) was engaged by AT&T to identify the extent of the problem and to evaluate alternatives for site remediation.

Over the years, various industrial solvents, etchants, and other industrial chemicals were used at the facility in the manufacturing and assembly process. These chemicals were stored on-site in both underground storage tanks (USTs) and in barrels. A site map showing the pertinent details of the facility is presented in Fig. 1.

CDM, in conjunction with AT&T, adopted a structured, phased approach for site remediation. This approach allowed for an efficient, cost-effective remediation consistent with the environmental and fiscal goals and objectives of AT&T and the regulatory agencies.

The programme consisted of the following elements:

- phased remedial investigation,
- execution of any required immediate remedial measures,
- feasibility study,
- design of selected remedial alternative,
- construction,
- start-up,
- post-construction monitoring,
- closure.

MULDOON, D. G., CONNOLLY, P. J., MAKOVITCH, A. W., HOLDEN, J. M. W. & TUNSTALL-PEDOE, N. 1998. Groundwater remediation of chlorinated hydrocarbons at an electronics manufacturing facility in northeastern USA. *In*: MATHER, J., BANKS, D., DUMPLETON, S. & FERMOR, M. (eds) *Groundwater Contaminants and their Migration*. Geological Society, London, Special Publications, **128**, 183–200.

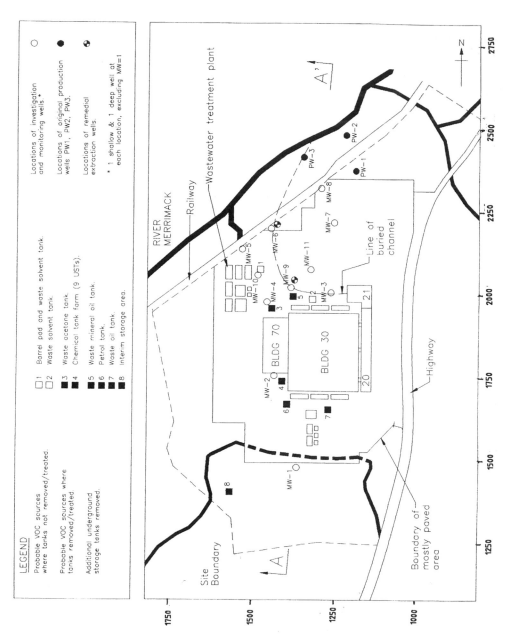

Fig. 1. Site plan. Scale in metres. The solid lines represent watercourses; dashed lines indicate where watercourses are culverted beneath paved areas.

Hydrogeological investigation

Phase I investigation

The objectives of the Phase I hydrogeological investigation were to evaluate the distribution of VOCs in the groundwater at the Merrimack Valley Works, and to identify potential VOC sources. The activities performed as part of the Phase I study were designed to achieve the following:

- develop an understanding of the hydrogeology at the site by installing monitoring wells, detailing the soil stratigraphy, and documenting the distribution of VOCs in the groundwater at the site, in nearby streams, and in the Merrimack River;
- assess the integrity of underground storage tanks at the Merrimack Valley Works to determine whether they may have released VOCs into the soil or groundwater;
- develop a plan for the removal of underground storage tanks, which are probable VOC sources;
- evaluate the potential for nearby industries to have released the VOCs detected in groundwater at the Merrimack Valley Works.

During the Phase I hydrogeological investigation, 14 soil borings were advanced and groundwater monitoring wells installed at selected locations. The general results of the Phase I hydrogeological study are summarized below.

Potential sources of VOC contamination. Soil and groundwater samples collected in the vicinity of four underground storage tank areas at the Merrimack Valley Works contained VOCs, suggesting that these tank areas might be VOC sources. The four tank areas (whose locations are identified on Fig. 1) comprise the following:

1. the barrel pad area consisted of an underground waste solvent storage tank and an above-ground barrel storage facility, both of which had been located approximately 150 m northwest of Building 70 (but which were closed by the time of the Phase I investigation);
2. the waste solvent tank, which had been located adjacent to the north face of Building 30 (this tank was removed in the year before the Phase I investigation);
3. the waste acetone tank, also previously used to store waste solvents, located along the north face of Building 70;
4. the tank farm, comprising nine underground storage tanks located along the south side of Building 70, adjacent to Building 30.

The most significant VOC source appears to have been at the tank farm where volatile organics were discovered in the groundwater and in the soil surrounding the tanks. The concentrations of total VOCs in groundwater at the tank farm measured as high as $300 \, mg \, l^{-1}$. Lower concentrations of volatile organics were measured in groundwater in the vicinity of the waste solvent tank, near the waste acetone tank, and near the barrel pad area.

Potential subsurface migration pathways. Based on historical information and the results of the Phase I subsurface investigation, three aquifers were identified at the site (see Fig. 2). One is a shallow silty-sand aquifer that is up to 12 m thick and underlies most of the site. Low levels of volatile organics were generally encountered in this formation. A localized deep aquifer, comprising sand and gravel in a buried channel, runs northward from Building 30 toward the river (Fig. 1). The full lateral extent of this aquifer was not defined in the Phase I study. The third aquifer is the bedrock, which is weathered and fractured siltstone and sandstone.

The deep aquifers were identified as pathways for VOCs to travel in groundwater. Water quality data indicated that the sand and gravel formation in the deep buried channel was hydraulically connected with the surficial aquifer; however, the available data were not sufficient to determine whether the connection is through the typically 1.5–15 m thick layer of silt and clay separating the water-bearing formations, or whether the deep channel aquifer tilts upward and meets the surficial sand formation. Except where the deep aquifer is present, the shallow aquifer is separated from the bedrock by a relatively impermeable layer of glacial till up to 15 m thick.

The transmissivity of this system was not determined in the Phase I study, but the bedrock did not appear to be a likely pathway for VOC movement in the ground since relatively small quantities of volatile organics were detected in the bedrock throughout the site. The groundwater in the deep aquifers was being extracted by pumping of AT&T's Production Well 3, which at the time of the Phase I investigation in 1986 had been in almost continuous operation at rates of up to $281 \, s^{-1}$ since its construction the previous year.

VOCs in the shallow aquifer are migrating towards the Merrimack River. Some of the VOCs in the shallow aquifer are being intercepted by pumping Production Well 1 which abstracts water from this formation, and some of the VOCs are migrating down to the deep

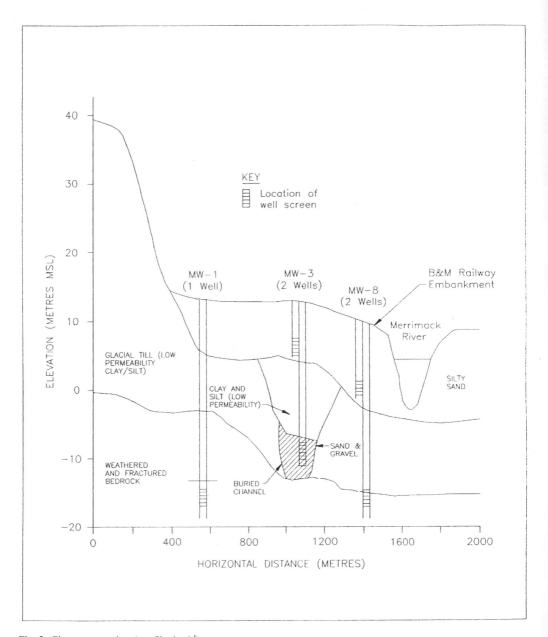

Fig. 2. Site cross-section (profile A–A').

aquifer. Groundwater from both the shallow and deep aquifers that is not intercepted by the production wells probably discharges to the Merrimack River. No VOCs were detected in any surface waters leaving or bordering the site, nor were VOCs detected in the sediments of the Merrimack River either immediately upstream or downstream of the AT&T site. There are no known municipal or industrial wells nearby which might be affected by groundwater from the AT&T site.

Phase II hydrogeological investigation

In the Phase I investigation, CDM developed an understanding of the magnitude of the impact of VOCs on the groundwater and the complexity of the hydrogeology and stratigraphy at the Merrimack Valley Works. It was agreed that more fieldwork and analysis was needed to further quantify groundwater flow and VOC transport mechanisms before remedial measures could be evaluated and implemented. Two principal tasks of the Phase II investigation were as follows:

- field investigation activities: installation of additional monitoring wells, sampling, and *in situ* tests of soil characteristics;
- hydrogeological study: analysis of field data and laboratory tests, refinement of the conceptual understanding of groundwater flow and VOC transport at the site, and computer simulation of groundwater flow and VOC transport using CDM's DYNFLOW and DYNTRACK three-dimensional models (Riordan *et al.* 1984).

Additional wells at seven locations were installed to complement and refine the information on stratigraphy, hydrology and groundwater quality obtained in Phase I. A detailed discussion of these results together with the results of the Phase I study are given in the following sections.

Results of detailed hydrogeological investigations

Stratigraphy

During the Phase I hydrogeological investigation, 14 monitoring wells were installed at the Merrimack Valley Works by CDM and Hydro Group, Inc., a drilling contractor. The borehole logs from these wells were useful in characterizing site stratigraphy. Five principal formations were encountered:

- Fine silty sand extends 6–12 m below ground surface across the site.
- Glacial till (comprising clay and clayey silts) lies directly above bedrock, and either outcrops at the ground surface, or is covered by the fine sand layer.
- There is a weathered and fractured siltstone and sandstone bedrock. Nine borings were extended into presumed bedrock, and in no case was relatively unweathered bedrock encountered.
- A relatively narrow band of coarse sand and gravel traverses the site, and is believed to be a buried post-glacial channel. It is located 18–27 m below ground surface, is approximately 4.5–9 m thick, and lies directly above bedrock.

- A relatively impermeable layer of silts and clays, which varies in composition and thickness across the site, lies directly below the fine sand and above either the buried channel or bedrock.

These principal formations are shown in cross-section A–A′ (Fig. 2).

Fourteen wells were installed at eight locations during the Phase I study, most of them as pairs, with one well screened in the shallow fine sand layer (designated S) and an adjacent well screened in either bedrock or the buried channel (designated D). Seven additional wells were installed in four locations in June 1987 to better quantify the stratigraphy in the middle of the AT&T site, and to provide additional locations for water quality sampling. Their locations are also shown on Fig. 1.

Generalized subsurface profiles using the stratigraphic information obtained during the installation of 21 monitoring wells in Phases I and II were evaluated. This information indicates that a highly permeable buried channel traverses the site. It is this critical subsurface feature that governs migration of groundwater contaminants at the Merrimack Valley Works, particularly in areas with a relatively thin confining layer between the buried channel and the surficial permeable fine sand aquifer.

Although the subsurface stratigraphy for this site is quite complex, with changes over relatively short distances, the monitoring wells installed during the Phase I and Phase II investigations provided sufficient subsurface stratigraphy information to characterize the subsurface conditions at the site.

Distribution of volatile organic compounds in groundwater

There are 21 monitoring wells in 11 locations, in addition to three production wells. Groundwater was sampled from each well at least once during the study. The magnitude of VOC concentrations in groundwater varied considerably across the site, but the concentrations of total VOCs measured in the wells on site fall into the following three broad categories:

source range:	200 000–300 000 µg l^{-1} total VOCs
transition range:	600–1100 µg l^{-1} total VOCs
low range:	10–100 µg l^{-1} total VOCs

The distribution of VOCs at the monitoring wells and production wells is provided in Table 1.

Table 1. *Distribution of VOCs at monitoring and production wells*

Category	Range of total VOCs ($\mu g\,l^{-1}$)	Well identification
Source	200 000–300 000	MW-2S
Transition	600–1100	MW-3S
		MW-4S
		MW-6D
		MW-10S
		MW-10D
Low	10–100	MW-2D
		MW-3D
		MW-5S
		MW-6S
		MW-7D
		MW-8S
		MW-9S
		MW-9D
		MW-11S
		PROD WELL 2
		PROD WELL 3
Absent	Below detection limit	MW-1D
		MW-4D
		MW-7S
		MW-8D
		MW-11D
		PROD WELL 1

Source range concentrations. Relatively high concentrations were measured at only one location: in the shallow well at the tank farm.

Transition range concentrations. Concentrations of VOCs ranging from 600 to 1100 $\mu g\,l^{-1}$ were measured in three shallow wells and two deep wells that were installed in the vicinity of a number of waste solvent tanks at the Merrimack Valley Works. The shallow wells exhibiting between 600 and 1100 $\mu g\,l^{-1}$ of total VOCs are MW-3S (at the location of the waste solvent tank), MW-4S (at the location of the waste acetone tank), and MW-10S (at the location of the barrel pad).

Deep wells MW-10D and MW-6D showed VOC concentrations in the transition range of 600–1100 $\mu g\,l^{-1}$. MW-10D is located in the bedrock and the most likely source of the VOCs in this well is the downward migration of waste solvents from the barrel pad facility.

Monitoring well MW-6D is screened in the buried channel, down-gradient of the four potential VOC sources mentioned previously. Because the shallow well at this location, MW-6S, shows only low levels (10–100 $\mu g\,l^{-1}$) of VOCs, it is not likely that there was a VOC source at this location. VOCs from any of the four potential

sources identified may have migrated in groundwater to MW-6D.

Low concentrations of VOCs. Low levels of VOCs were found in the shallow sand formation, in the buried channel, and in the bedrock. This is indicative of general migration of the VOCs, and of the relatively high degree of hydraulic communication between the shallow and deep aquifers.

Discussion

It was difficult to relate specific sources (i.e. tanks) to the VOCs found in specific wells. All of the VOCs found in the groundwater are industrial solvents (or their degradation products), which could be traced to any number of on-site sources.

Trichloroethene is the most prevalent VOC in groundwater at the Merrimack Valley Works. It is present in 14 of the 24 wells sampled. Tetrachloroethene (PCE), trichloroethene (TCE) and trichloroethane (TCA) were in the past stored in underground storage tanks or were used in manufacturing operations at the Merrimack Valley Works. Biotic and/or abiotic transformations of these parent compounds in groundwater generally produce dichloroethene (DCE) isomers, dichloroethane (DCA), vinyl chloride (VC), and chloroethane (CA). The transformation sequences are shown in Table 2. The presence of parent compounds (PCE, TCE and TCA) found in groundwater at the Merrimack Valley Works can probably be attributed to inadvertent leaks or spills at underground solvent storage areas. Because the daughter compounds (DCE, DCA, VC and CA) are not known to have been used at the plant or stored in underground facilities, their presence in groundwater at the site can probably be attributed to abiotic or biotic transformation of the parent compounds in the subsurface environment.

Table 3 shows the ratios of parent to daughter VOCs for the monitoring wells where total VOCs were measured in the transitional range (i.e. 600–1100 $\mu g\,l^{-1}$). Daughter compounds were present in concentrations over 400 $\mu g\,l^{-1}$ in wells MW-4S (near the waste acetone tank area), MW-6D, MW-10S (at the barrel pad area), and MW-10D. Unless the daughter compounds were present in the original source, their presence at these concentrations implies that VOCs had been in the ground for many years, provided transformation rates were similar to typical values for such compounds.

Table 2. *Biotic and/or abiotic chemical transformations (Vogel et al. 1987)*

Parent compounds	1,1,2,2-Tetracholoroethene (PCE)	
	Trichlorethene (TCE)	1,1,1-Trichorethane (TCA)
Daughter compounds	*trans*-1,2-Dichloroethene	1,1-Dichloroethene
	cis-1,2-Dichloroethene	1,1-Dichloroethene
	Vinyl chloride	Vinyl chloride
	Chloroethane	Chloroethane

Groundwater system: conceptual model

An initial understanding of the groundwater system at the Merrimack Valley Works was developed in the Phase 1 hydrogeological investigation. The Phase II hydrogeological investigation refined that 'conceptual model' of the groundwater system by better defining the hydraulic characteristics of each water-bearing formation, and by establishing the magnitude and location of hydraulic connections between formations.

Surface water and groundwater features. Groundwater flow at the Merrimack Valley Works is governed by recharge from Lake Cochichewick to the south, recharge from rainfall, discharge to the Merrimack River to the north, and the hydraulic properties of the soil and bedrock through which the water is conveyed. Water infiltrates from Lake Cochichewick into the ground and flows toward the Merrimack River. Precipitation within the Merrimack Valley Works watershed also enters the ground and flows toward the Merrimack River. Small surface streams provide water to the groundwater system in some places and elsewhere carry water from the ground to the Merrimack River.

Groundwater is conveyed through three water-bearing formations. Fine sands in the upper 6–12 m of the site contain an unconfined aquifer 3–9 m thick. Surface water bodies such as the local streams and the Merrimack River

occur where the water-table aquifer intersects the ground surface. Directly beneath the surficial sand aquifer is a relatively impermeable layer composed of glacial till, silt and clay. This layer varies in thickness from 1.5 to 15 m across the site. The large variations in vertical permeability in this layer are important to the groundwater flow system because there are two water-bearing formations directly beneath it. Water flowing vertically between the shallow and deep aquifers would be likely to convey VOCs present in the shallow aquifer to the relatively uncontaminated deep aquifer.

One of the deep aquifers is a narrow post-glacial channel that lies directly above bedrock and is covered with 18–27 m of clay, silt and sand. The buried channel is up to 9 m thick and approximately 15–61 m wide. It is present beneath Building 30 and the parking lots on the northern half of the Merrimack Valley Works. The buried channel is very transmissive; AT&T's Production Well 3, which is screened in this formation, produces up to 28 l s^{-1}. Before Production Well 3 was in operation, it is probable that water in the buried channel discharged slowly to the Merrimack River due to the natural upward vertical gradient.

The second deep aquifer is the highly weathered and fractured siltstone and sandstone bedrock. CDM drilled 11 boreholes up to 6 m into this bedrock. In no case was competent bedrock encountered. The transmissivity of the bedrock is anisotropic and quite variable across

Table 3. *Ratios of parent to daughter VOCs in wells*

Well identification	Total VOCs (ppb)	Parent VOCs (ppb)*	Daughter VOCs (ppb)†	Ratio of parent to total VOCs	Ratio of daughter to total VOCs
MW-3S	769	696	73	0.91	0.09
MW-4S	677	91	586	0.13	0.87
MW-6D	1068	595	473	0.56	0.44
MW-10S	683	243	440	0.36	0.64
MW-10D	895	102	793	0.11	0.89

* Parent compounds: 1,1,2,2-tetrachloroethene, trichloroethene, and 1,1,1-trichloroethane.
† Daughter compounds: 1,1-dichloroethene, 1,1-dichloroethane, trans-1,2-dichloroethene and vinyl chloride.

the site. The Merrimack River appears to be a regional groundwater sink, and probably receives discharge from the bedrock aquifer.

Anthropogenic pathways. When the underground storage tanks were excavated from the tank farm it was observed that the tanks were surrounded by relatively low-permeability soil. However, transmissive pathways may have existed leading from the tank farm to other parts of the site, e.g.

- the backfill around the foundation of Building 70;
- the backfill of a 150 mm (increasing to 375 mm) diameter sanitary drain under the tank farm, leading to the sanitary wastewater treatment plant on site.

There are many other underground utilities traversing the site; however, only the sanitary drain and the foundation backfill appeared to be deep enough to intersect the water-table and provide a contaminant migration pathway.

Piezometric head distribution. Contours of piezometric heads in the groundwater were evaluated for both the shallow and deep aquifers based on water level measurements taken over a two-year period. Horizontal and vertical gradients did not change significantly. The direction of the groundwater flow is generally towards the Merrimack River in both the shallow and deep aquifers.

The effects of extracting water from AT&T's production wells are apparent in the shallow aquifer, and even more pronounced in the deep aquifers. Production Well 3, screened in the buried channel, presently pumps from 12.6 to $28 \, l \, s^{-1}$ almost continuously; while Production Wells 1 and 2, screened in the surficial sand, are only pumped intermittently.

Horizontal groundwater gradients are approximately 1% (i.e. a 1 m drop in head over a 100 m horizontal distance, in the direction of flow) in each of the three water-bearing formations. However, horizontal gradients in the buried channel are artificially increased due to pumping of Production Well 3.

There are very strong vertical gradients across the site, ranging from an 11% downward gradient at MW-3, to a 5% upward gradient at MW-6. Zones of upward and downward vertical gradients are shown in Fig. 3.

Several conclusions can be drawn from the distribution of vertical gradients at the site. It shows that the Merrimack River is a point of groundwater discharge, even for the deep aquifers; downward gradients on the southern

half of the site yield to upward gradients near the Merrimack River. Production Well 3, when pumping, artificially creates a downward gradient within its zone of influence. The strong downward gradient at MW-3 does not necessarily indicate that a significant quantity of water is flowing downward at this location. A 9 m thick layer of clay separating the surficial sand aquifer from the buried channel at MW-3 most likely strongly inhibits vertical flow. The markedly decreased vertical gradients at nearby wells MW-11 and MW-9 imply that a break or change in composition of the layer separating the shallow and deep aquifers permits vertical flow and tends to equalize the piezometric heads in that vicinity. This assumption is consistent with changes in the soil stratigraphy observed during the installation of monitoring wells MW-3D, MW-4D and MW-9D, and pumping test data for the site.

Chloride concentrations in groundwater. During the summers of 1986 and 1987, groundwater from all monitoring and production wells was sampled to determine anion concentrations.

Chloride concentrations measured in the shallow sand formation were, on average, 100 times higher than chloride concentrations in the weathered bedrock, and approximately four times higher than chloride concentrations in the buried channel. The most likely source of chlorides in the shallow sand formation is road salt used in winter on roads and parking lots. Chlorides in such high concentrations do not occur naturally in groundwater in the area. As chloride is a conservative parameter, the road salt can be regarded as a tracer of groundwater movement from the shallow sand aquifer.

Table 4 shows the ranges and average chloride concentrations measured in the shallow sand formation, the buried channel and the bedrock. The bedrock is essentially free of chlorides (except at MW-10D and MW-5D) while the buried channel contains chloride concentrations approaching those of the shallow sand formation. Although MW-10D and MW-5D monitor bedrock groundwater, their chloride concentrations are more characteristic of the buried channel aquifer. They are referred to as deep transitional wells in Table 4.

The relatively high chloride levels in the buried channel imply that groundwater flows from the shallow sand formation to the buried channel, probably along much of its length. This hypothesis is supported by the presence of downward hydraulic gradients over much of the site, and by variations in the thickness and composition of the silt and clay layer separating the shallow

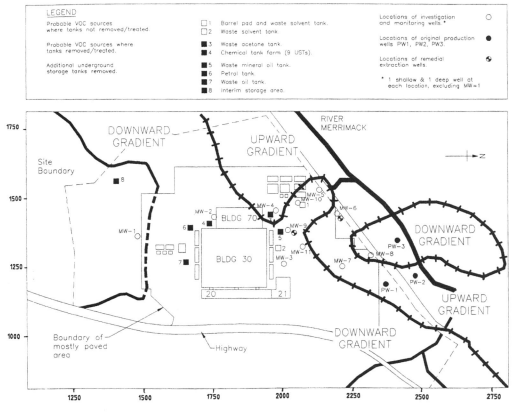

Fig. 3. Direction of hydraulic gradients. Scale in metres. The solid lines represent watercourses; dashed lines indicate where watercourses are culverted beneath paved areas. The crossed solid lines represent the boundaries between zones of upward and downward hydraulic gradient.

sand formation from the buried channel. The low chloride concentrations in much of the bedrock groundwater have the following implications:

- the till overlying the bedrock restricts vertical groundwater flow;
- the bedrock does not receive water from the buried channel;

- the bedrock is recharged primarily from south of the Merrimack Valley Works;
- the bedrock has low vertical permeability.

Pumping test. A seven-day pumping test was conducted at the site to investigate the conductivity and location of hydraulic connections

Table 4. *Chloride concentrations in groundwater*

Sampling time	Chloride concentration (mg l^{-1})			
	Surficial fine sand	Buried channel	Deep transition	Weathered bedrock
Summer 1986				
Range	67–2700	100–350	—	2.3–7.5
Average	726	197	43	5.1
Summer 1987				
Range	46–2600	110–550	55–500	2.4–25
Average	900	246	278	10.3

between the shallow sand aquifer and the deep aquifers. The well selected, MW-11D, draws water from the transmissive buried channel and contained no VOCs when first sampled and analysed.

MW-11D was pumped at an average rate of $41s^{-1}$ for seven days. Daily water levels were measured manually in wells MW-3 to MW-11, and electronic loggers collected data hourly in wells MW-3D, MW-4D and MW-9D. VOC concentration was measured in well MW-11D.

Measured drawdowns in the deep wells ranged from over 1. 2 m at well MW-9D to only 0.3 m at MW-5D. All shallow wells showed 0.15 m or less of drawdown and no shallow well registered more than 0.03 m of drawdown after two days of pumping. Drawdowns in the shallow wells are thus much smaller than drawdowns in the deep wells, indicating that the buried channel aquifer is confined but is subject to some leakage to and from the shallow aquifer.

Drawdowns in the deep wells were generally greater along an axis parallel to the alignment of the buried channel, supporting the hypothesis that the channel is a narrow, transmissive band. Wells in the weathered bedrock zone also show this preferential drawdown pattern, indicating that water may flow through the bedrock along fractures parallel to the buried channel.

The hydraulic properties of the deep aquifer were calculated from drawdown versus time plots. The hydraulic parameters calculated at MW-3D (in the buried channel) are

transmissivity: $130\,m^2\,day^{-1}$
hydraulic conductivity: $27\,m\,day^{-1}$
storativity: 2×10^{-4}

In summary, the test indicates that well MW-11D is drawing water from a leaky confined aquifer with preferential flow along the alignment of the buried channel. Since the shallow well registering the most drawdown was MW-4S, it is postulated that vertical leakage occurs in the area bounded by MW-4, MW-9 and MW-10.

Borehole permeability tests. Rising head, constant head and recovery tests were performed in the monitoring wells on site and the data were analysed to provide estimates of hydraulic conductivity in the vicinity of the well screens. The resulting values of hydraulic conductivity were plausible, but lower than anticipated. The average results, listed in Table 5, were used as conservative estimates of hydraulic conductivity.

Table 5. *Results of borehole permeability tests*

Formation	Hydraulic conductivity $(m\,day^{-1})$	
	Range	Average
Surficial fine sand	0.03–2.7	0.92
Buried channel	2.3–6.9	5.2
Weathered bedrock	0.06–0.92	0.31

Numerical modelling of groundwater flow

DYNFLOW computer program

A mathematical model of groundwater flow at the site was developed to improve our understanding of the groundwater flow regime and to provide a tool for evaluating contamination remedial alternatives.

The modelling was performed using DYNFLOW (Riordan *et al.* 1984), a computer program developed at CDM that simulates three-dimensional groundwater flow using the finite-element method. DYNFLOW solves the conventional equations of flow in porous media to simulate the response of groundwater flow systems to several types of natural and artificial stresses. It solves both confined and unconfined groundwater flow equations, handles the transition from confined to unconfined status and can accept anisotropic aquifer parameters.

Finite-element grid

The model includes the 68 ha AT&T Merrimack Valley Works property and neighbouring parts of North Andover and Methuen, including a 1.6 km stretch of the Merrimack River. The areal extent of the model is shown in Fig. 4. The model boundaries were set by sensitivity analysis sufficiently far from the site so that flow conditions at the boundaries do not affect gradients at the site.

A finite-element grid was created by discretizing the model area into a number of triangular elements, also shown in Fig. 4. The model grid has 258 non-uniformly spaced nodes (points of intersection among elements) and 479 elements. The high density of nodes at the site allows more refined representation of hydraulic gradients within the area of interest, since fluxes and piezometric heads are calculated at each node.

The vertical base of the model is unweathered bedrock, and the top of the model is the ground surface. The three water-bearing formations and the relatively impermeable layer that separates

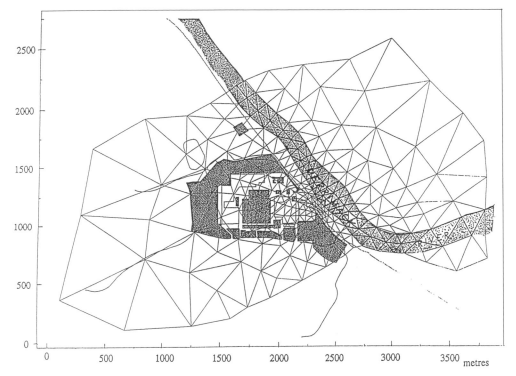

Fig. 4. Finite-element grid: plan view.

the shallow and deep aquifers were represented by four vertical layers in the model. The model layers are schematically shown in Fig. 5. The lowest layer (layer 1) represents a 10.6 m thick layer of weathered and fractured siltstone and sandstone lying above an impermeable unweathered bedrock base. The second layer contains the buried channel, where it is present at the site. Where there is no buried channel, layer 2 contains silty clay or glacial till. The third layer consists of silty clays and glacial till, while the uppermost layer contains the surficial sand aquifer and till outcrops. The elevations of each subsurface level were interpolated from well log data.

The backfill around utility lines and building foundations could represent highly transmissive conduits for groundwater flow. Of particular interest at the Merrimack Valley Works is the drainage potential of the backfill to a sewer under Building 70, and the backfill around the building's perimeter wall slab footing. These are believed to intersect the water-table and provide a pathway for water to flow from the tank farm to the rest of the site. Thus, a zone of transmissive material was included in the model under Building 70, with preferential flow along the axis of the sewer.

Boundary conditions

Boundary conditions in DYNFLOW describe externally imposed heads or fluxes (e.g. impermeable boundaries, constant flow or constant head).

The model's lateral boundaries were generally defined along ridgelines of hills of glacial till which probably behave as groundwater divides. Except in the vicinity of AT&T's production wells, the Merrimack River is a regional groundwater sink, bounding the groundwater system to the north. The model's northern boundary extends beyond the river in order to incorporate the full extent of drawdown around the AT&T production wells. The net flux of water in the model is thus determined primarily by recharge from precipitation, baseflow and pumping from AT&T's production wells.

Recharge was modelled by assuming infiltration rates of 0.4 m per year into surficial sands and 0.25 m per year in areas of glacial till. The areas of the AT&T site which are paved or covered by buildings (roughly 50% of the 68 ha site) were assumed to receive no recharge from rainfall.

Since both shallow and deep aquifers at the site discharge to the Merrimack River, the River's elevation largely governs piezometric

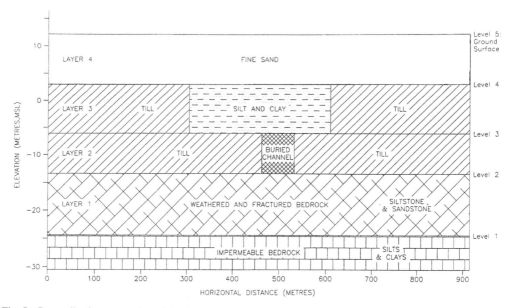

Fig. 5. Generalized cross-section of the finite-element grid.

heads in all aquifers at the site, especially in the unconfined silty sand aquifer. The water surface elevation of the Merrimack River was measured at 3 m above mean sea level (amsl), and was held constant in the model. The small streams draining the site were included in the model because they have a significant local effect on piezometric heads in both the shallow and deep aquifers. The water surface elevation in the marshy area in the hills south of the site was fixed at 41 m amsl.

AT&T's Production Wells 1, 2 and 3 are located within the model boundary. Pumping from these wells was included in the model by assigning discharge fluxes from the node nearest each well, at the elevation of the well screen. Average pumping rates were established from AT&T's hourly pumping records. During these periods, Production Well 3 was in use all of the time at rates up to $28 \, 1s^{-1}$, Production Well 2 came on line at water demands above $25-28 \, 1s^{-1}$ and Production Well 1 was turned on when the water demand exceeded approximately $50 \, 1s^{-1}$. The typical pumping rates used in the model are as follows:

Production Well 1 $4 \, 1s^{-1}$
Production Well 2 $9 \, 1s^{-1}$
Production Well 3 $24 \, 1s^{-1}$

Calibration of model

The model was calibrated to simulate average groundwater flow based on water levels measured in monitoring wells on-site. This was achieved by varying the hydraulic properties, layer thicknesses and principal axes of conductivity of the soils and bedrock.

Calibration of the model underscored the complexity of the geology at the AT&T site. It was necessary to incorporate 16 different element properties in the model, representing 16 different soil conditions. A wide range of values of hydraulic conductivity were tested for each soil formation. Table 6 lists the range of values tested. The adopted values of hydraulic conductivities, also shown in Table 6, fall within generally accepted ranges of values for the soil types at AT&T (e.g. values suggested in Freeze & Cherry 1979), and are consistent with the analyses of the pumping tests performed by CDM at the Merrimack Valley Works.

The seven-day pumping test was also used to calibrate the groundwater flow model. Drawdowns generated by DYNFLOW match measured drawdowns reasonably well (Table 7).

Summary

Several hypotheses previously formed were supported by the DYNFLOW model results:

- the weathered and fractured bedrock shows preferential flow along a NW–SE axis (perpendicular to the river);
- there is vertical groundwater flow between the shallow and deep aquifers;

Table 6. *Hydraulic conductivities used during model calibration*

Subsurface formation	Common range[*]	Hydraulic conductivities (m day^{-1})			
		Range tested		Range adopted	
		Lateral	Vertical	Lateral	Vertical
Shallow fine sand	0.03–30.7	0.02–61.4	0.003–15.4	0.03–30.7	0.02–0.08
Silt and clay	3×10^{-8}–3.1	0.003–15.4	3×10^{-4}–7.7	0.02–0.03	0.002–0.2
Glacial till	3×10^{-8}–0.31	0.003–15.4	3×10^{-4}–7.7	0.02	0.002
Buried channel (coarse sand and gravel)	3.1–3070	15.4–154	1.5–30.7	38–92	3.8–9.2
Bedrock	3×10^{-6}–30.7	0.003–61.4	0.002–61.4	0.02–61.4	0.02–61.4

[*] Adopted from Freeze & Cherry (1979).

- the hydraulic characteristics of all soil and rock formations are highly variable across the site.

Numerical modelling of contaminant transport

The migration of volatile organic compounds (VOCs) in the groundwater was simulated using DYNTRACK, CDM's groundwater mass transport computer program. The immediate objective of this modelling effort was to investigate the pathways of VOC movement from the underground storage tanks on-site through the subsurface system.

DYNTRACK uses a 'random walk' technique to simulate mass transport in groundwater flow. A mass of solute is represented by a cloud of discrete particles. Each particle is displaced in successive steps to represent advective processes in the groundwater flow field, with dispersion represented by imparting a random deflection. The mass represented by each particle can be reduced to simulate biodegradation or other decay processes.

The results (groundwater fluxes and piezometric heads) of the DYNFLOW simulations are used directly by the mass transport model DYNTRACK. Average groundwater flow conditions were simulated for the period when

VOCs may have entered the groundwater at the Merrimack Valley Works.

Four underground storage tank areas at the site – the tank farm, the waste solvent tank, the waste acetone tank and the barrel pad area – are suspected to have led to leakages or spills in the past. The release of VOCs from each source was simulated in separate model runs.

The tank farm

DYNTRACK was used to identify the paths that VOCs would follow if released into the groundwater from the tank farm. A plume of particles was generated by releasing particles from the tank farm continuously for 25 years. The 25-year simulation is used to estimate the migration pathway that the bulk of VOCs released from the tank farm would follow over time, under present well pumping conditions.

As would be expected, this simulation showed that the bulk of the particles travel north toward the Merrimack River. The existing monitoring wells are well placed to intercept any VOCs that may travel from the tank farm to the river.

Waste solvent tank

A 25-year simulation for this source shows that particles released from this location generally migrate toward the buried channel and the majority of particles are drawn down into the buried channel within a few hundred feet of the source.

Waste acetone tank

With the same modelling assumption as above, particles released from this location follow a

Table 7. *Comparison of computed and measured drawdowns*

Well type	Total number	Average of measured drawdown (m)	Average of computed drawdown (m)
Shallow	11	5.94	6.00
Deep	11	6.95	7.04

northerly flow path towards the Merrimack River. This plume appears to migrate gradually downward through the silt and clay, towards the buried channel and bedrock.

Barrel pad

Simulations for this source under the same conditions show that the plume does not follow a direct path toward the Merrimack River; rather, it migrates northward towards the buried channel. Some of the particles are drawn down through the silt and clay towards the buried channel and bedrock.

Summary of DYNTRACK modelling results

The particle migration pathways predicted by the DYNTRACK model generally support the conceptual model of VOC migration. The migration pathways from the four potential VOC sources appear to be predominantly in the surficial fine sand formation, but with a component being drawn down into the buried channel aquifer.

AT&T's Production Well 3 appears to capture the bulk of VOCs in the buried channel, and draws VOCs in from the surficial sand and bedrock formations. Those VOCs that remain in the surficial sand formation are either captured by AT&T's Production Wells 1 or 2, or migrate slowly to the Merrimack River.

Remediation

Immediate remedial measures

Based on the Phase I hydrogeological study, it was determined that the tank farm and other underground storage tanks at the site may have contributed to groundwater and subsurface soil contamination. These tanks were targeted for removal as an immediate remedial measure.

In all, nine solvent tanks in the tank farm (storing acetone, toluene, trichloroethene and trichloroethane), a gasoline tank and a waste oil tank were removed. In addition, a waste acetone tank was cleaned, filled with cement grout and abandoned in place.

During the tank farm closure over $230 \, m^3$ of contaminated soil from the excavations was removed from site. In addition, $190 \, m^3$ of groundwater contaminated with VOCs was treated and disposed of during excavation dewatering.

Design of permanent remedial system

Based on the results of the phased hydrogeological study, a feasibility study to evaluate six potential remedial alternatives was conducted:

A. No action/minimal action (no extraction wells).
B. A single deep extraction well at MW-6D.
C. Deep extraction wells at MW-6D and MW-10D.
D. Deep extraction wells at MW-6D and MW-9D.
E. A deep extraction well at MW-6D and shallow extractions wells at MW-5S and MW-6S.
F. Deep extraction wells at MW-6D and MW-9D and shallow extraction wells at MW-5S and MW-6S.

Because source removal was accomplished as part of the immediate remedial measures (IRM) programme, the only remedial alternatives considered were variations of a number of pump-and-treat scenarios. The effectiveness of various extraction well locations and recovery rates were evaluated using computer simulation models calibrated in this study.

The selected alternative D takes advantage of the unique hydrogeological characteristics of this site. Two extraction wells were installed in the buried channel between the site and Production Well 3 (see Fig. 1). The purpose of these extraction wells is twofold. First, contamination in the surficial aquifer would be drawn into the buried channel as was clearly demonstrated in the seven-day pumping test. This will eventually effect a long-term remediation of the surficial aquifer because the principal source of the contamination (underground storage tanks) has been removed. Secondly, groundwater contamination will be intercepted before it reaches Production Well 3. Elements of the proposed remediation programme included the following:

- two groundwater extraction wells located in the buried channel;
- total pump rate $6.3–19 \, l \, s^{-1}$;
- treatment of all extracted groundwater in an air stripper;
- discharge of the air stripper effluent to the AT&T outfall;
- sampling of selected monitoring wells quarterly;
- monitoring of the air stripper effluent as specified in the plant operating licence;

Table 8. *Design parameters for the groundwater treatment system*

Air stripper	
Design influent concentration:	1–$2\,mg\,l^{-1}$ VOCs
Design effluent concentration:	$<100\,\mu g\,l^{-1}$ VOCs
Air to water ratio:	40 to 1
Packing material:	25.4 mm tripacks, surface areas $276\,m^2\,m^{-3}$
Tower dimensions	
Diameter:	0.92 m
Height:	4.6 m
Packing height:	2.6 m
Influent water temperature:	10 °C
Blower:	1 Hp forced draft
Nozzle:	non-aspirated shower
Air emissions:	$0.005\,kg\,min^{-1}$ total VOCs

- treatment of all gases from the air stripper using vapour phase activated carbon;
- beginning groundwater extraction and treatment as soon as possible;
- discontinuing groundwater extraction and treatment when total VOC concentrations in extraction wells are below $100\,\mu g\,l^{-1}$ in two consecutive samples.

Design parameters for the groundwater treatment system are shown in Table 8. The system is shown schematically in Fig. 6.

As an added benefit, the treated effluent from the pump-and-treat system was used to augment the on-site supply. To accomplish this, a previously decommissioned water storage tank was incorporated into the final design.

Operation

System start-up. System start-up is a crucial period when initial operating parameters and equipment functions are monitored to assess the actual system operation in relation to the designed or estimated operation. In addition to verifying proper operation of the equipment, parameters such as groundwater elevations, VOC concentrations and anion concentrations in the monitoring wells and removal efficiencies in the air stripper tower and carbon adsorption system are closely monitored to establish baseline values. These baseline values are used to predict and evaluate future trends with respect to the remediation of the contaminated aquifer.

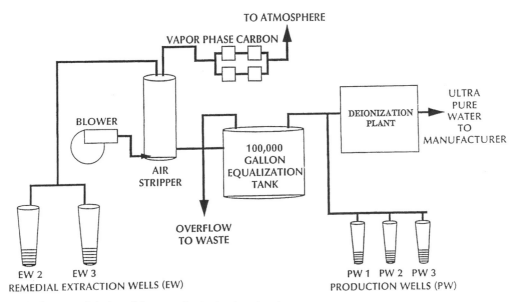

Fig. 6. Conceptual design of the groundwater treatment system.

Performance. The groundwater remediation system was started up on 6 November 1990 and has operated with minimal interruption except for a shutdown between February and October 1991 due to poor performance of the activated carbon units being used to treat off-gases from the air stripper. As required by the monitoring plan approved by the regulatory agencies involved, samples for VOC analysis of the air stripper tower influent and effluent groundwater streams were collected daily for the first week of operation and weekly for the following month. Sampling has continued on a monthly basis.

Results of the monthly influent and effluent air stripper tower sampling and associated removal efficiencies are presented in Table 9. Monthly sampling of the individual extraction wells EW-2 and EW-3 was also recommended for analysis of VOCs, to identify a reduction of VOCs in an individual well. A reduction in VOCs to below $100 \mu g l^{-1}$ in one well would mean that the water from this well would not require further treatment.

Analysis of both extraction wells for chloride and sulphate anions was also conducted to monitor migration of road-salt contaminated groundwater from the shallow aquifer (a surficial fine sand formation) to the deeper buried channel aquifer. As expected, chloride concentrations increased proportionally to the VOC concentration increase, indicating a drawdown of groundwater from the more highly VOC contaminated shallow aquifer.

The mass removed of VOCs and TCE (the highest individual VOC concentration) has been monitored and is presented in Table 10. By December 1993, approximately 0.64 million m^3 of groundwater had been extracted and treated, resulting in the approximate removal of 500 kg (350 l) of total VOCs, of which an estimated 330 kg (230 l) comprised TCE.

Initially the pumping rates of extraction wells EW-2 and EW-3 were $8.8 l s^{-1}$ and $6.3 l s^{-1}$, respectively. In the course of a month, the

Table 9. *Air stripper tower performance data** *

	Range	Mean
Influent water flows $(l s^{-1})$	291–901	520.0
Air flows in tower $(m^3 s^{-1})$	33–57	39.0
Treated effluent	(VOC) ND–25.7	7.6
concentration $(\mu g l^{-1})$	(TCE) ND–13	4.3

* Operational period: 6 November 1990 to date (1996), except for a shutdown during February to October 1991. Period for which monitoring data are available: 6 November 1990–15 December 1993.

Table 10. *Treatment of groundwater: mass removal of VOCs and TCE*

Month	Litres treated (litre $\times 10^6$)	VOC removed (kg)	TCE removed (kg)
November 1990	28.4	5.2	2.7
December	34.1	21.8	14.1
January 1991	20.8	21.8	17.0
February	14.8	18.9	12.4
October	25.4	5.2	3.0
November	32.9	37.6	27.0
December	30.7	35.0	24.6
January 1992	18.5	9.7	8.0
February	21.6	16.4	10.0
March	22.3	24.4	15.2
April	19.7	22.0	15.4
May	23.5	14.4	10.0
June	23.1	12.5	8.7
July	23.8	10.4	7.3
August	22.0	28.3	18.4
September	23.1	20.3	12.2
October	21.2	23.3	14.3
November	19.3	21.8	13.8
December	18.2	21.9	14.6
January 1993	14.8	8.5	6.3
February	13.6	7.4	5.7
March	14.8	5.7	4.4
April	14.4	6.8	4.9
May	17.4	14.7	9.1
June	15.1	10.5	6.4
July	18.5	13.2	7.7
August	15.5	15.9	9.6
September	11.4	7.1	4.6
October	12.9	17.7	9.0
November	22.3	8.4	5.9
December	28.8	18.4	11.0
Total to date	642.7	505.4	333.0

combined flow rate decreased to $12.5 l s^{-1}$, and one year from start-up the combined flow rate decreased to $9.5 l s^{-1}$. The decrease in flow rate is attributed to the low aquifer recharge rate.

The initial air stripper tower influent sample contained $41 \mu g l^{-1}$ of TCE and $120 \mu g l^{-1}$ of total VOCs. A gradual increase in the influent TCE concentration to $340 \mu g l^{-1}$ and the total VOC concentration to $551 \mu g l^{-1}$ by 4 December 1990 was observed. The TCE and total VOC concentrations increased more rapidly over the next two months to concentrations of $840 \mu g l^{-1}$ and $1277 \mu g l^{-1}$ in February 1991. At this time, operational problems forced the temporary shutdown of the system.

Following the shutdown the system was restarted in October 1991, and has operated consistently since. TCE/VOC concentrations $(120 \mu g l^{-1}$ and $206 \mu g l^{-1})$ after start-up in October 1991 were similar to the concentrations

Table 11. *Maintenance and operational problems and resolutions*

Problem	Symptom	Resolution
Inefficient vapour phase carbon system operation	Rapid breakthrough of carbon canisters	Installation of a drop-out leg and air/water separator equipped with demister pads to remove excess water droplets prior to entering the heat exchanger. In addition, all ductwork located outside the building and the carbon canisters were insulated to minimize condensation
Prolonged shutdown of wells due to operational problems allowed metal oxides to accumulate and inhibit the flow of groundwater	Decrease in pumping	Removal of pumps for cleaning of intake screens. Chemical cleaning and resurging of wells to remove oxidized metals and iron bacteria from well screen and discharge pipe
Metal oxide build-up in flow meters	Inaccurate flow measurement	Removal and cleaning of flowmeters on a consistent basis
Leak present at flange connection between two pieces of air stripper tower	Pooling of untreated groundwater on roof	Leak patched with fibre-reinforced plastic

detected in the initial start-up. The TCE/VOC concentrations increased rapidly to the levels reached prior to shutdown in February 1991.

Carbon adsorption system. The monitoring plan also requires monitoring of the air stripper tower off-gas stream before and after treatment through vapour phase carbon. This requirement is fulfilled through bi-monthly sampling of the off-gas stream at several strategic locations and on-site analysis utilizing a portable gas chromatograph.

Upon exiting the air stripper tower, the off-gas stream is directed through ductwork to two sets of carbon adsorption units. Each carbon set consists of two carbon canisters arranged in series configuration. The primary unit captures the majority of the VOCs. The secondary unit polishes the stream prior to discharge to atmosphere.

Analytical results obtained from the portable gas chromatograph are used to determine 'breakthrough' of the canisters. 'Breakthrough' is defined in the monitoring plan as occurring when the vapour phase concentration of TCE (reported as ppmv) in the discharge stream of the primary carbon canisters is equal to or greater than 75% of the vapour phase TCE concentration (ppmv) in the carbon adsorption system influent stream.

Since modifications to the off-gas stream ductwork were completed (as discussed in the following section), the average time period between 'breakthrough' events has been approximately four months.

Maintenance and operational problems. Over the course of the operation of the groundwater remediation system, several problems have surfaced that affected the efficiency of the system. A summary of the problems, the symptoms and the resolution is given in Table 11.

Conclusions

The paper describes a phased approach to the investigation and treatment of a complex heterogeneous aquifer system contaminated by chlorinated solvents with a density greater than that of water (DNAPLs). Numerical modelling of groundwater flows and contaminant transport has been of significant help in developing conceptual models and designing the remediation system.

Remedial treatment of contaminant sources included removal or grouting of solvent tanks and the excavation of heavily contaminated soils surrounding the tanks. Treatment of the dissolved plume involved pump-and-treat techniques with two remedial wells being installed in a deep aquifer beneath the site. Surface treatment of the abstracted water employed an air stripper with treatment of the off-gases from the air stripper by vapour phase activated carbon units.

Progress towards the treatment standard for the groundwater ($100 \mu g l^{-1}$ of dissolved total VOCs) has been fairly slow. In the three years of operation of the system for which results are available, a significant volume of total VOCs has been extracted from the groundwater. However, no trend has yet been discerned for a significant decrease in total VOCs in the groundwater being extracted by the treatment wells. This reflects the difficulties reported on many sites of treating groundwater polluted by chlorinated solvents in heterogeneous aquifers, particularly where difficulties are experienced in identifying and treating all source areas.

Nevertheless, the system has operated successfully as a hydraulic control, preventing the dissolved contaminant plume from migrating down-gradient in the deep buried aquifer to the production well located in this aquifer and which draws off a large proportion of water for use in the manufacturing plant. Water from this production well has remained unpolluted.

References

FREEZE, R. A. & CHERRY, J. A. 1979. *Groundwater*. Prentice Hall, Englewood Cliffs, New Jersey.

RIORDAN, P. J., HARLEY, B. M. & SCHREIBER, R. P. 1984. Three dimensional modelling of flow and mass transport processes in groundwater systems. *In: The National Water Well Association Conference on Practical Application of Groundwater Models*. Columbus, Ohio, August 1984. National Water Well Association, USA.

VOGEL, T. M., CRIDDLE, T. S. & McCARTHY, P. L. 1987. Transformation of halogenated aliphatic compounds. *Environmental Science Technology*, **21**(8), 722–736.

Groundwater pollution by chlorinated solvents: the landmark Cambridge Water Company case

B. D. R. MISSTEAR[1], R. P. ASHLEY[2] & A. R. LAWRENCE[3]

[1]*Department of Civil, Structural and Environmental Engineering, Trinity College, Dublin 2, Ireland (formerly with Mott MacDonald, Demeter House, Station Road, Cambridge CB1 2RS, UK)*
[2]*Mott MacDonald, Demeter House, Station Road, Cambridge CB1 2RS, UK*
[3]*British Geological Survey, Wallingford, Oxfordshire OX10 8BB, UK*

Abstract. The case of *Cambridge Water Company v Eastern Counties Leather plc and v Hutchings and Harding Ltd* is believed to be the first case this century in which common law principles have been applied to groundwater pollution in England. The case concerned the pollution of a public water supply borehole at Sawston, Cambridgeshire, by the chlorinated solvent tetrachloroethene, and the attempts by the water company to recover damages from the alleged polluters. Following discovery of the pollution in 1983 and subsequent investigations, the case was heard in the High Court in 1991, the Court of Appeal in 1992, and was finally decided by the House of Lords in 1993. This paper sets out the background and chronology of the case, reviews the investigations into the nature and extent of groundwater pollution, and then considers the legal arguments and findings of the courts and the House of Lords. The paper concludes with a discussion of some of the technical implications of this case, especially in relation to the collection of hydro-geological evidence.

In 1983, a water supply borehole operated by Cambridge Water Company (CWC) near Sawston, about 7 km south of Cambridge (Fig. 1), was found to contain the solvent tetrachloroethene (also known as perchloroethylene or PCE) at concentrations generally in the range $150–300 \,\mu\text{g}\,\text{l}^{-1}$, together with lesser concentrations of other solvents such as trichloroethene (TCE). The borehole source, Sawston Mill, was taken out of supply and investigations into likely sources of pollution were initiated. The investigations concentrated at sites where large quantities of solvents were known to have been used, namely the tanneries of Eastern Counties Leather plc and Hutchings and Harding Ltd. Sawston Village has a long history of leather processing, although tanning, as such, was discontinued before any of the events described in this paper occurred. The investigations and subsequent legal case concentrated mainly on the Eastern Counties Leather plc (ECL) site, and this is the main focus of this review.

Much of the information given here was obtained from the various parties to the case in the course of preparation for the legal hearings, including ECL itself, the former Anglian Water Authority (AWA) and the British Geological Survey (BGS). Specific references will be given where relevant.

The sequence of events related to the case, together with certain other relevant dates, is set out in Table 1. Of particular note are the relative dates when the presence of trace amounts of PCE and TCE in the aquatic environment became of concern to the scientific and regulatory community, compared to the dates of upgrading by ECL of its cleaning and effluent treatment systems, and compared to the dates of purchase and use of the Sawston Mill borehole by CWC.

Characteristics of solvents

The behaviour of chlorinated solvents in the Chalk and other UK aquifers has been studied in detail since the mid-1980s (see Rivett *et al.* 1990; Lawrence *et al.* 1992; Lerner 1993). However, at the time of the events described in this paper (the late 1970s and early 1980s), the behaviour was predicted on the basis of their properties, rather than on the results of field investigations. TCE and PCE share the common characteristics of all related chlorinated solvents, such as carbon tetrachloride and 1,1,1-trichloroethane. They are volatile, of low viscosity and denser than water; they are also only poorly soluble in water (Table 2).

These properties cause a solvent release to behave in a characteristic way in the subsurface.

MISSTEAR, B. D. R., ASHLEY, R. P. & LAWRENCE, A. R. 1998. Groundwater pollution by chlorinated solvents: the landmark Cambridge Water Company case. *In*: MATHER, J., BANKS, D., DUMPLETON, S. & FERMOR, M. (eds) *Groundwater Contaminants and their Migration*. Geological Society, London, Special Publications, **128**, 201–215.

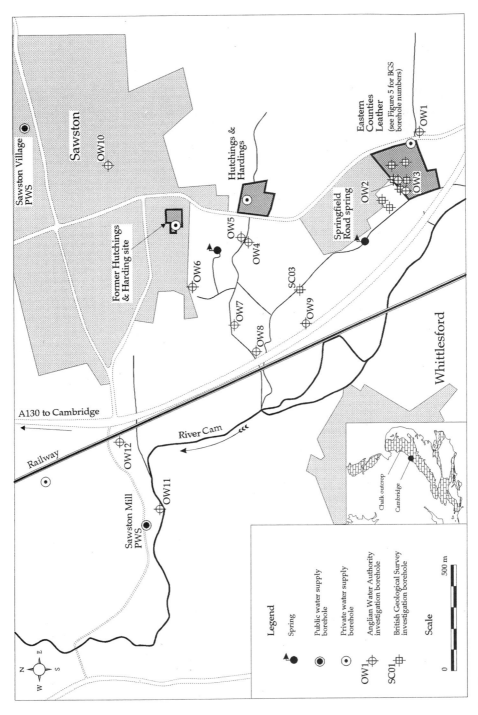

Fig. 1. Location map.

Table 1. *Chronology*

Date	Event
1879	ECL founded. No chlorinated solvents in existence.
1958	ECL introduces TCE as a solvent for cleaning leather.
1960s	Early 1960s: ECL introduces PCE as a replacement for TCE.
	Mid to late 1960s: improvements to drainage and waste systems at ECL site.
1975	US Environmental Protection Agency (US EPA 1975) cites TCE and PCE as drinking water contaminants. First major public recognition of potential hazards.
1976	CWC agrees to purchase Sawston Mill borehole from local paper factory: purchase completed in 1977.
1970s	ECL introduces bulk storage for solvent (to replace 40-gallon drums) in about 1976, and purchases new cleaning machinery in 1978, with improved solvent distillation equipment.
1976	Closure of Thomas Evans Ltd, leather processor in Sawston. Hutchings and Harding Ltd (HHL), another leather processor, moves from central Sawston to former Thomas Evans site (Fig. 1).
1979	CWC starts abstracting from Sawston Mill. No analysis for PCE carried out.
1980	European Community drinking water quality directive (80/78/EEC) makes no specific reference to TCE or PCE, other than as generic organochlorine compounds. A 'guide level' for organochlorine compounds of $1\,\mu g\,l^{-1}$ was given, but no 'maximum admissible concentration' was set.
1983	PCE detected in water supply by a local laboratory developing testing methods for TCE and PCE. Sawston Mill was pinpointed as the source ($125\,\mu g\,l^{-1}$) by CWC and by Anglian Water Authority as the then regulatory body, and closed down in October. The borehole was subsequently pumped to waste to control PCE migration in the aquifer.
1984	AWA detects high levels of PCE and TCE in ECL site borehole. AWA conducts investigations into contaminant migration routes in Sawston area: boreholes at 12 sites.
1984	World Health Organisation (WHO 1984) sets 'tentative guideline values' for TCE ($30\,\mu g\,l^{-1}$) and PCE ($10\,\mu g\,l^{-1}$).
1986	CWC commences proceedings against ECL and HHL.
1987	British Geological Survey commences research investigation around ECL site.
1989	UK drinking water standards (DoE, 1989) specify 'maximum concentrations' of TCE ($30\,\mu g\,l^{-1}$) and PCE ($10\,\mu g\,l^{-1}$).
1991	CWC action against ECL and HHL fails in High Court.
1992	CWC appeal against decision in respect of ECL succeeds in Appeal Court.
1993	ECL appeal against Appeal Court decision succeeds in House of Lords.

It tends to migrate rapidly downwards, much more rapidly than water in soil or rock of identical permeability. On reaching the water-table, it continues to migrate downwards until it reaches an effectively impermeable bottom to the aquifer. A formation that is normally regarded as an aquiclude may be permeable to a liquid of this type, and care must be taken in interpreting hydrogeological data to predict the behaviour of chlorinated solvents.

In the saturated zone of the aquifer, the solvent will slowly dissolve in the passing groundwater

Table 2 *Comparison of solvent properties with water (after Verschueren 1983)*

	Specific gravity	Absolute viscosity (cp)	Solubility in water (mg/l)
Water	1	1	–
PCE	1.63	0.9	200
TCE	1.46	0.57	1100
Carbon tetrachloride	1.59	0.97	785
1,1,1-Trichloroethane	1.35	0.84	720

flow. Because the solvents are poorly soluble in water, and because the accepted water quality standards are set very low, a moderate quantity of solvent is capable of contaminating a large area of aquifer over a long period of time.

Research (Freedman & Gossett 1989) has shown that bacteria are capable of degrading aliphatic chlorinated solvents such as PCE under anaerobic conditions, albeit slowly. Under aerobic conditions, degradation appears to occur only by co-metabolism in the presence of other contaminants, such as hydrocarbons from mineral oils. The degradation proceeds (in simplified terms) from PCE by the removal of a chlorine atom to yield TCE, which degrades in turn to *cis*- and *trans*-1,2-dichloroethene and 1,1-dichloroethene, then to vinyl chloride, to ethene, and eventually to carbon dioxide and water.

Hydrogeology

The Sawston area is underlain entirely by Chalk bedrock, dipping gently from northwest to southeast (Fig. 2). The base of the Chalk and

B. D. R. MISSTEAR *ET AL.*

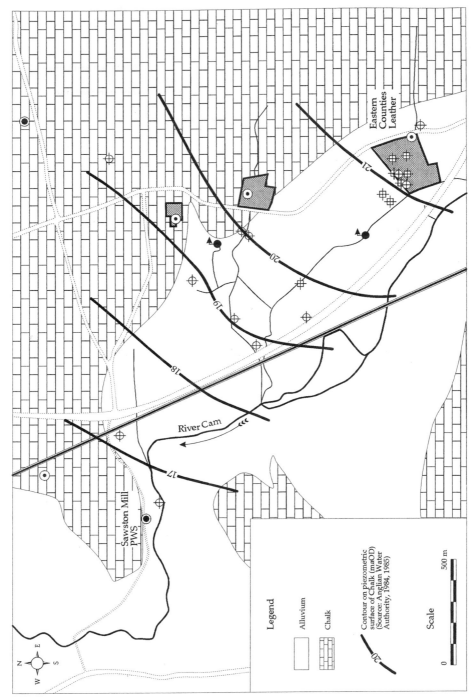

Fig. 2. Geology and Chalk piezometry. (Adapted from Anglian Water Authority 1984, 1985.)

the underlying Gault Clay outcrop in Cambridge north of the area. The Gault Clay is believed to be at a depth of about 70 m below the ECL site, and about 50 m below the Sawston Mill site. There is a strip of recent alluvium along the valley of the River Cam, extending to the ECL site where it is about 7 m thick.

The Chalk is a 'dual-porosity' aquifer: the main groundwater flow is through highly transmissive fractures and solution-fissures, which occupy a small volume of the strata in percentage terms. However, the low-permeability intergranular pores, typically of 1 mm diameter, may occupy up to 50% of the bulk volume, and thus contain a large amount of water. Wells drilled in the Chalk aim to penetrate one or more fissure zones, thereby creating a highyielding source. Wells which do not penetrate such a zone are very low-yielding as water cannot be extracted from the pores by normal pumping.

Groundwater flow in the Sawston area is from southeast to northwest along the valley of the Cam. Formerly, the groundwater discharged into the Cam and from local springs, such as the one at Springfield Road near the ECL site. Latterly, a considerable proportion of the flow was intercepted by water supply wells such as Sawston Mill and others in the area.

Although the Chalk is generally unconfined, there is some evidence that the main body of the aquifer is partly isolated from water in the shallow drift, as a result of the development in its upper layers of low-permeability 'putty chalk', i.e. soft weathered chalk, formed by freezing and thawing at the end of the last ice age.

Pollution investigations

Initial investigations

The initial investigations into the sources of contamination at Sawston are summarized in Mott MacDonald (1991). The first investigations were conducted by AWA in its capacity as regulator of drinking water quality and environmental water quality. Boreholes were drilled at 12 locations, at the ECL site, near Sawston Mill and in between, including the HHL site (Fig. 1). In recognition of the distinction between the shallow and deep groundwater systems, some of the sites contained two boreholes, one shallow and one deep. Water samples from all the boreholes and from local surface water sources were analysed for PCE. Figure 2 shows the piezometric surface in the deep Chalk; a similar pattern was found for the shallow groundwater system.

Figure 3 shows the results of analysis of the deeper Chalk groundwater samples, and Fig. 4 shows the results from the surface water samples. It is noticeable that, although there were high concentrations of PCE in the boreholes at the ECL site, in the Springfield Road spring, and in boreholes and springs near the HHL site, the boreholes between these locations and Sawston Mill failed to detect any major plume of contamination. Even in the borehole (OW11) about 100 m upstream of Sawston Mill there were no detectable levels of PCE contamination. These results are evidence of the difficulty of tracing contamination in fissured aquifers.

BGS investigations

The BGS investigations were conducted as a part of a research project studying synthetic organic contaminants in UK aquifers (Lawrence & Foster 1991; Lawrence et al. 1992). The investigations included the following:

- logging and depth sampling of Sawston Mill and other boreholes;
- drilling ten boreholes, mainly on and adjacent to the ECL site (see Fig. 5) to determine the vertical distribution of PCE in core samples and in the fissure water;
- packer-testing of cored borehole SC01;
- a tracer test in boreholes OW11 and OW8 near the Sawston Mill site, to determine actual contaminant migration rates;
- a soil gas survey of part of the ECL site.

The data are mainly of relevance to the ECL site. Figure 5 shows the maximum concentrations of PCE found in boreholes below 20 m depth, which indicate that the PCE is migrating along a very narrow plume to the northwest. The soil gas survey in the same area (Fig. 6) indicated two plumes (Stuart et al. 1990). First, a broad plume which appeared to emanate from the direction of the former effluent settling tanks but which did not extend much beyond 200 m. Second, a narrow plume which could be traced for about 1 km. The latter soil gas plume coincided with the high PCE concentrations observed in groundwater samples from the cored boreholes SC04, SC07 and SC08.

Figure 7 shows the vertical distribution of PCE in pore water samples from the three boreholes situated along the line of the plume. The data suggest that the PCE entered the aquifer upstream of SC07, and was initially held up by the shallow layer of 'putty chalk' before eventually penetrating to the top of the Chalk Marl at about 50 m. The highest concentrations of

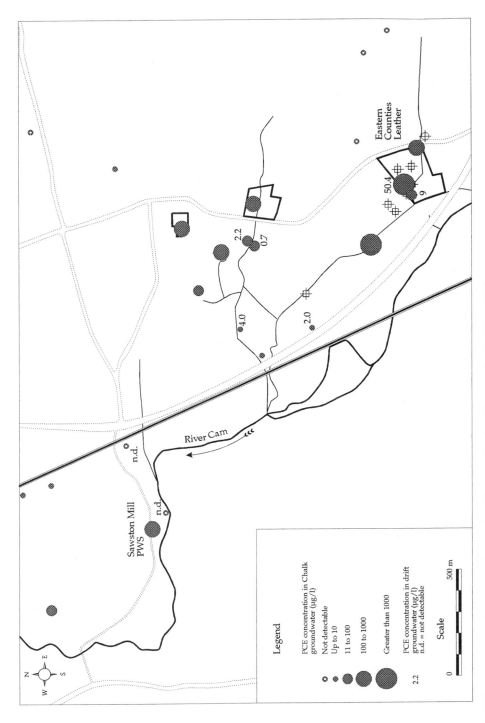

Fig. 3. PCE concentrations in groundwater samples. (Based on Anglian Water Authority 1984, 1985.)

Fig. 4. PCE concentrations in surface water samples. (Based on Anglian Water Authority 1984, 1985.)

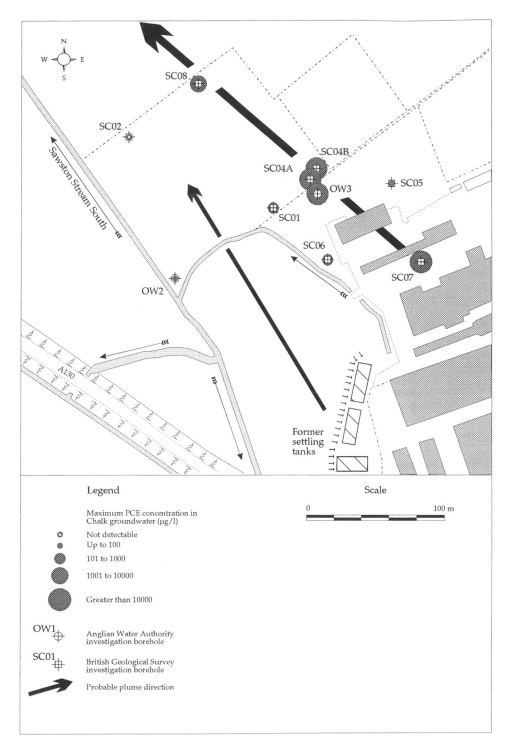

Fig. 5. Maximum PCE concentrations in Chalk groundwater at ECL site. (Source: Lawrence *et al.* 1992.)

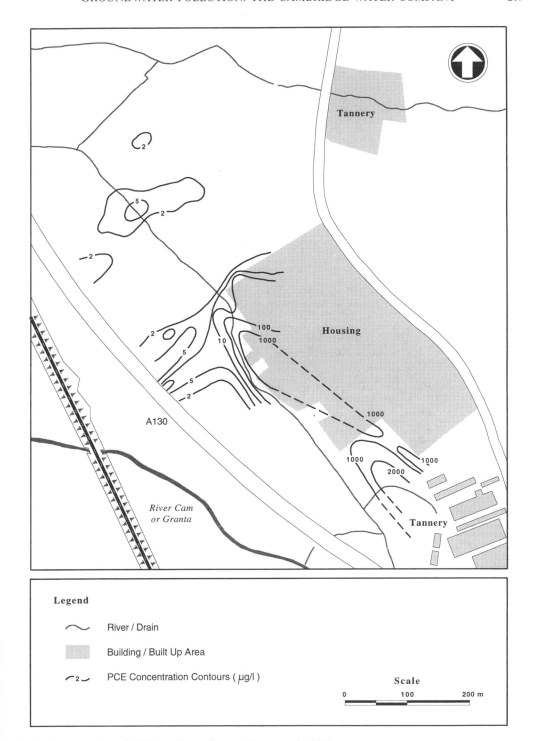

Fig. 6. Concentration of PCE in soil gas. (Source: Stuart *et al.* 1990.)

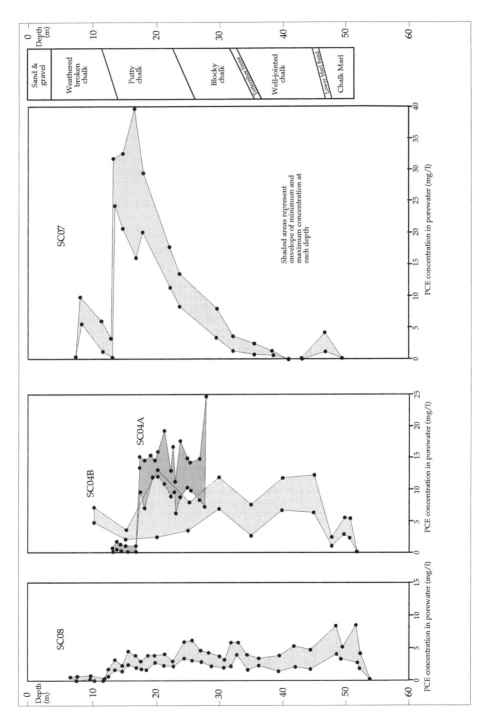

Fig. 7. Profiles of PCE concentrations in pore waters. (Source: Lawrence *et al.* 1992.)

PCE on top of the Chalk Marl are observed in borehole SC08, some 200 m downstream of where PCE probably initially entered the aquifer. The concentrations of PCE found in SC07 ($40\,\mathrm{mg\,l}^{-1}$ in some samples) are consistent with the presence of neat PCE in the rock; an observation that may also be true for SC04A ($25\,\mathrm{mg\,l}^{-1}$).

The borehole logging and packer testing showed that the highest permeability zones in the Chalk occurred within the depth interval 25–50 m below the ECL site. The tracer tests showed that the groundwater velocity near Sawston Mill averages $8\,\mathrm{m\,day}^{-1}$. It is worth noting as evidence of the difficulty of such tests, that the tracer injected into OW11 was not detected in Sawston Mill, about 100 m distant.

The results of the investigations of AWA and BGS, although unsuccessful in demonstrating a clear flow path between the ECL and HHL sites on the one hand and the Sawston Mill borehole on the other, provided sufficient information to conclude that the ECL site, at least, was a potential source of the PCE in the water supply.

Age and travel time of the pollutant

It was clearly of importance to establish, if possible, the travel time for PCE from the ECL site to Sawston Mill, so that, for example, the date of a polluting incident could be compared with the implementation of relevant legislation.

The BGS tracer tests imply a travel time of about 250 days from the ECL site, although a longer time may be more realistic, considering that the flow rates were measured in the centre of the Cam valley near to the abstracting Sawston Mill borehole. However, the period for PCE to migrate from the surface down to the main aquifer is much more difficult to estimate, as there are no data pertaining to migration rates through the low-permeability putty chalk. Periods in excess of several years are quite conceivable, but remain speculative.

The contaminant concentrations in Sawston Mill borehole remained roughly steady since first detection at about $150–300\,\mu\mathrm{g\,l}^{-1}$, which suggested that PCE was present in the Sawston Mill water for some time before first detection.

From indirect evidence, it can be concluded that spillages at the ECL site would be more likely to have occurred in the earlier years of PCE/TCE use, when the solvent was delivered and handled in drums rather than in bulk storage piped to the cleaning machines. Similarly, losses from the aqueous effluent system were more

likely in the 1960s than in the late 1970s, by which time ECL had remodelled its effluent treatment system and was using more modern cleaning machines.

Estimates of the quantity of PCE that must have been released to the ground vary considerably, but, based on the quantities pumped from Sawston Mill since 1984, and on the flow rate from the Springfield Road spring, at least 5770 kg (3600 litres) had been removed from the aquifer by 1989. The BGS data showed that considerable quantities remained in the ground. These considerations are consistent with the finding of the High Court that the PCE losses occurred as a series of small leaks.

The Case in Law

The High Court

The case was heard in the High Court in January 1991, and the judgement delivered by Ian Kennedy Justice some six months later (and published by the Queen's Bench Division in 1992). ECL admitted, on the balance of probabilities, that the pollution derived, at least in part, from its works. Kennedy J found that this pollution occurred through accidental spillages in the years prior to 1976 when the solvent was delivered in 40-gallon drums and moved around the site using forklift trucks (this practice ceased with the introduction of bulk storage in 1976). Kennedy J found that pollution also originated from the HHL works, but he could find no evidence of this pollution having a measurable effect on the Sawston Mill borehole. The legal arguments therefore focused mainly on the case against ECL.

The plaintiff brought the case on three grounds: strict liability under *Rylands* v *Fletcher*, negligence and nuisance. The Victorian case of *Rylands* v *Fletcher* (1866) concerned the escape of impounded water from a reservoir which flooded a neighbour's land. Over the years, the application of this ruling has been generally limited to situations where the use of the land has been considered to be 'non-natural'. In the present case, Kennedy J concluded that the storage of solvents at a tannery of long standing such as ECL did not constitute a non-natural use of the land:

> In reaching this decision I reflect on the innumerable small works that one sees up and down the country with drums stored in their yards. I cannot imagine that all those drums contain milk and water, or somelike

innocuous substance. Inevitably that storage presents some hazard, but in a manufacturing and outside a primitive and pastoral society such hazards are part of the life of every citizen.

The case therefore failed on *Rylands* v *Fletcher*.

Negligence and nuisance were essentially considered together, and the main issue here was one of foreseeability of the damage caused by the escape of solvent from the site. Kennedy J found that a responsible site supervisor at ECL in the early 1970s would not reasonably have foreseen that small spills of solvent would lead to pollution of the aquifer downstream. The case therefore also failed on the grounds of negligence and nuisance.

The fact that actual damage only occurred on the introduction of the relevant water quality regulations in the 1980s clearly influenced the decision. Kennedy J added:

> There must be many areas within England and Wales where activities long ceased still have their impact on the environment, and where the perception of such impact depends on knowledge and standards which have been gained or imposed in more recent times. If it is right as a matter of public policy that those who were responsible for those activities, or their successors, should now be under a duty to undo that impact (or pay damages if a cure is impractical), that must be a matter for Parliament. The common law will not undertake such a retrospective enquiry.

The Court of Appeal

CWC appealed against the decision of the High Court, and the Court of Appeal gave its judgement in November 1992 (Court of Appeal, Civil Division 1993). The lead judgement was given by Lord Justice Mann.

Whereas much of the legal argument at the Court of Appeal was again on the issue of strict liability under *Rylands* v *Fletcher*, the judgement centred on the grounds of nuisance and, in particular, on another Victorian case, *Ballard* v *Tomlinson* (1885). The latter case concerned the pollution of a brewer's well by sewage effluent disposed of down his neighbour's well, the court finding that the defendant had interfered with the natural right of the plaintiff to abstract unpolluted groundwater from beneath his land.

The Court of Appeal concluded that the present case was 'not distinguishable' from *Ballard* v *Tomlinson*. The appeal was therefore allowed and damages of approximately £1 million plus

legal costs were awarded to CWC. The question of quantum (the allowed amount of damages) had been decided in the High Court by Kennedy J in the event that his judgement might be overturned on appeal. Kennedy J supported the plaintiff's case on quantum, which was to claim the costs of developing a replacement borehole source, as opposed to the (cheaper) alternative of providing suitable water treatment facilities at Sawston Mill. The plaintiff argued successfully that the treatment technology was not well proven at the time when the decision on the future of the water supply was made.

The House of Lords

Owing to the importance and potential ramifications of the case in terms of no-fault liability for historic pollution, ECL successfully petitioned the House of Lords to decide on the case. The appeal was heard in October 1993, and the leading judgement was given by Lord Goff in December (House of Lords, 1994).

The House of Lords decided that the recovery of damages in nuisance depended on foreseeability, and that the same principle should be applied to strict liability under *Rylands* v *Fletcher*. Lord Goff agreed with the earlier conclusion of Kennedy J that ECL could not have reasonably foreseen the consequences of the escape of solvent from its property. Therefore ECL's appeal was allowed.

The House of Lords also considered the argument that ECL should be strictly liable for damages on the grounds that pollution was still occurring from the pools of solvent deep within the Chalk aquifer beneath its site, and the consequences of this were now all too foreseeable. Lord Goff rejected this argument on the basis that the solvent was clearly beyond the control of ECL.

Although not a deciding factor in this ruling, Lord Goff did not agree with Kennedy J's earlier decision on the issue of non-natural use of the land:

> Indeed I feel bound to say that the storage of substantial quantities of chemicals on industrial premises should be regarded as an almost classic case of non-natural use.

Lord Goff, however, agreed with the opinion of Kennedy J that strict liability in such cases should be a matter for Parliament and statute:

> Like the judge in the present case, I incline to the opinion that, as a general rule, it is more appropriate for strict liability in respect of

operations of high risk to be imposed by Parliament, than by the courts. If such liability is imposed by statute, the relevant activities can be identified, and those concerned can know where they stand.

In reaching this decision, the House of Lords considered the CWC case in the context of current European developments, referring specifically to the Council of Europe's *Convention on Civil Liability for Damage Resulting from Activities Dangerous to the Environment* (the Convention), published in March 1993. A major aim of the Convention is to make operators of dangerous activities liable for the costs of environmental damage. However, this liability is not to be retrospective. In this respect, Lord Goff argued that it would 'be strange if liability for such (historic) pollution were to arise under a principle of common law'. It should be added that the UK has not ratified the Convention at the time of writing.

Implications

The legal implications of the case have been described in numerous environmental and legal journals (e.g. ENDS 1993; Bryce 1994) and discussed at legal and scientific seminars following the House of Lords ruling (including those organized by Berrymans 1994, and IBC Technical Services and Dames and Moore 1994). The remainder of this paper will mainly consider the implications of the case for the scientist responsible for collecting hydrogeological evidence.

In this civil case the hydrogeological evidence was judged on the 'balance of probabilities' and not the 'burden of proof' (beyond all reasonable doubt) which would have been necessary under criminal law. In the view of the authors it seems open to question whether the hydrogeological evidence would have been sufficient if the case had been brought under a criminal law, such as the Control of Pollution Act.

For the purpose of making a decision on the basis of balance of probabilities it has been shown that it is sufficient for there to be evidence that groundwater has become contaminated at a source site, that contaminated groundwater is present at another, remote, site, and that, although a migration path between the two cannot be found, no other more likely source can be identified. Thus in similar cases in the future, careful consideration should be given to the level of detail and the location of any investigations designed to obtain data for a civil law suit.

For the purpose of making a decision on the basis of burden of proof, it appears that a clearer relationship between cause and effect would have to be identified, implying a better knowledge of the migration pathway. For example, fissure flow aquifers such as the Chalk are notorious for transmitting a high proportion of their flow through narrow pathways relating to palaeo-water-tables, stream channels, glacial features, fault zones and primary joints; contaminants may follow a tortuous path, or may originate in an unexpected quarter of the catchment. In the present case, for example, no PCE was detected in borehole OW11 located upstream of the polluted Sawston Mill borehole, and only about 100 m from it (see Fig. 1). In such cases, a considerable amount of detailed investigation is clearly necessary to prove the connection between a pollution source and the point of detection.

A useful comparison may be made with practice in the USA, where the formal investigation procedures are much more rigorously defined. The level of effort needed to conduct an adequate investigation is affected by the legal end point. Superfund projects require very strict adherence to the procedures of the Environmental Protection Agency (EPA), and the preparation of extensive planning and quality control documents (see, for example, Anon (1990), for a summary of US environmental regulations). Hazards and clean-up levels are dictated by calculated risks of contaminants to human health and the environment, so that assessments must address details of exposure pathways along with the current nature and extent of contamination. Organizations leading clean-ups require extensive documentation of data collection and analysis procedures so that the results are admissible and valid in court.

As well as Superfund projects, groundwater clean-ups may also be carried out under the Resource Conservation and Recovery Act (RCRA) or individual State regulations. Application of Superfund, RCRA or State clean-up criteria depends on which agency maintains authority over the specific site. However, in the majority of cases, clean-up criteria are likely to be based on detailed risk assessment and potential receptor identification.

The level of effort needed to complete an assessment conducted as part of a civil law suit in the USA depends on the level of risk perceived by the legal counsel, and this is similar to the UK situation for civil cases. Lawyers are concerned with the potential loss of real estate value, as well as human health and the environment, so that clean-up levels tend to be more stringent than those required for a Superfund or RCRA clean-up. At one such US site of which the authors are aware, the clean-up level for a solvent was one-fifth of the current regulatory maximum contaminant level (MCL). The justification for this was the possibility that the MCL would be lowered during the period of the investment, thereby causing potential future liability.

UK and European practice in environmental matters tends to follow a similar route to that in North America, and we should be aware of the greater emphasis on formal procedures and rigorous documentation that pertain there. This will almost inevitably mean higher costs for pollution investigations, but should reduce the scope for subsequent disputes in court and at preliminary stages over the validity and implications of hydrogeological data and interpretations.

It is beyond the scope of this paper to discuss groundwater remediation options in the Sawston case, and only a few general points about remediation will be made. First, for any remedial programme the criteria for clean-up need to be clearly stated. Issues that should be addressed include the following:

- *concentration*: what residual concentrations must be attained before a site can be accepted as 'clean'?
- *location*: does the groundwater need to meet these criteria within the site of the polluting activity, or outside the site boundary only?
- *sampling method*: it is important that the method of sampling to ascertain whether an aquifer is 'clean' should be clarified, e.g. should samples be of flowing groundwater, porewater, or both?

Finally, recent practice in North America and increasingly also in the UK, is to recognize that full and rapid clean up of groundwater contamination by chlorinated hydrocarbons on the one hand is often not feasible and, on the other, may not justify the very high costs involved. Attention is focusing more on a combination of different approaches, including containment at source, treatment at the point of abstraction, and monitoring (and, where possible, enhancing) the natural degradation processes ('intrinsic remediation').

This paper is an updated and revised version of the paper first presented at the Second Annual Groundwater Pollution Conference, held in London on 16–17 March 1993. The authors would like to thank the organizers of that event for permission to publish this paper. This paper is also published by permission of the Director of the British Geological Survey, a component institute of the Natural Environment Research Council. We would like to thank the contribution made by colleagues at Mott MacDonald, especially the late D. Milne. The data provided by the former Anglian Water Authority and by ECL is also gratefully acknowledged. Thanks also to S. Shemmings for reviewing the legal aspects of the original manuscript, and J. Kennedy and M. P Sherrier for advice on relevant American regulations.

References

ANGLIAN WATER AUTHORITY 1984. *Tetrachloroethylene pollution of Cambridge Water Company source at Sawston Mill*. Report No. 1.

—— 1985. *Tetrachloroethylene pollution of Cambridge Water Company source at Sawston Mill: Water Quality Investigation*. Report No. 2.

ANON 1990. *Selected environmental law* statutes. West Publishing Co., St Paul.

BERRYMANS 1994. *Cambridge Water Company v Eastern Counties Leather: the future*. Proceedings of seminar, 13 January 1994, London.

BRYCE, A. 1994. Environmental liability in the UK and under European law. *Institute of Wastes Management Proceedings*, January 1994.

COUNCIL OF EUROPE 1993. *Convention on Civil Liability for Damage Resulting from Activities Dangerous to the Environment*. Council of Europe Press, Strasbourg.

COURT OF APPEAL, CIVIL DIVISION 1993. *Cambridge Water Company Ltd v Eastern Counties Leather plc*. Environmental Law Report 287.

DEPARTMENT OF THE ENVIRONMENT 1989. *Water supply (water quality) regulations 1989*. HMSO, London.

ENDS 1993. *Key ruling on civil liability in House of Lords*. ENDS Report 227.

EUROPEAN ECONOMIC COMMUNITY 1980. Directive relating to the quality of water intended for human consumption 80/78/EEC. *Official Journal of the European Community*, **L129**.

FREEDMAN, D. L. & GOSSETT, J. M. 1989. Biological reductive dechlorination of tetrachloroethylene and trichloroethylene to ethylene under methanogenic conditions. *Applied Environmental Microbiology*, **55**(4), 1009–1014.

HOUSE OF LORDS 1994. *Cambridge Water Company Ltd v Eastern Counties Leather plc*. Environmental Law Report 105.

IBC TECHNICAL SERVICES and DAMES and MOORE 1994. *Controlling Environmental Pollution*. Proceedings of conference, 22 June 1994, London.

LAWRENCE, A. R. & FOSTER, S. S. D. 1991. The legacy of aquifer pollution by industrial chemicals: technical appraisal and policy implications. *Quarterly Journal of Engineering Geology*, **24**, 231–240.

——, STUART, M. E., BARKER, J. A., CHILTON, P. J., GOODY, D. C. & BIRD, M. J. 1992. *Review of groundwater pollution of the chalk aquifer by the halogenated solvents*. R&D Note 46 prepared by the Hydrogeology Research Group of BGS for National Rivers Authority.

LERNER, D. N. (ed.) 1993. Coventry groundwater investigation: sources and movement of chlorinated hydrocarbon solvents. *Journal of Hydrology*, **149**(1–4), Special volume

MOTT MACDONALD 1991. *Report on alleged tetrachloroethene pollution of groundwater*. Mott MacDonald, report.

QUEEN'S BENCH DIVISION, HIGH COURT 1992. *Cambridge Water Company Ltd v. Eastern Counties Leather plc and v. Hutchings and Harding Ltd*. Environmental Law Report 116.

RIVETT, M. O., LERNER, D. N. & LLOYD, J. W. L. 1990. Chlorinated solvents in UK aquifers. *Journal of the Institute of Water & Environmental Technology*, **4**, 242–250.

STUART, M. E., CHENEY, C. S. & BOYES, S. E. 1990. *Soil gas survey to map contaminant plume migration in the Chalk aquifer*. British Geological Survey Technical Report WD/90/48C.

UNITED STATES ENVIRONMENTAL PROTECTION AGENCY 1975. *Preliminary assessment of suspected carcinogens in drinking water*. Report to Congress.

VERSCHUEREN, K. 1983. *Handbook of environmental data on organic chemicals*. Van Nostrand Reinhold.

WORLD HEALTH ORGANISATION (WHO) 1984. *Guidelines for drinking water quality*. WHO, Geneva.

Detection of point sources of contamination by chlorinated solvents: a case study from the Chalk aquifer of eastern England

B. D. R. MISSTEAR[1], P. W. RIPPON[2] & R. P. ASHLEY[2]

[1] Department of Civil, Structural and Environmental Engineering, University of Dublin, Trinity College, Dublin 2, Republic of Ireland (formerly with Mott MacDonald, Demeter House, Station Road, Cambridge CB1 2RS, UK)

[2] Mott MacDonald, Demeter House, Station Road, Cambridge CB1 2RS, UK

Abstract. Groundwater contamination by chlorinated solvents such as trichloroethene is of increasing concern in the UK. This paper describes a case study of chlorinated solvent contamination of the Chalk aquifer at a groundwater supply source in eastern England. The investigations comprised two phases: the first phase focused on the extent and nature of the contamination within the main Chalk aquifer (both in the fissures and in the Chalk matrix); and the second phase targeted the sources of contamination through detailed sampling of the unsaturated drift at suspect sites. The volatile nature of the organic contaminants required special hygiene procedures during drilling and sampling to avoid cross-contamination between boreholes, or loss of contaminants in the samples collected. Four potential sources of contamination were identified within an industrialized area about 1 km up hydraulic gradient from the contaminated well. However, the contamination was found to be extremely localized, with closely spaced investigation boreholes at individual sites giving highly varied results. To be successful, investigation boreholes had to target precise locations where these chemicals were known or suspected to have been used or stored. Therefore, a key aspect of the investigation was a solvent usage survey prior to each phase of drilling.

Chlorinated solvents have been widely used in the UK since the 1930s. Whereas the toxicity of these volatile compounds in liquid and vapour form has long been recognized, hazards associated with their presence as trace contaminants in drinking water supplies (as potential carcinogens) were only perceived during the 1970s. Drinking water standards introduced as a response to these concerns (notably EEC Directive 80/778/EEC of 1980) have required routine monitoring for chlorinated solvents and other organic compounds. Not surprisingly, there has been an increasing number of cases reported in recent years of chlorinated solvents being found as groundwater contaminants, affecting major aquifers such as the Sherwood Sandstone in the English Midlands (Rivett et al. 1990; Lerner, 1993), and the Chalk in eastern England (Lawrence et al. 1992). One occurrence of contamination of the Chalk aquifer in East Anglia by chlorinated solvents resulted in the recent landmark pollution case involving Cambridge Water Company (Bishop et al. 1998; Misstear et al. 1998).

The behaviour of chlorinated solvents in porous and fractured media is described by Schwille (1988), amongst others. These compounds are volatile, denser than water, of low viscosity and are only poorly soluble in water. They can occur in the gaseous, immiscible or dissolved phases in the subsurface. Owing to their comparatively high density and low viscosity, the immiscible solvents tend to be more mobile than water in porous media. On entering the ground they migrate downwards and through the water-table until they reach the base of the aquifer or some other impermeable layer. They degrade slowly in groundwater and hence are very persistent contaminants. Given their persistence and widespread occurrence, therefore, chlorinated solvents are a major groundwater quality issue in the UK, as in other industrialized countries.

The case study

The case study which is the subject of this paper involved contamination of the Chalk aquifer by chlorinated solvents at an important groundwater supply source in eastern England. The paper describes the investigations carried out to identify the sources of the contamination. It highlights some of the problems of detecting contaminant sources and tracing chlorinated hydrocarbon migration in the ground, problems which are the natural consequence of the unusual physical and chemical properties of the contaminants.

MISSTEAR, B. D. R., RIPPON, P. W. & ASHLEY, R. P. 1998. Detection of point sources of contamination by chlorinated solvents: a case study from the Chalk aquifer of eastern England. *In*: MATHER, J., BANKS, D., DUMPLETON, S. & FERMOR, M. (eds) *Groundwater Contaminants and their Migration*. Geological Society, London, Special Publications, **128**, 217–228.

During routine monitoring of raw water quality by the water undertaker in 1987, the groundwater source was found to be contaminated by the chlorinated solvents trichloroethene (TCE) and tetrachloroethene (also known as perchloroethylene or PCE). The concentrations of dissolved TCE in pumped samples were generally between 40 and 80 µg l^{-1}, with dissolved PCE present at lower levels of between 5 and 10 µg l^{-1}.

The source-works were taken out of supply immediately after detection of the contamination, and a water treatment system was installed based on air stripping (Burley *et al.* 1990).

The source-works are located on the outskirts of a city near the confluence of two rivers (Fig. 1). There is a long-established, light industrial area about 1 km upstream of the groundwater abstraction. Within this area there are metal

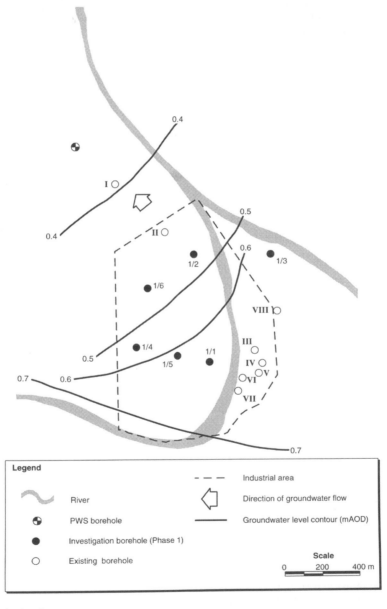

Fig. 1. Schematic site plan.

plating works, rail yards, electrical engineering works and other sites where industrial solvents such as TCE and PCE might be expected to have been used.

The groundwater supply source comprises two boreholes, 67 m deep and 10 m apart, which abstract from the Cretaceous Upper Chalk aquifer. The Chalk is several hundred metres thick in this area, but the main water-producing zones are located in the upper strata within 40 m of ground level where there is extensive, solution-enhanced fissuring. The uppermost bedrock layer (above the main flow horizons) has been degraded to a structureless, low-permeability putty-like texture (herein referred to as putty chalk), which has a maximum thickness of about 10 m. The Chalk is overlain by drift deposits, which typically consist of 5–10 m of sands and gravels and Boulder Clay, but attain greater thickness in buried channels. Some of the channels are of glacial or sub-glacial origin and are up to 30 m deep and steep-sided. Where saturated, the sands and gravels within the drift also form an aquifer, which has a variable hydraulic connection with the underlying Chalk aquifer. Along the river valleys, the water-table is generally close to the ground surface.

Planning the investigation

After discovery of the contamination at the source-works, preliminary investigations into the extent of the contamination were carried out by the water undertaker, including sampling of boreholes close to the source-works, and a survey of solvent usage in the industrial area nearby. The sampling programme detected dissolved TCE and PCE at two sites (boreholes I and II on Fig. 1) and PCE at an additional site (borehole VIII). The initial survey of solvent usage by the water undertaker confirmed that TCE and PCE were widely used within the industrial area, and further investigations, involving the authors, were therefore carried out to identify the source or sources of pollution.

The hydrogeology of the area suggested that there would be four potential reservoirs for contaminants in the subsurface: the unsaturated and saturated drift, the Chalk matrix and the Chalk fissure system. The subsurface investigations were undertaken in two distinct phases:

- Phase 1, to determine the characteristics and extent of contamination in the Chalk aquifer, both within the matrix and the fissures; and

- Phase 2, to determine the sources of contamination more accurately, by identifying contaminants within the saturated and, especially, the unsaturated drift deposits overlying the Chalk aquifer.

Although in certain cases an initial sampling programme of the shallow unsaturated zone might be useful prior to carrying out deeper groundwater investigations, in this case the deeper investigations were necessary in Phase 1 to identify the contamination source areas more clearly before the more localized investigations could be pursued in Phase 2. Before describing the two phases of investigation in turn below, it should be added that the investigations were carried out at a time when the potentially useful technique of soil gas surveying was only just being introduced in the UK. Although soil gas survey trials were held in Phase 2, it was not a principal investigation tool.

Phase 1 investigations

Methodology

The first phase included the drilling, test pumping and sampling of six exploratory boreholes (Fig. 1), which were subsequently completed as observation wells. In view of the volatile nature of the chlorinated solvents, cable-tool was the preferred method for the drilling of the boreholes since this would not require the use of drilling fluids such as air or mud which might affect the concentration of contaminants in rock or water samples. It had originally been anticipated that a rotary rig might be needed to drill through hard rock, but the Chalk proved so soft that this was unnecessary.

Special hygiene precautions were taken during drilling to prevent accidental contamination of samples, cross-contamination between boreholes or loss of volatile compounds during storage. Before use, and in every borehole, all temporary casing, bailers and trip hammers were cleaned with a high-pressure water hose. Immediately prior to running in, all drill tools and temporary casing were flame-cleaned to remove all volatile organic compounds such as degreasing agents, grease and oil.

The boreholes were completed in Chalk bedrock at depths of between 50 m and 62 m. Short pumping tests, using a stainless steel pump and rising main, and geophysical logging, were carried out on the open-hole sections of Chalk before constructing the observation wells with 75 mm diameter PVC casing and screen. A fully screened borehole design was adopted as the cost of more expensive installations was not considered to be justified at this stage of the investigations. The potential problems of using continuous screens were recognized, however, and the analytical results of the bulk groundwater samples taken from the completed holes (see below) were treated with due caution.

During the investigation, a large number of samples were collected, including the following:

- bailed water samples collected inside the temporary casing that was used to stabilize the drift and weak Chalk during drilling;
- water samples collected by a depth sampler from open-hole sections of Chalk to determine the vertical chemical profile in groundwater flowing in fissured Chalk;
- chalk core samples, analysed for solvents absorbed into the pore structure or dissolved in pore water; analysis was carried out by disaggregation followed by steam distillation, liquid–liquid extraction and gas chromatography;
- bulk samples of water collected from the entire screened section of the completed boreholes during the pumping tests to determine the chemistry of the main groundwater flow zones.

It was recognized that the collection of water samples during drilling operations was open to the risk of loss of volatile constituents by agitation of the sample, but in this case no alternative method was available to obtain a depth profile of contaminant concentrations.

Almost 100 samples were analysed at the laboratory of Resource Consultants Cambridge for volatile organic compounds. The sampling procedures incorporated a large number of duplicate samples as quality controls. The analytical techniques used were standard methods (Anon 1985).

Hydrogeology

Five of the six investigation boreholes encountered Chalk bedrock at depths of between 5 and 10 m, the overlying drift deposits comprising mainly sands and gravels. The exception was borehole 1/3, which encountered a much thicker, and finer, sequence of mainly fluvio-glacial and glacial sand and clay deposits, before reaching Chalk bedrock at a depth of 41.5 m. The Chalk at this borehole site is about 10 m deeper than at any other reported site in this area, and indicates the presence of a deep buried channel running along the approximate line of one of the modern-day rivers (Fig. 2). The upper 10–15 m

of Chalk in most of the boreholes was found to be highly weathered, putty chalk.

The sand and gravel deposits form an unconfined aquifer. Water level measurements made during drilling indicated that water levels in the drift were generally slightly higher than those in the underlying Chalk aquifer, indicating a downward hydraulic gradient; however, the data are not conclusive in this respect. Although no hydraulic testing was carried out on the drift, its coarse lithology at most sites implied a relatively high permeability, indicating that the drift would not present a significant barrier to the downward migration of dense non-aqueous phase liquids such as TCE and PCE.

Pumping tests on the open-hole sections of Chalk confirmed a highly productive aquifer, with estimated transmissivity values of between 250 and 11 000 m^2 day^{-1}, and an average value of 2700 m^2 day^{-1}. Geophysical logs of fluid temperature and conductivity obtained during four of the tests indicated that the main flow horizons occur in the upper horizons of relatively fresh chalk rock, below the weathered putty chalk.

Groundwater level measurements indicated a shallow hydraulic gradient of about 0.2 m km^{-1} in the Chalk aquifer, with flow directions running along the main river valley, from the industrial area towards the contaminated supply well (Fig. 1). The depth to water in the completed boreholes (for winter conditions) ranged from 0.9 m in borehole 1/2, close to the river, to 7.6 m in borehole 1/4, located about 8 m above the river valley.

Groundwater contamination

The Phase 1 investigations showed the Chalk and drift aquifers to be widely contaminated by chlorinated solvents in the dissolved phase. The compounds most commonly detected were TCE, PCE and 1,1,1-trichloroethane (TCA). The results of the chlorinated hydrocarbon analyses on the different types of sample collected at each borehole are summarized in Tables 1–3. The extent of the main areas of contamination is shown in Fig. 3.

Maximum concentrations of dissolved TCE in the groundwater of the main Chalk aquifer were generally in the range 10–100 µg l^{-1}. However, very high concentrations of TCE were measured in the upper, low-permeability putty chalk at one site (borehole 1/6: 19 850 µg l^{-1}) and in the permeable drift above this putty chalk at a second site (borehole 1/1: 25 300 µg l^{-1}). Maximum concentrations of PCE and TCA detected were 225 µg l^{-1} and 536 µg l^{-1} respectively. A

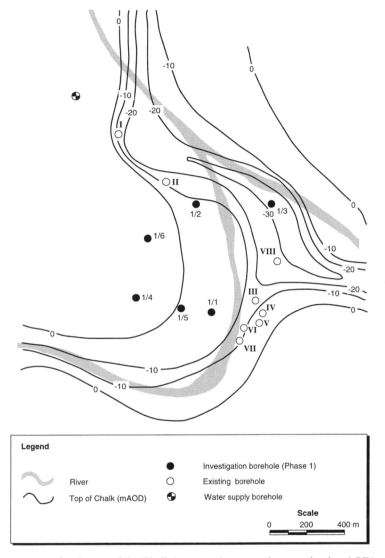

Fig 2. Structure contours for the top of the Chalk (contours in metres above sea level, mAOD).

fourth chlorinated solvent, carbon tetrachloride (CTC), was found in boreholes 1/1 and 1/2 at levels of up to $9\,\mu g\,l^{-1}$, as well as in samples from the existing boreholes III, IV and V nearby.

The main concentrations of chlorinated solvents in the subsurface, therefore, proved to be in the sand and gravel drift deposits overlying the low permeability putty chalk above the main Chalk aquifer. Contaminant concentrations in the fissured Chalk, as indicated by the pumped samples, and in the Chalk matrix, as indicated by the pore water analyses, were orders of magnitude less: the maximum pore

water concentration was $24.8\,\mu g\,l^{-1}$ of TCE at 25 m depth in borehole 1/6 (Table 3).

The distribution of the different contaminants suggested that there had been multiple slugs of chlorinated solvents entering the groundwater system at a number of sources over several years. Positive identification of the sources of contamination was not possible during Phase 1, although it was considered that two of the investigation boreholes with the highest levels of shallow contamination (boreholes 1/1 and 1/6) were probably at, or very close to, sources of contamination.

Table 1. *Chlorinated solvent analyses for Phase 1 boreholes: bailed and depth samples*

Borehole	Depth (m)	Sample type	Geology	TCE ($\mu g l^{-1}$)	PCE ($\mu g l^{-1}$)	TCA ($\mu g l^{-1}$)
1/1	7	B	Drift	25 300	123	nd
	31	D	Chalk	nd	2	52
	36.5	D	Chalk	nd	nd	2
	44	D	Chalk	nd	1	38
1/2	3.7	B	Drift	6.4	0.9	nd
	8	B	Base of drift	70	0.5	nd
	15	B	Putty chalk	67	nd	nd
	31	D	Chalk	5	190	nd
	35	D	Chalk	3	108	nd
	41	D	Chalk	nd	28	nd
	45	D	Chalk	nd	33	nd
	49	D	Chalk	nd	14	nd
1/3	3	B	Made ground	11	nd	nd
	5.6	B	Drift	12	nd	nd
	23	B	Drift	nd	nd	130
	32	B	Drift	nd	nd	13
	40	B	Drift	nd	nd	24
	50	B	Chalk	nd	nd	536
	60	D	Chalk	nd	nd	4.3
1/4	20	B	Putty chalk	nd	0.5	1.3
	26	B	Putty chalk/ Chalk	nd	0.5	1.7
	31.5, 37, 40.5, 47	D	Chalk	nd	nd	nd
1/5	3.5	B	Drift	42	2.5	nd
	8, 18	B	Putty chalk	nd	nd	nd
	24	D	Chalk	1.5	nd	nd
	29	D	Chalk	nd	nd	nd
	34	D	Chalk	nd	nd	1
	39	D	Chalk	nd	nd	4
	44.5	D	Chalk	nd	nd	nd
1/6	12	B	Putty chalk	19 850	225	434
	23	B	Putty chalk	1 543	20.8	33.3
	27	B	Putty chalk	1 714	22.2	38.5
	30	D	Chalk	14.9	16.5	6.5
	35	D	Chalk	19.1	21.3	7.5
	41	D	Chalk	27.6	7.8	6.5
	47	D	Chalk	9.6	4.8	3.7

Notes: B, bailed sample from inside temporary drill casing; D, depth sample from specific sections of Chalk open hole; nd, not detected.

Phase 2 investigations

Methodology

Phase 2 commenced with a more detailed survey of past and present solvent usage within the areas of pollution identified during Phase 1. Only one site was found where large quantities of solvents were still in use: in this case it was TCA for degreasing electrical equipment. However, the survey confirmed the earlier findings that solvents had been widely used in the past at several sites. Wherever possible, the locations of buildings and tanks where solvents had been used or stored were identified in the follow-up survey, and this information was used to target precise locations for the Phase 2 drilling investigations.

Thirteen shallow boreholes were drilled at five sites of known solvent usage, wherever possible within a maximum of 10 m of the likely sources of leaks or spills. The aim was to identify the presence of contaminants in the unsaturated zone above the water-table in the drift, thereby confirming sources of pollution from above. The drilling, sampling and analytical techniques were similar to those adopted in Phase 1. The

Table 2. *Chlorinated solvent analyses for Phase 1 exploratory boreholes and existing borehole II: pumped samples*

Borehole	Time of pumping (min)	TCE ($\mu g\,l^{-1}$)	PCE ($\mu g\,l^{-1}$)	TCA ($\mu g\,l^{-1}$)
1/1	80	nd	nd	5
	180	nd	nd	5
	360	nd	nd	4
1/2	105	25	165	nd
	240	37	215	nd
	350	24	153	nd
1/3	40	nd	nd	nd
	140	nd	nd	nd
1/4	60	nd	nd	1
	220	nd	nd	1
	345	nd	nd	nd
1/5	80	3	nd	nd
	220	4	nd	nd
	340	1.6	nd	2.8
1/6	80	23.4	5.8	3.1
	345	43.6	11.6	11.1
II	220	289	348	19.5

Note: nd, not detected.

Table 3. *Chlorinated solvent analyses for Phase 1 exploratory boreholes: whole-chalk porewater samples*

Borehole	Depth (m)	TCE ($\mu g\,l^{-1}$)	PCE ($\mu g\,l^{-1}$)	TCA ($\mu g\,l^{-1}$)
1/1	8.6	2.99	2.51	na
	21	3.68	1.05	na
	31	9.95	0.65	na
	39.5	1.91	0.39	na
	49.5	1.51	0.22	na
1/2	20	nd	0.15	na
	30	5.62	6.83	na
	40	nd	2.50	na
	50	0.56	0.09	na
	57	1.98	0.53	na
1/3	10	2.1	0.24	0.87
	33.5	nd	nd	nd
	45	3.7	nd	0.65
	60	0.66	nd	0.33
1/4	26	1.6	0.22	0.55
	39	0.86	0.14	0.68
	49.5	1.44	0.14	0.41
1/5	19.5	4.13	nd	nd
	29.5	5.96	0.18	nd
	39	0.82	nd	nd
	49	0.80	0.22	nd
1/6	25	24.8	0.25	nd
	35	5.3	0.25	nd
	45	5.1	nd	nd
	49.5	2.6	0.22	nd

Notes: depths are in metres below ground level to top of U100 core; nd, not detected; na, not analysed.

boreholes ranged in depth from 3.75 m to 11.25 m , the majority being terminated in weathered chalk. About 60 drift and putty chalk core samples were collected and analysed for TCE, PCE, TCA and CTC. Upon completion of sampling, the boreholes were backfilled.

Soil and groundwater contamination

The Phase 2 shallow drilling programme identified significant contamination by chlorinated solvents at four of the five sites investigated, with the contamination at two of these sites being particularly severe. Profiles of contaminant concentrations with depth at the two most contaminated sites are shown in Figs 4 and 5, where they are referred to as Site B and Site D respectively. In constructing these profiles, it has been assumed that all the solvent in the solid samples was in a dissolved form, and hence concentrations have been converted from micrograms per kilogram to micrograms per litre using laboratory measurements of moisture content. This was considered to be a reasonable assumption, since even the highest concentrations were well below saturation concentrations for the individual solvents (Verschueren 1983). However, it should be remembered that high

concentrations in the sample moisture may result from low total amounts in the solid material if the moisture content is low.

At Site B, the four Phase 2 boreholes encountered between about 5 m and 11 m of sands and gravels overlying weathered Chalk, with a water-table at a depth of between 4 and 5 m. The top of the bedrock here is formed of low-permeability putty chalk, and it is assumed that the groundwater is hydraulically partially or fully isolated from the Chalk aquifer. The boreholes were all located within 100 m of the Phase 1 borehole 1/6 which had detected high concentrations of TCE, and to a lesser extent PCE and TCA, in the upper putty chalk. One of the Phase 2 boreholes (borehole 2/3), sited adjacent to a building in which solvents had been used, detected very similar maximum concentrations of TCE ($18\,100\,\mu g\,l^{-1}$), PCE (1720) and TCA (1740), at least two metres above the water-table. CTC was also detected at higher concentrations (430) than had been found at any site during Phase 1.

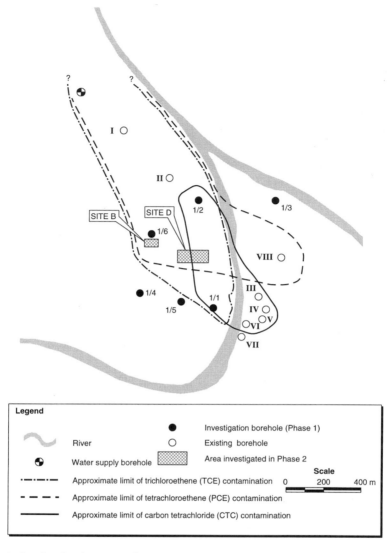

Fig 3. Schematic site plan showing areas of contamination.

The three remaining boreholes at Site B (2/4, 2/5 and 2/6) failed to detect high concentrations of contaminants. Since the severely contaminated borehole 2/3 was located between boreholes 2/4 and 2/5, and only about 25 m away from either, this serves to confirm how localized the high concentration hotspots can be.

Two boreholes were drilled at Site D, and encountered between 6 and 7 m of sands and gravels overlying weathered Chalk, with a water-table at about 2 m depth. Both boreholes were sited adjacent to a building in which TCA was in use at the time of the investigation.

Borehole 2/10 detected high concentrations of all four solvents above the water-table (the maximum values were TCE, $87\,500\,\mu g\,l^{-1}$; PCE, $354\,\mu g\,l^{-1}$; TCA, $6875\,\mu g\,l^{-1}$; CTC, $164\,\mu g\,l^{-1}$), although the concentration of PCE is within that which could be expected as a manufacturing impurity in TCE or as an impurity in recycled solvents. Borehole 2/11 showed high maximum concentrations of TCE above the water-table (up to $27\,700\,\mu g\,l^{-1}$). Unfortunately, these high levels of TCE masked the gas chromatography peaks for PCE, TCA and CTC which therefore could not be quantified.

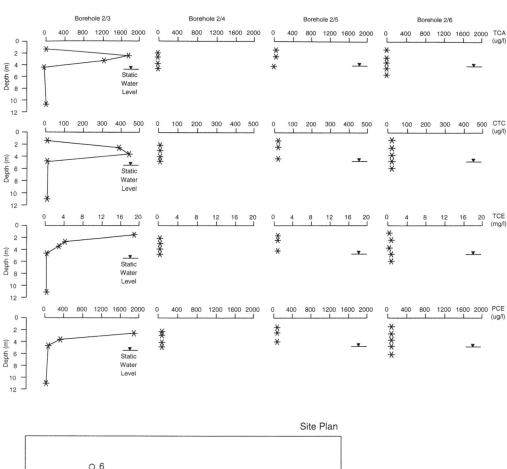

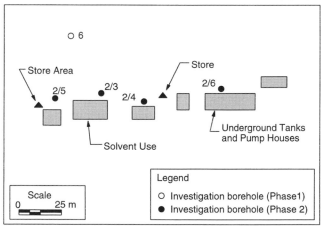

Fig. 4. Contaminant profiles – Site B.

The investigations led to the development of a conceptual model for the contamination, whereby organic solvents spilled or leaked from the ground surface at a number of sites, migrated rapidly downwards through the permeable drift to accumulate at the top of the low-permeability putty chalk in small depressions in its irregular surface. The solvents then seeped slowly through this layer into the fissured chalk below, where flow rates are rapid. This is shown schematically in Fig. 6. A similar conceptual model was formulated to analyse contaminant behaviour in the

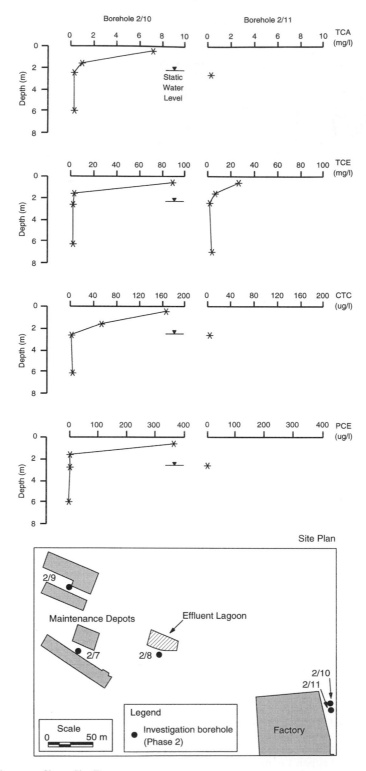

Fig. 5. Contaminant profiles – Site D.

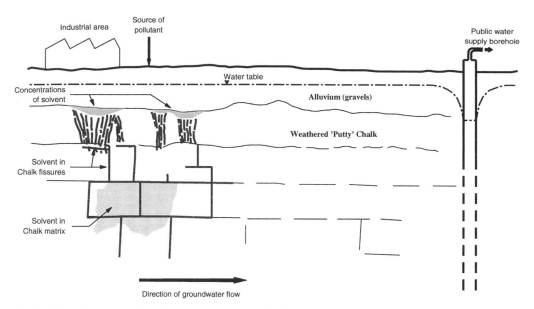

Fig. 6. Schematic cross-section of solvent migration in Chalk.

Sawston pollution case in Cambridgeshire (Lawrence *et al.* 1992; Bishop *et al.* 1998). Attempts were made to determine the time of travel of the contaminants from the source sites to the polluted supply well using a numerical flow model. Reasonable estimates could be made of travel times within the sand and gravel drift deposits and within the fissured Chalk aquifer (although fissure orientation could influence rates of migration in the Chalk). A wide range of possible travel times was estimated in the putty chalk, however, owing to the wide range of possible effective porosity values (5–35%). Consequently, it was not possible to determine the age of pollution based on first discovery in the supply well and predicted travel times.

Conclusions

The investigations showed that within even a lightly industrialized urban area, groundwater contamination by chlorinated hydrocarbons can be widespread. The identification of at least four sites of contamination, and the range of chemicals found within the drift and Chalk aquifers, suggest that this type of contamination should be more properly referred to as multiple point source rather than point source contamination, similar to those at other locations on the Chalk aquifer investigated by Longstaff *et al.* (1992).

The investigations also showed that the sources of contamination at individual industrial sites were very localized. Boreholes drilled only 25 m apart found extremely varied levels of contamination. Therefore, an important feature of planning the drilling investigations was the need to know which chlorinated solvents had been used in the area, and where and how they had been used and stored. Without such detailed information from the solvent usage surveys, it would not have been possible to site the exploratory boreholes sufficiently accurately to identify major sources of contamination.

Finally, the investigation also highlighted the difficulty of determining the age of the groundwater contamination in cases where low-permeability layers such as the putty chalk have a major influence on contaminant movement. The putty chalk would be a suitable topic for the kind of detailed research programme that is beyond the scope of a case study such as that presented here. Further research is needed into the properties of the putty chalk, how these vary spatially and vertically, and the controls which affect these properties.

The authors wish to thank their colleagues at Mott MacDonald who worked on this study, especially P. Ravenscroft and J. Dottridge. The authors would also like to acknowledge the staff of the water undertaker and regulatory authority who were involved in the project, and to thank the water undertaker for permission to publish this paper.

References

ANON 1985. *Methods for the Examination of Waters and Associated Materials – Halogenated Solvents and Related Compounds in Sludge and Waters.* HMSO, London.

BISHOP, P. K., LERNER, D. N. & STUART, M. 1998. Investigation of point source contamination by chlorinated solvents in two different geologies: a multi-layered Carboniferous sandstone-mudstone sequence and the Chalk. *This volume.*

BURLEY, M., MISSTEAR, B. D. R., ASHLEY, R. P., FERRY, J. & BANFIELD, P. 1990. Groundwater pollution by volatile organic solvents – source identification and water treatment. *In: Proceedings of the International Conference 'Advances in Water Treatment and Environmental Management'*, Lyon, 27–29 June 1990.

EUROPEAN ECONOMIC COMMUNITY 1980. Directive relating to the quality of water intended for human consumption 80/78/EEC. *Official Journal of the European Community*, **L129**.

LAWRENCE, A. R., STUART, M. E., BARKER, J. A., CHILTON, P. J., GOODY, D. C. & BIRD, M. J. 1992. Review of groundwater pollution of the chalk aquifer by the halogenated solvents. R&D Note 46 prepared by the Hydrogeology Research Group of BGS for National Rivers Authority.

LERNER, D. N. (ed.) 1993. Coventry groundwater investigation: sources and movement of chlorinated hydrocarbon solvents. *Journal of Hydrology*, **149**(1–4), Special volume.

LONGSTAFF, S. L., ALDOUS, P. J., CLARK, L., FLAVIN, R. J. & PARTINGTON, J. 1992. Contamination of the Chalk aquifer by chlorinated solvents: a case study of the Luton and Dunstable area. *Journal of the Institution of Water and Environmental Management*, **6**(5), 541–550.

MISSTEAR, B. D. R., ASHLEY, R. P. & LAWRENCE, A. R. 1998. Groundwater pollution by chlorinated solvents: the landmark Cambridge Water Company case. *This volume.*

RIVETT, M. O., LERNER, D. N. & LLOYD, J. W. L. 1990. Chlorinated solvents in UK aquifers. *Journal of the Institution of Water & Environmental Technology*, **4**, 242–250.

SCHWILLE, F. 1988. *Dense Chlorinated Solvents in Porous and Fractured Media – Model Experiments.* Lewis.

VERSCHUEREN, K. 1983. *Handbook of Environmental Data on Organic Chemicals.* Van Nostrand Rheinhold.

Investigation of point source pollution by chlorinated solvents in two different geologies: a multi-layered Carboniferous sandstone–mudstone sequence and the Chalk

P. K. BISHOP[1], D. N. LERNER[2] & M. STUART[3]

[1] *Mott MacDonald Group Limited, Demeter House, Station Road, Cambridge CB1 2RS, UK*
[2] *Groundwater Protection and Restoration Group, Department of Civil and Structural Engineering, University of Sheffield, Mappin Street, Sheffield S1 3JD, UK*
[3] *British Geological Survey, Maclean Building, Crowmarsh Gifford, Wallingford, Oxfordshire OX10 8BB, UK*

Abstract: Two investigations of point source chlorinated solvent pollution of groundwater from very different settings are described: an industrial site in Coventry underlain by a multi-layered Carboniferous sandstone–mudstone sequence, and a village site (Sawston) underlain by Chalk in rural eastern England. The investigations are probably two of the most comprehensive studies of point source pollution by solvents in the UK. Both followed a phased approach with key elements being soil gas surveys, and definition of vertical pollution profiles through development of multi-level groundwater sampling and pore-water extraction techniques. The studies identified very significant contamination by chlorinated solvents, trichloroethene and 1,1,1-trichloroethane in the case of the Coventry site, and tetrachloroethene in the case of the Sawston site. The non-aqueous solvent phase was not detected at either site. The main similarities and differences between the studies are brought out, and the successes and failures of the studies discussed. The main successes are considered to be the development of effective soil gas and pore-water sampling methods, and sampling methods for determining vertical variations in groundwater quality. The main failures are considered to be the occurrence of cross-contamination at the Coventry site and the failure to accurately define pollution plumes. An ideal methodology is presented for future investigations, involving a phased approach maximizing the quality of information gained in each phase and preventing cross-contamination.

This paper describes and compares two investigations of point source pollution of groundwater by chlorinated solvents from very different settings: an industrial site underlain by a multi-layered sandstone–mudstone sequence from Coventry in the industrial English Midlands, and a leather processing factory in Sawston, a village site underlain by the Chalk in Cambridgeshire (Fig. 1), both in the UK.

The study of the Coventry site formed a key element of a three-year research investigation by the University of Birmingham into chlorinated solvent groundwater pollution in the city. The study was initiated in 1988 and has been reported in a series of papers (Lerner 1993). At about the same time as the Coventry study, a major investigation into chlorinated solvent pollution of groundwater was also being performed at the Eastern Counties Leather factory in Sawston. This investigation was prompted following pollution of the Cambridge Water Company borehole at Sawston Mill by tetrachloroethene (TeCE), resulting in eventual closure of the source. The Sawston investigation formed part of a six-year research project undertaken by the British Geological Survey (BGS), initiated in 1986. The legal aspects of the Eastern Counties Leather case have been widely reported (e.g. House of Lords 1994; Misstear *et al.* 1998) and various technical papers have also been published regarding aspects of the site investigation (Chilton *et al.* 1990; Lawrence *et al.* 1990; Lawrence & Foster 1991).

These two investigations are perhaps the most comprehensive studies of point source pollution by solvents in the UK. The studies were both performed at a time when the profile of chlorinated solvent pollution of groundwater was high. They followed similar investigation procedures and both had notable successes and failures. They revealed many useful aspects which the authors feel will be of use to practising hydrogeologists in future investigations. By comparing the studies in this paper and by drawing attention to the mistakes made and the lessons learned, it is hoped that future investigations can be optimized.

BISHOP, P. K., LERNER, D. N. & STUART, M. 1998. Investigation of point source pollution by chlorinated solvents in two different geologies: a multi-layered Carboniferous sandstone–mudstone sequence and the Chalk. *In*: MATHER, J., BANKS, D., DUMPLETON, S. & FERMOR, M. (eds) *Groundwater Contaminants and their Migration*. Geological Society, London, Special Publications, **128**, 229–252.

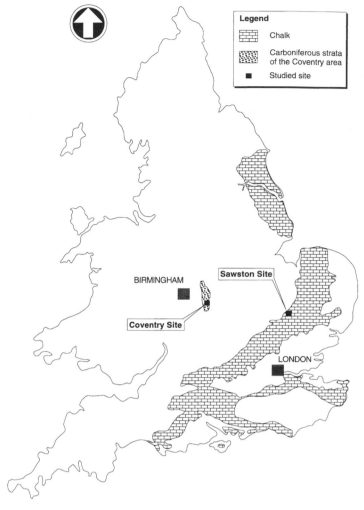

Fig. 1. Location map showing the two sites studied.

Description of sites

The site selected for study in Coventry (Fig. 1) was a major automotive plant with a long industrial history. Chlorinated solvents had been used for many years for degreasing metal components. As with many of the larger industrial sites in Coventry, the site had an abstraction borehole supplying water for heating and some site production processes. An initial survey of Coventry's groundwater indicated that the abstraction borehole (in common with most other abstractions in the city) was contaminated with solvents, most notably trichloroethene (TCE) at around $250\,\mu g\,l^{-1}$.

The Eastern Counties Leather site is located on the eastern part of the flood plain of the River Cam in Sawston (Fig. 1). The Sawston Mill public water supply borehole is located approximately 2 km to the northwest of the site. Sawston has a long history of leather processing and Eastern Counties Leather was founded in 1879. In the early 1960s, Eastern Counties Leather introduced TeCE as a solvent for leather cleaning. It was TeCE contamination of the Sawston Mill borehole, first detected in 1983 ($150\,\mu g\,l^{-1}$), that led to investigation of the Eastern Counties Leather site as a suspected groundwater pollution source. Full details of the chronology of events surrounding the history of the site and investigation (including the BGS investigation) are given by Misstear *et al.* (1998).

Geology and hydrogeology

The geology and hydrogeology of the two sites played a crucial role in the planning and execution of the two studies. A schematic depiction of the geology and hydrogeology of the sites is given in Fig. 2.

The Coventry site is underlain by the Carboniferous Coventry Sandstone Formation, a multi-layered sandstone–mudstone sequence comprising the main aquifer in the area. The sandstone and mudstone layers vary in thickness, typically from a few centimetres up to several metres. The units are known to be laterally discontinuous, owing to their lenticular shape, and the mudstones are relatively soft and prone to slumping. There is little superficial cover apart from a metre or so of made ground. The rest water level at the site was approximately 20 m below surface.

The Carboniferous aquifers of the Coventry area are classified by the Environment Agency

(EA) as 'Minor Aquifers' (National Rivers Authority 1992). They have historically been exploited for industrial water supply and, to a lesser extent, for public water supply. The hydrogeology of the system has been described by Lerner *et al.* (1993). Owing to the multi-layered nature of the aquifer, strong vertical hydraulic gradients are observed. Pumping tests indicate very low permeabilities for mudstone units and average sandstone permeabilities of 1.5–3 m day^{-1}, with overall formation transmissivities of generally between 100 and 300 m^2 day^{-1}. There is evidence of important fracturing in the sandstones, but no evidence of such secondary porosity development in the mudstones.

In contrast to the Coventry site, the Sawston site is underlain by the most important aquifer in the UK, the Chalk, with limited cover by thin sandy gravels. The Chalk is a relatively homogenous soft white micro-porous limestone of Cretaceous age and is classified by the EA as a 'Major Aquifer'. The detailed Chalk stratigraphy beneath the

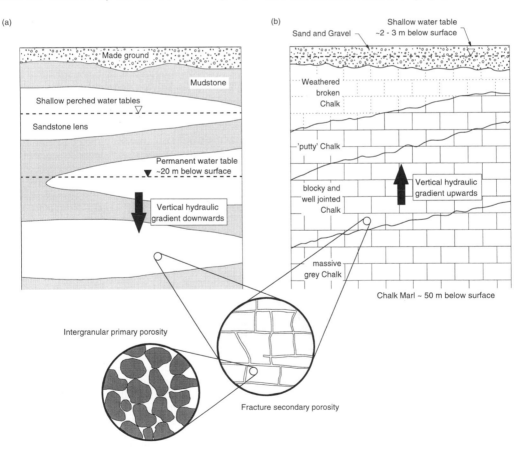

Fig. 2. Schematic representation of the geology and hydrogeology of the two sites: (a) Coventry site; (b) Sawston site.

site comprises weathered broken Chalk and 'putty' Chalk, underlain by blocky and massive Chalk to a depth of approximately 50 m where the Chalk Marl is encountered (Fig. 2). There is a shallow water-table within the overlying gravels some 2–3 m below ground level. The vertical direction of groundwater flow is upwards at the Sawston site.

The hydrogeology of the Chalk is well documented and has been summarized for the Sawston site by Lawrence *et al.* (1990). In common with the sandstone units underlying the Coventry site, the Chalk is a dual porosity aquifer, with the bulk of groundwater movement occurring through fractures and fissures. The presence of enlarged fissures results in enhanced permeabilities and local transmissivity values can exceed $1000 \, m^2 \, day^{-1}$.

The two sites therefore presented geological and hydrogeological settings with some similarities, but more differences. The main similarities between the two sites were the lack of significant Drift cover and the dual porosity nature of the aquifers (the latter applies to most of the important UK aquifers). The main differences were the multi-layered nature of the Coventry Sandstone aquifer compared to the relatively uniform Chalk and the hard cemented nature of the Coventry sandstone units compared to the soft Chalk. The geological/hydrogeological environments of the sites were carefully assessed in planning the site investigations.

Methodology

Similar phased investigations were undertaken at both sites, as described below.

Site walkover and discussions with site personnel

A preliminary site walkover and discussions with site personnel comprise an important initial phase of any groundwater pollution investigation (useful guidance on site walkovers is given by the Association of Groundwater Scientists and Engineers (1992), although this publication is geared to the USA). In our experience, it is not always possible to gain good quality information owing to qualms site owners may have regarding disclosure and potential liability issues. It is also frequently difficult to find site personnel who have been employed for a sufficient period of time and have good memories. When information is forthcoming it is often confused and inconsistent between different parties.

Fortunately, good co-operation was received from the owners of the Coventry site. Discussion with the works manager led us to focus on the area of the site where bulk storage of solvent had historically occurred, and poor housekeeping and spillage were evident. The background research into the Sawston site was somewhat different to that at Coventry, since some previous investigations had been performed by the former Anglian Water Authority and these were used to focus the investigations described in this paper.

Soil gas surveying

Sampling soil gas for chlorinated solvents is a rapid method of detecting contamination and identifying pollution 'hotspots' (see Devitt *et al.* 1987; Marrin & Thompson 1987). The soil gas sampling apparatus designed for the Coventry investigation has been described in detail by Bishop *et al.* (1990) and is depicted schematically in Fig. 3(a). The system comprised hollow stainless steel probes hammered into the ground, with gas sucked into glass bulbs using a hand pump. The glass bulbs were then transported to the laboratory for analysis by gas chromatography (GC).

The soil gas method employed in the Sawston investigation was similar to the Coventry method in that it was a 'dynamic' method using driven probes. However, in contrast to the Coventry investigation, soil gas analysis at Sawston was performed using an on-site laboratory, and soil gas surveying was performed after the main phase of investigation borehole drilling. The equipment used for the Sawston investigation is shown in Fig. 3(b). Soil gas was extracted using silicone tubing attached to a probe and a rubber hand pump. After a known volume of gas had been pumped, samples were taken using a gas-tight syringe for analysis by GC.

Drilling and borehole completion

The soil gas surveys were successful in defining pollution hotspot areas where drilling efforts could be concentrated, particularly at the Coventry site. At both the Coventry and Sawston sites, investigation boreholes were drilled, core was taken and boreholes were completed as monitoring points. However, because of the very different geologies at the two sites there were significant differences in the drilling programmes.

At the Coventry site, the budget for drilling and borehole construction was approximately £36 000. The initial plan had been to drill several

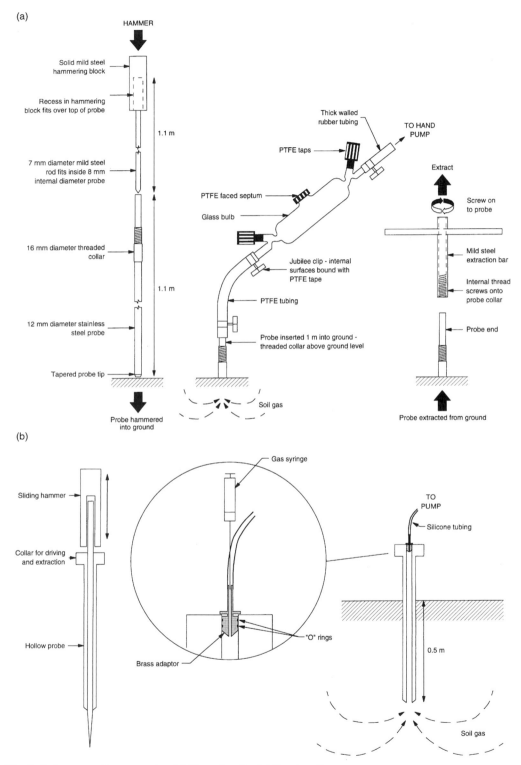

Fig. 3. Soil gas sampling apparatus: (a) Coventry site; (b) Sawston site.

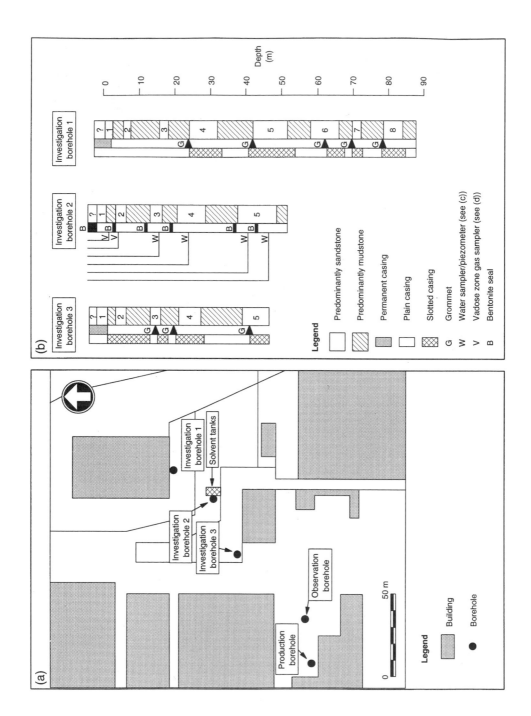

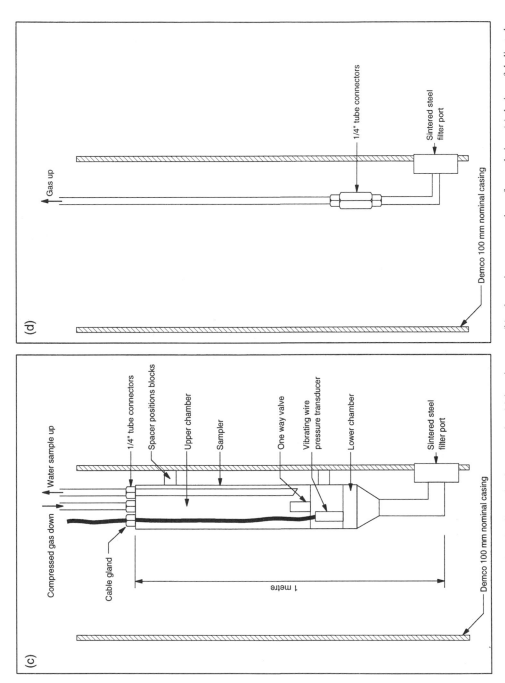

Fig. 4. Investigation borehole completion details at the Coventry site: (a) location map; (b) schematic representation of completion; (c) design of dedicated groundwater samplers; (d) design of gas samplers.

boreholes around the site, but since the soil gas survey indicated that subsurface pollution was concentrated in the area around the bulk solvent storage tanks, it was decided that efforts should be focused here. The designs and locations of the three deep boreholes completed at the Coventry site are depicted schematically in Fig. 4. Drilling was performed using a down-the-hole air hammer (a method anticipated to give rapid penetration rates, and thereby minimize the opportunity for mudstone collapse), apart from the upper 30 m of borehole 3 which was cored using rotary air flush. Selected core material was examined for mineralogy, porosity and organic carbon content, as well as being used for laboratory experiments (Bourg *et al.* 1993; Mouvet *et al.* 1993). However, the hard consolidated nature of the sandstones resulted in it not being possible to analyse pore waters for solvent concentrations; a literature search yielded no satisfactory methods.

On completion, all boreholes were gamma logged to differentiate sandstone and mudstone units. To compare and assess various level-determined sampling techniques, two methods of borehole completion were used (Fig. 4(b)). Boreholes 1 and 3 were cased throughout their entire length, with slotted casing adjacent to major sandstone units (referred to as 'open' boreholes in this paper). Casing diameters were 125 and 100 mm, respectively. It was anticipated that slumping of the mudstone layers against plain casing sections would create hydraulic seals in the borehole annulus between individual aquifers. To preclude the downward collapse of mudstone adjacent to slotted screen sections, short sections of plain casing incorporating dual or triple rubber collars (shown as 'grommets' in Fig. 4(b)) were placed directly above the slotted sections. The collars were slightly oversized with respect to the borehole diameter, forcing them to flex upwards and form a tight seal in the annulus.

Investigation borehole 2 was completed with a dedicated sampling system, four gas-driven water samplers (incorporating pressure transducers; Fig. 4(c)) being placed adjacent to major saturated sandstone units, and two gas samplers (Fig. 4(d)) being placed adjacent to sandstone units in the vadose zone (see Bishop *et al.* 1991). Bentonite layers (1 m thick) were emplaced in the annulus using a tremmie pipe to form the hydraulic seals between sampler ports (Fig. 4(b)). The tubes and wires emanating from the dedicated samplers installed in borehole 2 did cause some logistical problems during installation (Fig. 5) and it is to the credit of the contractor that his patience resulted in the

system being installed without too many hitches.

In contrast to the Coventry investigation, some drilling at the Sawston site was performed prior to the soil gas surveys in an area of suspected contamination. Unlike the Coventry site, the geological situation at Sawston was relatively simple and the choice of drilling method was not so critical. Eight boreholes were drilled in all, using a cable tool percussion rig to drive 0.5 m lengths of 100 mm diameter steel tube (U-100 tubes) into the Chalk. Core was extracted from the U-100 tubes on site, using a hand-operated hydraulic extruder, and analysed on site. At the Sawston site it was pore-water profiles that yielded information on vertical solvent distribution. Therefore, unlike the Coventry site, specialized completion of the boreholes was not necessary. Boreholes were cased in the upper softer sections and left as open holes below. All relevant borehole locations are shown in Fig. 6.

Testing, sampling and analysis

The primary objective of the drilling at the two sites was to gain good quality information on aquifer properties and contaminant distribution and behaviour. Thus the programme of testing and sampling at the two sites was broadly similar, comprising:

- hydraulic testing (pumping and packer tests);
- tracer tests;
- groundwater sampling from discrete horizons in the boreholes.

At the Coventry site, the testing and sampling programme benefited substantially from the involvement of the Geological Survey of Denmark, a collaborator in the EC-sponsored project. The Geological Survey had developed some sophisticated techniques for multi-level groundwater sampling and these were applied to the two open investigation boreholes (boreholes 1 and 3) and the existing observation borehole. The methods generally involved the use of several small-diameter submersible pumps in a single borehole together with packers and various gas-driven sampling devices. The methods have been described in detail by Bishop *et al.* (1992) and can be divided into two broad categories:

(i) *Category 1* where a narrow interval above a packer was sampled using a low-capacity pump, while high-capacity pumping from the zone above eliminated vertical mixing in the sampling interval (Fig. 7(a)).

Fig. 5. Installation of dedicated sampler system in borehole 2.

(ii) *Category 2* where a principle called 'separation pumping' was used, whereby flow in a borehole is diverted to an upper and lower pump and either (a) a third pump samples laterally flowing groundwater from the water divide created, or (b) a calculation scheme is used to evaluate solvent distribution (Fig. 7(b)).

Samples were also returned from below packers using gas-driven and submersible pump arrangements. All sampling relied heavily on flow logging methods to identify the flow regimes in the boreholes.

Meaningful hydraulic testing at the Coventry site was severely hampered by the cyclic pumping of the site's abstraction borehole. However, some pumping tests were performed using the newly created observation borehole network, and the transducer system installed in borehole 2 was successfully used to monitor groundwater heads in individual sandstone layers. The most successful hydraulic test performed was a tracer test where lithium chloride tracer was injected into borehole 3 with pumping and monitoring of borehole 1 (the site abstraction borehole was not operational at the time). Tracer breakthrough was successfully

monitored yielding useful information on the dual porosity nature of the aquifer, such as solute retardation characteristics and coefficients of diffusion.

Groundwater sampling at the Sawston site centred on pore-water extraction from Chalk core obtained from the drilling (Lawrence *et al.* 1990). Each 0.5 m core sample was sub-sampled six times (i.e. three pairs of samples, near the top, centre and bottom of the tube). Bulk groundwater sampling was a less important part of the Sawston site investigation than that at Coventry. However, some bailed samples were taken during drilling to compare with pore-water samples. Hydraulic testing at the Sawston site comprised packer testing of borehole SC01, and a tracer test in two boreholes (OW11 and OW8) located 50 and 1000 m respectively from the pumping station.

Standard analytical methods were used for soil gas and groundwater solvent analysis in the Coventry study, as described by Bishop *et al.* (1990) and Burston *et al.* (1993). At Sawston, the soft friable nature of the Chalk allowed a method to be developed to analyse solvent profiles in pore water. Solvent was extracted from small sub-core samples using a test-tube

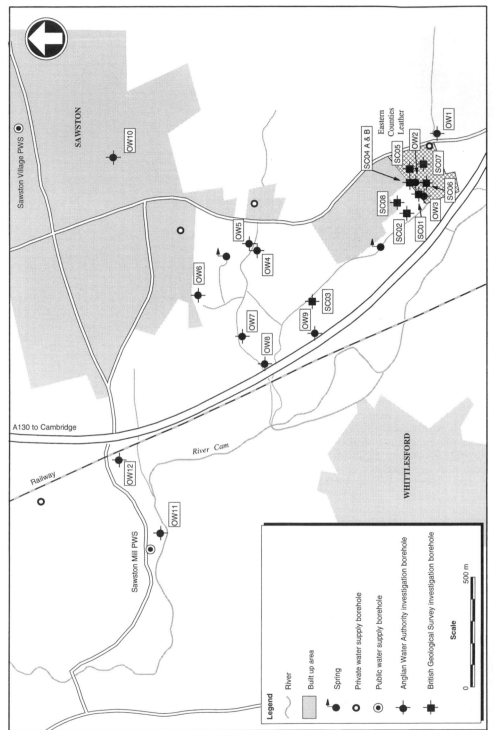

Fig. 6. Map showing locations of investigation boreholes at the Sawston site (based on Misstear *et al.* 1998).

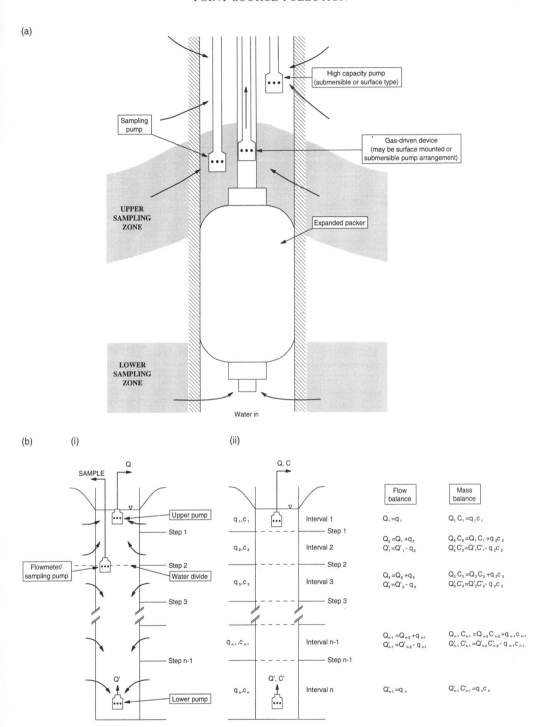

Fig. 7. Level-determined sampling methods employed at the Coventry site: (a) category 1 – packer method; (b) category 2 – separation pumping; (i) combined flow logging and sampling system to sample directly from water divide, and (ii) calculation of flow and solute concentrations from flow/mass balance calculations.

vortex mixer to break down the sample and release the pore water directly to a pentane solvent (Lawrence *et al.* 1990). Analysis was performed on-site. The analysis required one member of staff to be present throughout the drilling, resulting in an analysis cost of about £3000 per borehole (at 1989 prices).

Results and interpretation

This section gives a brief resume of the main results from the two studies, with limited interpretation since the detailed findings are published elsewhere (Chilton *et al.* 1990; Lawrence *et al.* 1990; Lawrence & Foster 1991; Lerner 1993).

Coventry site

The main results gained from the Coventry site investigation were information on solvent distribution and the hydraulic properties of the system. The study also allowed an assessment to be made of the effectiveness of various sampling methods.

The soil gas survey confirmed the bulk solvent storage tanks as the most likely source of groundwater pollution at the site. An initial reconnaissance survey identified very high concentrations of both 1,1,1-trichloroethane (TCA) and TCE in this area (maximum values of $2000 \mu g \, l^{-1}$ and $45 \mu g \, l^{-1}$ in soil gas, respectively, compared with average background values of generally less than $1.0 \mu g \, l^{-1}$ for the rest of the site). A detailed follow-up soil gas survey in the solvent tanks area confirmed the findings of the reconnaissance survey and plumes were mapped for both TCA and TCE (Fig. 8). An interesting observation from the survey was the dominance of TCA in soil gas in the solvent tanks area (see Bishop *et al.* 1990). This contrasted significantly with the pollution characteristics of the site abstraction borehole, where TCE contamination was dominant (TCE and TCA concentrations in groundwater were around 240 and $7 \mu g \, l^{-1}$, respectively, during the period of the soil gas survey).

Vertical profiles of groundwater quality were similar for all three investigation boreholes and profiles from borehole 2 are shown in Fig. 9 as representative of the overall trends. Two main features were evident from the TCE and TCA profiles from all three boreholes:

(i) Maximum observed concentrations of both solvents in groundwater were around

$5000 \mu g \, l^{-1}$. Although very high with respect to drinking water standards, these values do not approach solubility limits (1100 and $4400 \, mg \, l^{-1}$ at 20 °C for TCE and TCA, respectively; see Verschueren 1983), nor values observed where immiscible phase solvents have been found (see Lawrence & Foster 1991). This corresponds with the lack of direct evidence for DNAPL (dense non-aqueous phase liquid) during subsurface investigations.

(ii) All three deep investigation boreholes indicated that TCA concentrations are greater than TCE concentrations at shallow depths, the converse generally being true below aquifer unit 3 (Fig. 9). This consistent trend is interpreted as a probable reflection of how long the two solvents have been in use on the site, rather than the different physical properties of each solvent. The use of TCE on the site between 1970 and 1980 meant that it has had time to penetrate to greater depths than TCA, storage and use of which superseded TCE (Bishop *et al.* 1993). This assumes that vertical migration occurred as the dissolved phase, rather than the DNAPL phase.

The vertical pollution profiles explain the different patterns of solvent pollution seen in soil gas and exploited groundwater at the site. Abstracted groundwater is influenced by solvent use practices at the site possibly ten or more years ago and is thus dominated by TCE. Soil gas reflects more recent practice and release of solvents, and hence is dominated by TCA. The investigation boreholes also indicated that significant spatial variations in solvent pollution existed, even within the small area of the site studied. The site abstraction borehole showed lower solvent concentrations than all three investigation boreholes, reflecting the fact that it draws on a wider area and a greater depth than the investigation boreholes.

Monitoring of solvent concentrations at the site abstraction borehole was conducted for a period of some two years, including through the drilling and testing programme (Fig. 10). It is immediately evident from the data that the drilling resulted in increased contamination of the abstraction borehole; a pathway was created for shallow, grossly contaminated groundwater (or possibly DNAPL) to reach deep exploited levels in the multi-layered aquifer. This is likely to have occurred mainly during the drilling, since packers were installed in boreholes 1 and 3 immediately after borehole construction to preclude further cross-contamination.

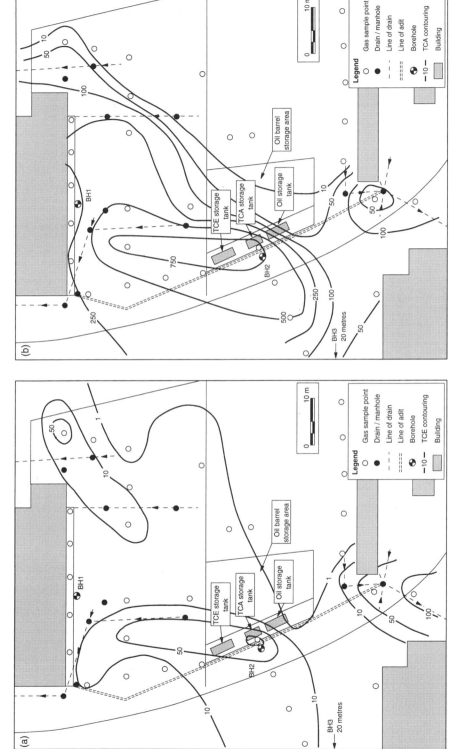

Fig. 8. Solvent distributions in soil gas at the Coventry site: (a) TCE concentrations; (b) TCA concentrations (both in micrograms per litre).

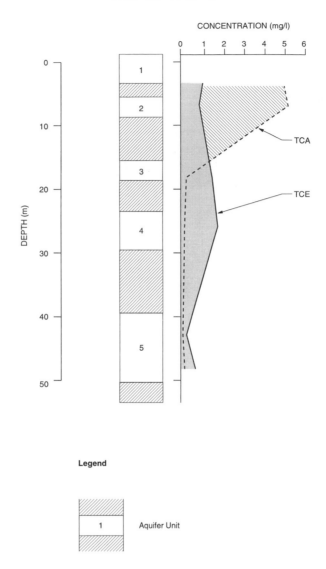

Fig. 9. Vertical profiles of solvent concentrations in groundwater from investigation borehole 2 at the Coventry site (averaged concentrations over an approximate 12-month sampling period; concentrations of solvents in perched groundwater in the upper two sandstone units were calculated from measured gaseous concentrations using partition coefficients derived by Eastwood *et al.* 1991).

Of the various pumping tests performed at the site, the monitoring of groundwater heads at various levels in the system using the permanent installation in borehole 2 proved of most value (Bishop *et al.* 1993), with hydraulic conductivities of 1.5 m day^{-1} and between 10^{-3} and 10^{-2} m day^{-1} obtained for sandstone and mudstone units, respectively. The tracer test yielded useful information on solute transport in the dual porosity aquifer. The asymmetry of breakthrough curves, coupled with laboratory diffusion experiments performed on core, indicated that diffusive exchange is an important process for solvent transport (as it is for inorganic tracers) and is likely to be the main process responsible for reducing solvent concentrations in the subsurface. By comparison, sorption and bacterial degradation of solvents were shown to be relatively unimportant (Bourg *et al.* 1993; Burston *et al.* 1993; Mouvet *et al.* 1993).

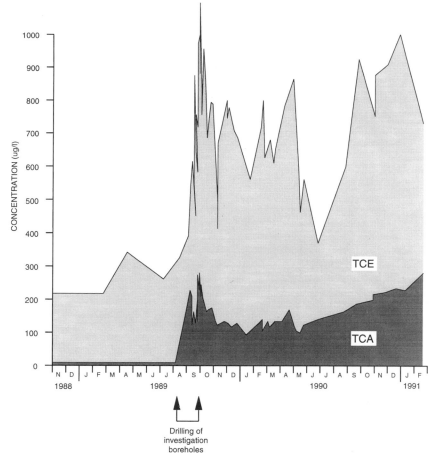

Fig. 10. Solvent concentrations in groundwater from the abstraction borehole throughout the testing programme at the Coventry site.

Sawston site

Drilling and soil gas surveying at the Sawston site suggested that two solvent pollution plumes were present in the system. The most important was a narrow plume with high solvent concentrations (up to 40 000 $\mu g\, l^{-1}$) in both fissure and pore waters from near surface to about 50 m depth. A secondary, shallow plume of much lower concentration (up to 100 $\mu g\, l^{-1}$) was also present, which did not appear to migrate much beyond about 200 m laterally. All further discussion in this paper relates to the primary main plume. The investigation boreholes were unable to demonstrate a clear groundwater flow path between the site and the Sawston Mill public supply source borehole, reflecting the problems investigating pollution in an anisotropic aquifer such as the Chalk. Other groundwater discharges

in the area (such as the spring shown in Figs 6 and 11) appear to have exerted important controls on plume migration.

Concentrations of solvent within pore water from cored boreholes within the main plume are shown and compared with the geological log in Fig. 11. Since an upward hydraulic gradient exists at this site, the solvent could only have penetrated to such depths if density driven as DNAPL (Chilton *et al.* 1990). Cross-contamination issues related to drilling were assumed to have had negligible impacts on relatively immobile pore-water concentrations. The absence of solvent within the Chalk Marl was attributed to the lack of jointing in this horizon; a fact confirmed by the exceedingly low permeability of the Chalk Marl determined during packer testing at the site. The profiles SC4A-B and SC7 (Fig. 11) were taken at locations

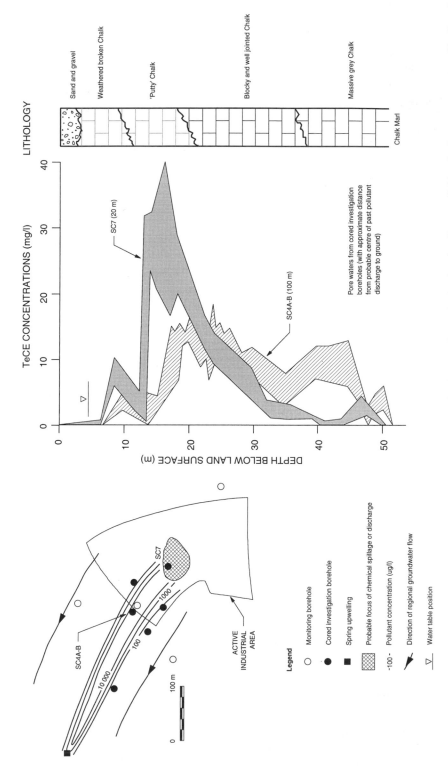

Fig. 11. Tentative contours (micrograms per litre) on dissolved solvent contamination in groundwater (main plume) at Sawston (left) and solvent pore-water profiles (based on maximum and minimum values obtained in the field) at the Sawston site from boreholes SC4A, SC4B and SC7 (right).

close to where the main spillage is believed to have occurred. Solvent concentrations in these profiles therefore result from dissolution of DNAPL (as it migrated downwards through the aquifer) and subsequent diffusion into the Chalk matrix. Thus, these profiles suggest a possible history of DNAPL migration as follows:

(i)   The DNAPL migrated rapidly through the gravels and upper part of the Chalk, allowing little opportunity for solvent retention, dissolution and diffusion into the matrix to occur. As a consequence, pore-water concentrations at these depths are relatively low (less than $1000 \,\mu g \, l^{-1}$).

(ii) The DNAPL ponded upon less permeable, soft 'putty-like' chalk at approximately 15 m depth before penetrating this horizon through the numerous but poorly developed joints. The greater residual saturation of this horizon and the considerable contact time allowed dissolution and diffusion of the solvent into the matrix to occur, producing the marked increase in the pore-water concentrations ($10\,000–40\,000 \,\mu g \, l^{-1}$) observed at this depth.

(iii) The DNAPL eventually migrated through the putty chalk layer and continued downwards until it reached the top of the Chalk Marl at about 50 m depth. Pore-water concentrations within the Chalk aquifer over the 25–50 m depth range are 5000–$15\,000 \,\mu g \, l^{-1}$.

(iv) DNAPL probably ponded upon the Chalk Marl, since no solvent was detected within this horizon above $50 \,\mu g \, l^{-1}$ (the limit of detection for the sampling method utilized). The absence of jointing within the Chalk Marl must have prevented further DNAPL migration. It could also be argued that solvent concentrations are lower in the high-permeability layers owing to increased groundwater flow causing more dissolution and removal of DNAPL

The contaminant plume was shown by the soil gas survey to extend for at least 1 km from the pollution source (Fig. 12), but almost certainly the plume extends for 2 km, which is the distance of the Sawston Mill public supply borehole from the site.

Similarities and differences between studies

The main similarities between the two studies, both methodological and in the results obtained, may be summarized as follows:

- both investigations adopted a sensible phased approach including soil gas surveying;
- both studies failed to identify a correlation between deep groundwater pollution and soil gas characteristics;
- both studies focused on defining and interpreting important vertical solvent profiles;
- both studies failed to accurately define the full spatial extent of dissolved solvent pollution plumes;
- DNAPL was not detected in either study, although its presence was inferred at Sawston.

The main differences between the studies were:

- the methods used for soil gas analysis;
- the methods used to define vertical solvent profiles.

Both studies adopted a phased approach to the investigations which seemed sensible; at neither site was a decision made prematurely regarding where the pollution was. At the Coventry site, in particular, discussion with employees and a site walkover proved invaluable.

Both studies used similar dynamic driven-probe methods for soil gas sampling (Fig. 3), but the analysis at Sawston was performed directly on-site whereas at Coventry samples were sealed in glass bulbs and transported to a laboratory for analysis. Both methods gave satisfactory results but it is probably fair to say that on-site analysis is the preferred method. The main advantages that on-site analysis offers are as follows:

- immediate checking for contamination of blanks;
- immediate checking of anomalies;
- the survey can be adapted as results are obtained.

Care has to be taken in interpretation of soil gas results in the deep aquifer situations typical of the UK. This was highlighted most significantly by the Coventry site study, where TCE and TCA contamination patterns in soil gas did not reflect those observed in deep groundwater. High results may also be obtained where solvent has been spilled onto the ground and, although present in the soil, has not reached groundwater; this may be partly responsible for the much higher solvent soil gas concentrations obtained at the Coventry site compared to the Sawston site (Figs 8 and 12). The overall conclusion, therefore, is that soil gas reflects shallow contamination whereas groundwater contamination patterns deep in the aquifer may be quite different. However, although care must be taken in the interpretation of soil gas data, this

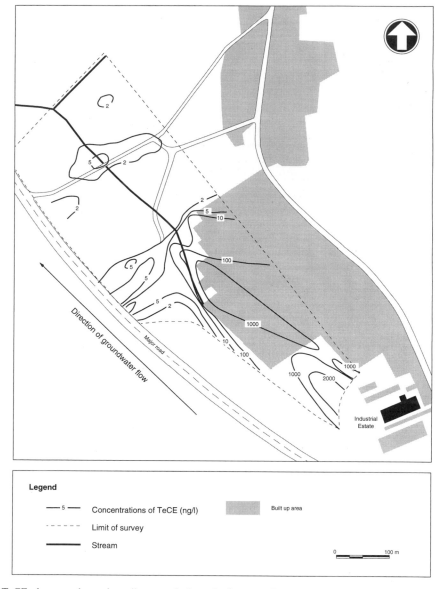

Fig. 12. TeCE plume as shown by soil gas analysis at the Sawston site.

does not detract from the value of the technique as a reconnaissance tool. These overall conclusions are supported by the results of recent field experiments from the Borden research site in Canada (Rivett 1995).

With regard to groundwater sampling and analysis, both studies focused on defining and interpreting vertical solvent profiles. The difference between the two studies was the methods used to achieve this. At the Sawston site, the soft nature of the Chalk enabled a suitable method to be developed for pore-water extraction and analysis. This was important as it was pore waters that held the key to solvent migration characteristics in the Chalk. However, a suitable method of pore-water extraction could not be applied to the hard, cemented Coventry sandstones and therefore specialized level-determined groundwater sampling methods had to be applied. The solvent profiles obtained at both sites provided information on migration rates and characteristics.

(a)

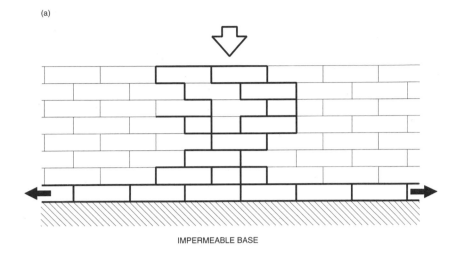

IMPERMEABLE BASE

(b)

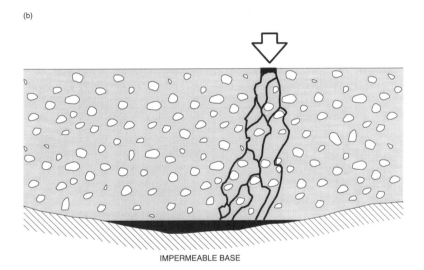

IMPERMEABLE BASE

Fig. 13. DNAPL accumulation at the base of (a) a fissured aquifer and (b) a granular aquifer.

The final significant similarity between the two studies is that DNAPL was not detected at either site, although it is almost certain to have been present at some stage in the subsurface at both sites. One of the reasons that DNAPL was not detected at either site may relate to the probable migration of DNAPL in fractured rocks. Where a spillage is sufficiently large that DNAPL migrates to the aquifer base (as inferred at the Sawston site), it will tend to spread out over a wide area and infill vertical joints (Fig. 13(a)); this contrasts with granular aquifers where a significant 'pool' may form, infilling hollows at the aquifer–bedrock interface (Fig. 13(b)). Recognition of DNAPL and the

identification of the solvent/water interface may therefore be practically impossible in the case of fissured aquifers given that the volume of DNAPL intercepted by any vertical borehole is small. An alternative explanation for the non-detection of DNAPL may be its absence from fractures owing to diffusion into the aquifer matrix (Parker *et al.* 1994).

Successes and failures of studies

The main successes of the studies are considered to be as follows:

Table 1. *Recommended procedure for site investigations*

Phase	Components	Details	Ranking
1. Preliminary study	Site history	Gives information on potential pollutants, likely areas of contamination[1]	1
	Geology, hydrogeology, groundwater quality	Allows a preliminary assessment of subsurface conditions, degree of pollution[2]	1
	Site walkover and interviews	May yield information on likely pollution hotspots; initial point of contact should be Works Manager, but search out employees with long employment history	1
2. Field reconnaissance survey	Groundwater and surface water sampling from existing boreholes (both site and surrounding area), drains, sumps, ditches, etc.	Yields information on actual pollutants, and magnitude and extent of pollution	1
	Soil gas survey and analysis	Reconnaissance tool for defining pollution hotspots[3]	1
	Soil survey/trial pitting and analysis	Soil samples can be collected and analysed for solvents[4]	3
	Trial drilling	Trial drilling outside the contaminated area may yield useful information on geological/hydrogeological conditions[5]	3
	Geophysical logging	Geophysical logging of existing boreholes and trial boreholes (if applicable)[6]	1
3. Detailed drilling and testing	Drilling of investigation boreholes	Choose drilling method appropriate to geology; where cross-contamination is perceived as a threat, drilling should be conducted layer-by-layer[7]	1
	Coring and core analysis	Important for Chalk, but solvent analysis methods are not likely to prove applicable to other commonly encountered rock types[8]	2^{8}
	Borehole completion	Dedicated, multi-level monitoring systems or completion as cased 'open' boreholes[9]	1^{9}
	Hydraulic testing	Important to define controls on groundwater movement and therefore controls on the advective transport of solvent[10]	2^{10}
	Groundwater sampling and analysis	Can be either sampling of bulk, mixed groundwaters or specialized level-determined sampling[11]	1
	Grouting of boreholes	Sealing of boreholes no longer required by grouting[12]	1^{12}

Ranking system: 1, considered essential; 2, useful, but not considered essential; 3, optional, depending on circumstances.

[1] A useful summary of background on industrial sites is given by the Department of the Environment (1994). Useful sources are likely include Ordnance Survey (topographical) maps, discharge consents, hazardous substances consents and local archives.

[2] Useful sources are likely to include geological maps, borehole records and logs, and Environment Agency groundwater quality data.

[3] Care should be taken in attempting correlation between soil gas and groundwater contamination; this may only be applicable in very shallow groundwater environments.

[4] Measuring solvent concentrations in bulk samples may prove useful if, for any reason, a soil gas survey cannot be performed.

[5] This step is only considered necessary if quality data are not available from the desk study stage. It is considered important to have a good knowledge of the geology and hydrogeology underlying the site prior to drilling in contaminated areas, owing to the need to preclude cross-contamination. Trial drilling will also determine the efficacy of the chosen drilling method.

Table 1. *Continued*

[6] Geophysical logging of existing boreholes is considered an important aspect of the reconnaissance investigation owing to its ability to identify argillaceous strata. The two studies described have proved the importance of argillaceous strata in controlling the movement of solvents.

[7] One of the most important points raised by the investigations described in this paper is the potential for cross-contamination when drilling boreholes. Where cross-contamination is perceived as potentially important, success is most likely if boreholes are drilled in stages, stopping at each low-permeability horizon. The borehole can then be cased and plugged, and drilling continued through the base plug. Such techniques have been used for landfill investigations (Harrison 1976) and have successfully prevented cross-contamination, albeit at significantly higher cost.

[8] Core sampling and analysis is considered an invaluable technique for the Chalk where accurate determination of solvent distribution with depth is required. However, as discussed in this paper, the analytical method is not readily applied to other rock types. Coring may also prove useful where rock material is required for organic carbon analysis or laboratory experiments to determine solvent retardation. However, such experiments are usually confined to large research investigations.

[9] Dedicated, multi-level monitoring systems (such as that described for the Coventry investigation) are considered the best option for long-term monitoring of groundwater quality. Once installed they prevent cross-contamination of the system, but they are relatively expensive. For multi-level sampling, nested piezometers may prove a cheaper option. The most basic option is completion as traditional monitoring boreholes with slotted and plain casing.

[10] Considered important where sufficient boreholes are drilled to yield meaningful results. Tracer tests may yield particularly important data on travel times and dual porosity transport characteristics.

[11] The basic sampling will involve pumping bulk, mixed groundwater samples from 'open' boreholes. Care needs to be given to purging the boreholes and the type of sampling equipment used. Techniques involving agitation of samples are not appropriate as solvent may be lost through volatilization; certain materials are also not appropriate for construction of sampling equipment as they may introduce bias through sorption or other phenomena. The best materials to use are stainless steel, Teflon and rigid PVC. Useful information on sampling methods is given by Pohlmann & Hess (1988), Clark & Baxter (1989), Pearsall & Eckhardt (1987), Pettyjohn *et al.* (1981); Reynolds *et al.* (1990), Schuller *et al.* (1981) and Sykes *et al.* (1986). Specialized level-determined sampling methods (such as those described in this paper) may be employed where information on solvent distribution with depth is considered important. However, 'open' boreholes will always be affected by cross-contamination and the quality of data obtained cannot be expected to match that from well-installed permanent dedicated systems.

[12] Sealing of 'open' boreholes at the end of an investigation is considered important if they are not required for monitoring. If left unsealed, such boreholes may always provide a vertical conduit for cross-contamination. An additional potential hazard is that they may be used for illicit disposal of chemicals, particularly if they are located on an industrial site. When planning such an investigation, costs associated with sealing boreholes should always be included in the budget.

- the development of soil gas sampling techniques and their use as a reconnaissance tool;
- the development and application of a method for analysing solvents in Chalk pore waters;
- the development and application of various methods for determining vertical variations in groundwater quality.

The main failures of the studies are considered to be as follows:

- the cross-contamination of the aquifer system that occurred at the Coventry site;
- the failure to define accurately the spatial distribution of pollution;
- the limitations of soil gas analysis in defining the characteristics of deep groundwater contamination.

Soil gas surveying was considered an important element of both studies. The authors consider the soil gas sampling method that was developed for the Coventry project to have been a notable success. It was novel and relatively cheap. The results obtained attest to its value as a reconnaissance tool. However, the method did suffer from some drawbacks; most notably bending of probes, and glass bulbs being prone to breakage in the field. It was necessary to wait for the results and therefore the sampling programme could not be adapted during a single site visit. Although suitable for the Coventry project (since samples could be analysed at Birmingham within an hour or two), the technique would probably not suit sites where transportation to a laboratory would take significantly longer (on-site analysis is recommended in such circumstances). The technique also suffers from severe limitations in correlation with deep groundwater pollution. The authors cannot envisage the technique being used to define groundwater pollution plumes in many of the hydrogeological scenarios typical of the UK.

The development and application of a method for analysing solvents in Chalk pore waters at the Sawston site was also considered a success. The extraction method and on-site analysis enabled accurate definition of pollution profiles and a conceptual model to be developed for solvent migration. There would have been no satisfactory method for defining such pollution profiles from bulk groundwater samples from completed boreholes.

At the Coventry site, groundwater sampling from completed boreholes was relied on to define vertical pollution profiles, since porewater extraction was impractical. However, by using some fairly specialized sampling techniques and installing a dedicated sampling system, it was possible to characterize the vertical distribution of solvents in the system and the results obtained were credible. Nevertheless, obtaining the data did prove costly. The dedicated sampling system cost approximately £6000 (1989 prices), and the level-determined sampling from open boreholes took somewhere between four and ten man-days to obtain good quality chemical data (therefore assume a cost of between £1000 and £2500 for labour, not including the cost of equipment).

The main failure of the Coventry study was in allowing cross-contamination of the system to occur. In practice, the prevention of this will always prove difficult and the process occurs to some degree in all drilling investigations. The next section presents a methodology whereby such cross-contamination may be minimized. The most obvious impact of the cross-contamination of the system was the increased levels of contamination mobilized to the site abstraction borehole (Fig. 10). In practice, this caused little concern as the groundwater was not for potable use. One could even stretch one's imagination to believing that a scavenger borehole remediation scheme had accidentally been created for the site. However, to minimize the effects, packers were installed for the duration of the project and the 'open' investigation boreholes were grouted up at the end of the project. The cross-contamination is also likely to have affected the profiles obtained from the investigation boreholes, but the controlled sampling performed immediately after borehole construction is likely to have minimized the effects.

The final failing of both studies was the inability to define the spatial extent of groundwater pollution. At the Sawston site, a total of 20 boreholes drilled by both BGS and the Anglian Water Authority failed to demonstrate a clear flow path between the site and the Sawston Mill source. The success rate of boreholes drilled to intercept and delineate the contamination plume was likely to be low owing to the probable narrowness of the solvent plume (Fig. 11). At the Coventry site, the budget

only allowed the drilling of three boreholes and therefore definition of the spatial distribution of solvent was always going to be problematic. Many more monitoring boreholes would be needed to define accurately the pattern of pollution beneath this single site, but problems of diminishing returns from additional boreholes are always likely to be encountered.

Implications for future investigations

The studies described in this paper have raised some important issues with regard to future investigations. The authors consider the most important of these to be as follows:

- the value of soil gas surveying and pore-water sampling in investigations of solvent pollution;
- the completion of permanent monitoring boreholes with dedicated samplers as opposed to completion as 'open' boreholes;
- the care needed to prevent cross-contamination;
- the fact that pollution plumes are hard to define and map accurately;
- the budget required for a detailed investigation.

The value and limitations of soil gas sampling when applied to investigations where volatile organic compounds are concerned have been discussed above. It suffices to say here that it can be considered an important reconnaissance tool in any such investigation and, given the choice, the authors would always choose on-site analysis.

Pore-water sampling is considered to be an important method applied to studies of the Chalk aquifer and it may be possible to extend it to other aquifers (Stuart 1991). It is recommended that pore-water sampling and analysis is always considered when performing solvent pollution investigations in the Chalk, although it is acknowledged that this will again be controlled by budgetary constraints.

With regard to borehole completion, it is considered that dedicated sampling systems represent the best technology available. Their main advantages are the relative ease of sampling and the preclusion of cross-contamination once they are installed. Their main disadvantage is cost, but another consideration is the inability to perform remedial measures should the system fail. A cheaper alternative to gas-driven water samplers would be stacked piezometers for multi-level completion in a single borehole, with bentonite seals preventing cross-contamination.

Dedicated sampling systems may reduce or eliminate cross-contamination when boreholes are complete, but an important consideration in any investigation is the cross-contamination that may be caused during drilling, as highlighted by the Coventry investigation. Before any drilling investigation is conducted, careful thought needs to be given to this issue and any adverse impacts. Groundwater pollution is almost always stratified and boreholes provide vertical pathways for contaminant migration.

The difficulty in defining spatial pollution plumes has been highlighted. This is particularly problematic in fissured, dual porosity aquifers where pollution is likely to be concentrated along narrow flow lines. Increasing the number of boreholes obviously increases the definition of any pollution plume: the plume was much better defined in the Sawston investigation (where 20 boreholes were drilled) than in the Coventry investigation (where only three boreholes were drilled). However, the number of boreholes is never enough and hydrogeologists should not anticipate accurate plume definition from such studies.

The budget for undertaking such detailed investigations should never be underestimated. It is estimated that the total cost of the investigation at the Coventry site was in the order of £70 000, whereas the total cost of the Sawston investigation was approximately £300 000. The costs associated with soil gas sampling and analysis, pore-water extraction and analysis, installation of dedicated sampling systems and specialized sampling are all relatively high. Accurately defining the subsurface distribution and behaviour of solvent at any site requires good quality data which are expensive to obtain.

The lessons learned from the two studies have made it possible to develop a recommended procedure for future site investigations (Table 1). This is based on a consolidated, layered geology typical of the UK. The key principles underlying this procedure are as follows:

- a phased approach to the investigation;
- maximizing the quality of the information gained in each phase;
- preventing cross-contamination.

It is clear that budgets will seldom allow all the work detailed in Table 1 to be performed. Therefore a ranking system has been devised, which gives an idea of which aspects of the investigation are considered essential, which are not so essential, and how compromises may be made.

This paper is published by permission of the Director of the BGS, a component institute of the Natural Environment Research Council. The Coventry study was funded by the Commission of the European Communities under Contract EV4V-0101-C(BA) and involved collaboration between the University of Birmingham, the Geological Survey of Denmark and the Bureau de Recherches Géologiques at Minières, France. Drilling at the Coventry site was performed by Wyatts of Whitchurch. The work at Sawston was undertaken as part of a study funded by the Department of the Environment and then by the NRA. The authors would like to thank the many BGS staff who contributed to this study, in particular A. Lawrence, J. Chilton and S. Foster. Drilling at Sawston was mainly carried out by FHV Hewson and Son, Shipdham, Norfolk, and part of the pore-water analyses by Clayton Environmental Consultants, Birmingham.

References

ASSOCIATION OF GROUND WATER SCIENTISTS AND ENGINEERS. 1992. *Guidance to Environmental Site Assessments*. National Ground Water Association, Dublin, Ohio.

BISHOP, P. K., BURSTON, M. W., LERNER, D. N. & EASTWOOD, P. R. 1990. Soil gas surveying of chlorinated solvents in relation to groundwater pollution studies. *Quarterly Journal of Engineering Geology*, **23**, 255–265.

——, ——, CHEN, T. & LERNER, D. N. 1991. A low-cost dedicated multi-level groundwater sampling system. *Quarterly Journal of Engineering Geology*, **24**, 311–321.

——, GOSK, E., BURSTON, M. W. & LERNER, D. N. 1992. Level-determined groundwater sampling from open boreholes. *Quarterly Journal of Engineering Geology*, **25**, 145–157.

——, LERNER, D. N., JAKOBSEN, R., GOSK, E., BURSTON, M. W. & CHEN, T. 1993. Investigation of a solvent polluted industrial site on a deep sandstone–mudstone sequence in the UK. Part 2. Contaminant sources, distributions, transport and retardation. *Journal of Hydrology*, **149**, 231–256.

BOURG, A. C. M., DEGRANGES, P., MOUVET, C. & SAUTY, J. P. 1993. Migration of chlorinated hydrocarbon solvents through Coventry sandstone rock columns. *Journal of Hydrology*, **149**, 183–207.

BURSTON, M. W., NAZARI, M. M., BISHOP, P. K. & LERNER, D. N. 1993. Pollution of groundwater in the Coventry region (UK) by chlorinated hydrocarbon solvents. *Journal of Hydrology*, **149**, 137–161.

CHILTON, P. J., LAWRENCE, A. R. & BARKER, J. A. 1990. Chlorinated solvents in chalk aquifers: some preliminary observations on behaviour and transport. *In: Chalk: Proceedings of the International Chalk Symposium*. Thomas Telford, London, 605–610.

CLARK, L. & BAXTER, K. M. 1989. Groundwater sampling techniques for organic micropollutants: UK experience. *Quarterly Journal of Engineering Geology*, **22**, 159–168.

DEPARTMENT OF THE ENVIRONMENT 1994. *Documentary research on industrial sites. Contaminated Land Research Report No. 3.* HMSO, London.

DEVITT, D. A., EVANS, R. B., JURY, W. A., STARKS, T. R., EKLUND, B., GNOLSON, A. & VAN EE, J. J. 1987. *Soil Gas Sensing for Detection and Mapping of Volatile Organics.* National Water Well Association, Dublin, Ohio.

EASTWOOD, P. R., LERNER, D. N., BISHOP, P. K. & BURSTON, M. W. 1991. Identifying land contaminated by chlorinated hydrocarbon solvents. *Journal of the Institution of Water and Environmental Management*, **5**, 163–171.

HARRISON, I. 1976. Construction in investigatory boreholes in landfill sites. *Surveyor*, 25 June, 28-37.

HOUSE OF LORDS 1994. Cambridge Water Company v. Eastern Counties Leather Plc. *Environmental Law Report 105.*

LAWRENCE, A. R. & FOSTER, S. S. D. 1991. The legacy of aquifer pollution by industrial chemicals: technical appraisal and policy implications. *Quarterly Journal of Engineering Geology*, **24**, 231–239.

——, CHILTON, J. P., BARRON, R. J. & THOMAS, W. M. 1990. A method for determining volatile organic solvents in chalk pore waters (southern and eastern England) and its relevance to the evaluation of groundwater contamination. *Journal of Contaminant Hydrogeology*, **6**, 377–386.

LERNER, D. N. (ed.) 1993. Coventry groundwater investigation: sources and movement of chlorinated hydrocarbon solvents in a consolidated sedimentary aquifer system. *Journal of Hydrology*, **149**, 111–272.

——, BURSTON, M. W. & BISHOP, P. K. 1993. Hydrogeology of the Coventry region (UK): an urbanised, multilayer, dual porosity aquifer system. *Journal of Hydrology*, **149**, 111–135.

MARRIN, D. L. & THOMPSON, G. M. 1987. Gaseous behaviour of TCE overlying a contaminated aquifer. *Ground Water*, **25**, 21–27.

MISSTEAR, B. D. R., ASHLEY, R. P. & LAWRENCE, A. R. 1998. Groundwater pollution by chlorinated solvents: the landmark Cambridge Water Company case. *This volume.*

MOUVET, C., BARBERIS, D. & BOURG, A. C. M. 1993. Adsorption isotherms of tri- and tetrachloroethylene by various natural solids. *Journal of Hydrology*, **149**, 163–182.

NATIONAL RIVERS AUTHORITY 1992. *Policy and Practice for the Protection of Groundwater.* NRA, Bristol.

PARKER, B. L., GILLHAM, R. W. & CHERRY, J. A. 1994. Diffusive disappearance of immiscible-phase organic liquids in fractured geologic media. *Ground Water*, **32**, 805–820.

PEARSALL, K. A. & ECKHARDT, D. A. V. 1987. Effects of selected sampling equipment and procedures on the concentrations of trichloroethylene and related compounds in ground water samples. *Ground Water Monitoring Review*, **7**, 64–73.

PETTYJOHN, W. A., DUNLAP, W. J., COSBY, R. & KEELY, J. W. 1981. Sampling ground water for organic contaminants. *Ground Water*, **19**, 180–189.

POHLMANN, K. F. & HESS, J. W. 1988. Generalized ground water sampling device matrix. *Ground Water Monitoring Review*, **8**, 82–84.

REYNOLDS, G. W., HOFF, J. T. & GILLHAM, R. W. 1990. Sampling bias caused by materials used to monitor halocarbons in groundwater. *Environment Science and Technology*, **24**, 135–142.

RIVETT, M. O. 1995. Soil-gas signatures from volatile chlorinated solvents: Borden field experiments. *Ground Water*, **33**, 84–98.

SCHULLER, R., GIBB, J. & GRIFFIN, R. 1981. Recommended sampling procedures for monitoring wells. *Ground Water Monitoring Review*, **8**, 90–96.

STUART, M. E. 1991. *Determination of chlorinated solvents in aquifer porewaters.* BGS Technical Report WD/91/37.

SYKES, A. L., MCALLISTER, R. A. & HOMOLYA, J. B. 1986. Sorption of organics by monitoring well construction materials. *Ground Water Monitoring Review*, **6**, 44–47.

VERSCHUEREN, K. 1983. *Handbook of Environmental Data on Organic Chemicals.* Van Nostrand Reinhold, New York.

Section 6: Groundwater pollution by radionuclides

Experienced hydrogeologists specialising in radionuclide transport in groundwater are seldom short of work, whether it be in the field of contaminant migration in connection with accidents such as Chernobyl (Bugai *et al.* 1996) or in the design of disposal sites. Many extremely costly and detailed investigations for radioactive waste repositories are being carried out in the world today (Chan 1992) but, as yet, few final disposal sites for high level waste are currently in operation. The Russian Federation is an exception, however, with major radwaste disposal operations being carried out in Siberia, at, for example, sites near Krasnoyarsk, Chelyabinsk and Tomsk (Bradley & Jenquin 1995). Of these disposal operations, some have been more successful that others: the disposal of large amounts of waste to surface water lakes and lagoons at Mayak, near Chelyabinsk has led to the area being characterised as one of the worst-polluted in the world (Solodov *et al.* 1994). The operation at Tomsk-7 appears, however, to be one of the more successful, at least in a short-to-medium term perspective, with wastes being injected to isolated, deep aquifer horizons.

We are privileged to be able to include two papers from the Tomsk area in this volume. Drs. **Lgotin** and **Makushin** give us an introduction to the hydrogeological setting of Tomsk and the radwaste disposal operation at Tomsk-7. They also describe the groundwater monitoring network forming the basis for civilian monitoring of groundwater quantity and quality. Dr. Igor **Solodov** of the Russian Academy of Sciences in Moscow is a veteran of radwaste disposal investigations at Mayak (Solodov *et al.* 1994). He has in recent years turned his attention to the hydro-geochemical processes occurring in connection with waste injection at Tomsk. In this paper he makes an interesting analogy with geochemical processes occurring during uranium leach-mining in Uzbekistan.

For readers interested in finding out more about Siberian nuclear facilities, the Norwegian environmental organisation, Bellona, maintains a comprehensive web site at:

http://www.bellona.no/e/russia/sibir/index.htm

BRADLEY, D. J. & JENQUIN, U. P. 1995. Radioactive inventory and sources for contamination of the Kara Sea. *In:* STRAND, P. & COOKE, Λ. (eds.) *Proc. Intl. Conf. on «Environmental radioactivity in the arctic», Oslo, Norway, 21st-25th August 1995*. Statens Strålevern (Norwegian Radiation Protection Authority), Østerås, Norway, 51–56.

BUGAI, D., SMITH, L. & BECKIE, R. 1996. Risk-cost analysis of strontium-90 migration to water wells at the Chernobyl nuclear power plant. *Environmental and Engineering Geoscience,* **2**, 151–164.

CHAN, C. Y. 1992. Radioactive waste management: an international perspective. *IAEA Bulletin*, **3/1992**, 7–15.

SOLODOV, I. N., VELICHKIN, V. I., ZOTOV, A. V., KOCHKIN, B. T., DROZHKO, E. G., GLAGOLEV, A. V. & SKOKOV, A. N. 1994. Distribution and geochemistry of contaminated subsurface waters in fissured volcanogenic bedrocks of the Lake Karachai area, Chelyabinsk, southern Urals. *Lawrence Berkeley Laboratory, University of California (Earth Sciences Division) Report* No. **LBL-36780 / UC-603**.

Groundwater monitoring to assess the influence of injection of liquid radioactive waste on the Tomsk public groundwater supply, Western Siberia, Russia

V. LGOTIN & Y. MAKUSHIN

Territorial Centre 'Tomsk Geomonitoring', Sovpartshkolny 13, Tomsk 634009, Russia

Abstract. This paper considers the potential impact of the deep injection of liquid radioactive waste (LRW) on the continued operation of the Tomsk public groundwater supply. The close proximity of the LRW injection polygons to the wellfield creates a geoenvironmental conflict, due to the possibility of migration of radionuclides and other toxic components of LRW from the Cretaceous aquifer system used for LRW injection to the Palaeogene aquifer complex used for water supply. Although a significant argillaceous aquitard and a non-utilized buffer aquifer horizon lie between the aquifer systems, there are indications that such inter-aquifer migration may take place, either due to sandy windows in the aquitard or to inadequately sealed well annuli. A major groundwater monitoring network, based on the co-operative use of automated information systems and GIS technologies, has been established. In order to simulate contaminant migration scenarios, the use of variably scaled numerical hydrogeomigrational models is recommended. These would allow prognostic estimations and form a basis for administrative decision-making to minimize damage to the geological environment.

Pollution of the natural environment is one of the most important problems of our modern time – a fact that is becoming increasingly recognized in the former Soviet Union. Such pollution has far-reaching sanitary and ecological consequences and threatens the population's health and economic activity. One of the main contributors to such pollution is industrial waste. The area of Western Siberia has to tackle the special dangers presented by a specific type of industrial waste, namely liquid radioactive waste (LRW), the disposal of which is, where possible, performed by borehole injection into deep water-bearing horizons. Additionally, wastes have formerly been disposed of to surface reservoirs, lakes and watercourses (this practice is, to some extent, still current).

Liquid radioactive waste is one of the by-products of enterprises manufacturing nuclear fuels and other materials and contains large concentrations of a variety of radionuclides (Bradley & Jenquin 1995). There are three main such enterprises in Western Siberia, namely at Mayak (also known as Ozersk, near Chelyabinsk; Solodov *et al.* 1994), Krasnoyarsk-26 (also known as Zhelenogorsk) and Tomsk-7 (also known as Seversk). These sites are located in the upper reaches of the catchments of the great Siberian rivers, the Ob and the Yenisey, both of which flow northwards (Fig. 1). The quantities of radionuclides transported to the Kara Sea in the period 1961–1989 are estimated to have included *c.* 650 TBq ^{90}Sr from the Ob, 450 TBq ^{90}Sr from the Yenisey and an estimated 100 TBq ^{137}Cs for the two rivers combined (NATO 1995) (1 TBq $= 10^{12}$ Bq $= 10^{12}$ disintegrations per second). The impact of these quantities of waste on the Arctic Ocean has raised some concern (Strand & Holm 1993; Strand & Cooke 1995), and the possible transport of radionuclides southwards in icebergs has been considered by Norwegian authorities (Edwards 1996).

One of the three enterprises, known as the Siberian Chemical Combine (SCC), is located at Seversk, in close proximity to Tomsk, the administrative centre of Tomsk Oblast'. (An Oblast' is a Russian administrative region.) The SCC was created as a uniform complex dealing with the entire nuclear technological cycle and manufacturing a wide variety of nuclear materials, except for the production and processing of raw nuclear material. The SCC is one of the largest nuclear manufacturing enterprises in the world and is comparable with similar operations at Hanford (USA) and Sellafield (UK) (Rikhvanov 1994). As a result of its activity, large amounts of liquid and solid waste are generated at Seversk (in addition to gaseous and aerosol emissions), which require storage and underground disposal.

In 1993, a well-publicized accident occurred after an explosion in a holding tank released approximately 4 TBq long-lived nuclides over an area of around 120 km^2 (NATO 1995). Fortunately, the wind direction was to the northeast, away from the cities of Seversk and Tomsk, which therefore escaped the worst consequences of the accident. Several smaller communities were, however, in the path of the plume. The

LGOTIN, V. & MAKUSHIN, Y. 1998. Groundwater monitoring to assess the influence of injection of liquid radioactive waste on the Tomsk public groundwater supply, Western Siberia, Russia. *In*: MATHER, J., BANKS, D., DUMPLETON, S. & FERMOR, M. (eds) *Groundwater Contaminants and their Migration*. Geological Society, London, Special Publications, **128**, 255–264.

Fig. 1. Map of Russia, showing locations of named sites.

purported ecological and human effects of the SCC's activities and of the accident are reported by Tomsk Oblast' State Committee for Ecology (1994, 1995) and Shvartsev *et al.* (1990).

Seversk contains several graphite-moderated, water-cooled reactors for the production of nuclear materials, three of which (commissioned in 1955, 1958 and 1961) have now been shut down. The reactors are dual purpose and produce electricity for the cities of Tomsk and Seversk. Cooling water is taken from, and released to, the Chernilshikov River. In the early days of SCC, LRW from uniflow nuclear reactors was directed to settling ponds and

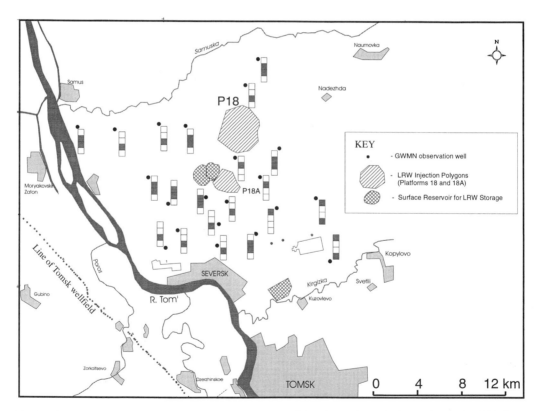

Fig. 2. (**a**) Map of the Seversk area, illustrating the locations of the selected Seversk observation wells which form part of the Tomsk Groundwater Monitoring Network (GWMN). The map also shows the locations of surface reservoirs and injection polygons for LRW. Platform 18 is used for the injection of low-level waste, platform 18a for medium- and high-level waste. The key for the observation well 'columns' is as for Fig 2(b).

thereafter was discharged to the River Tom (a major tributary of the Ob). Until 1982 (NATO 1995), LRW was disposed into at least two surface reservoirs, which are now being covered (and which contain an estimated 4.7 Ebq of long-lived nuclides: 1 Ebq = 10^{18} Bq). Radioactive contamination of the Tom can be traced downstream as far as its confluence with the Ob. Since 1963, LRW has been disposed of by borehole injection into Cretaceous strata at a depth of 280–400 m at a distance of about 10–20 km east of the Tom River. Around 40 EBq of long-lived nuclides are reported to have been disposed of by these means (Bradley & Jenquin 1995; NATO 1995), and the total volume of disposed wastes may reach tens of millions of cubic metres (Rikhvanov 1994).

The injection zones are located in relatively close proximity to the Tomsk groundwater abstraction system and also two abstractions for Seversk, the only reliable sources of drinking water in the Tomsk area. All these abstractions exploit groundwater from Palaeogene aquifer horizons, lying stratigraphically above the Cretaceous aquifer units used for LRW injection. The Tomsk groundwater abstraction system consists of long lines of abstraction boreholes to typical depths of 80–170 m. The lines lie approximately parallel to the Rivers Tom and Ob and within the triangle formed by their confluence (Fig. 2). The water is of acceptable quality, though basic treatment for high levels of dissolved iron is necessary.

It has thus become a matter of urgency to evaluate any potential conflict between the

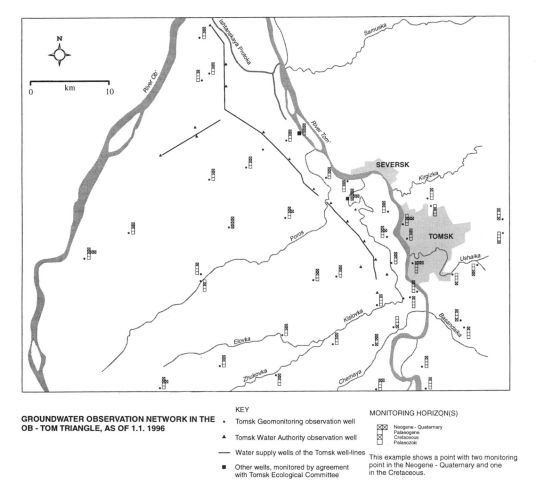

GROUNDWATER OBSERVATION NETWORK IN THE OB - TOM TRIANGLE, AS OF 1.1. 1996

KEY

• Tomsk Geomonitoring observation well

▲ Tomsk Water Authority observation well

— Water supply wells of the Tomsk well-lines

■ Other wells, monitored by agreement with Tomsk Ecological Committee

MONITORING HORIZON(S)

Neogene - Quaternary
Palaeogene
Cretaceous
Palaeozoic

This example shows a point with two monitoring point in the Neogene - Quaternary and one in the Cretaceous.

Fig. 2. (b) Map of the Ob–Tom triangle, illustrating the locations of the Tomsk wellfield and the observation borehole network.

injection zones and the Tomsk/Seversk abstractions and to establish an effective network for monitoring groundwater quality.

Geological and hydrogeological setting of the region

The study region is located near to the zone of conjunction of the Tom–Kolyvan terrane (plicate zone), dominated by Palaeozoic rocks, and the Meso-Cenozoic Western Siberian plate. The conjunction has a stepped character, and is characterized by a series of ruptures in the Palaeozoic basement. These discontinuities occur in various orientations, but are dominated by NE–SW and approximately E–W orientations.

The basement rocks outcrop at the surface some 20–25 km SSW of the zone of LRW injection. To the west and north of the outcrop, the basement dips below thick cover deposits which achieve a thickness of some 350–400 m at the injection zone. The Palaeozoic formations consist of sandstones, siltstones, argillaceous slates and other argillites, deformed into intricate folds and penetrated by numerous dolerite, basalt and diabase dykes. An argillaceous 'crust', the hydrogeological role of which is the subject of some debate, is found at the erosional unconformity at the top of the Palaeozoic. Rybalchenko et al. (1994) believe this crust to function as a low-permeability 'skin', although other authors (Rikhvanov 1994) are of the opinion that it can possess quite a significant permeability.

The Meso-Cenozoic cover consists of rhythmically alternating sandy and argillaceous rocks. The clays frequently contain sandy windows, which have been reliably detected within the triangle formed by the Ob and Tom Rivers. The facies of the cover rocks is reflected in its location close to an ancient zone of transition of continental sedimentary rocks with marine and littoral-marine rocks and again with continental deposits. Such conditions of sedimentation have led to abrupt facies variation, changes in thickness, and *outclining* in a southeasterly direction.

In the alternating sandy–argillaceous sequence, an aquifer complex of Upper Cretaceous deposits has been defined, consisting of permeable sandy horizons (I, II, III, IV) and low-permeability argillaceous layers (A, B, C, D). Overlying this is a not dissimilar complex of Palaeogene and Neogene–Quaternary deposits again consisting of sandy (IVa, V, VI) and argillaceous (E, F, G) horizons (Fig. 3). The boundary between Cretaceous and Palaeogene complexes

corresponds to the horizon defined by Lower Eocene marine deposits, and separating horizons IV and IVa.

LRW is injected into sand horizons II and III, underlain by aquitards B and C, and overlain by D. The Upper Cretaceous (Senomanian) aquifer horizon IV is essentially a buffer aquifer separating the injection horizons from the main water supply aquifer horizon (IVa) of Oligocene age.

The sandy horizons have rather variable grain size and are frequently somewhat argillaceous. Mineralogically, they consist of quartz, orthoclase, microcline and plagioclase feldspars, micas and hydromicas, and argillaceous minerals of the kaolinite and montmorillonite groups, as well as carbonates and organic matter. Aquitard horizons are composed partly of multicoloured, dense, swelling clays, and partly of sideritized sandy silts. Within the clays, fracturing is frequently observed (Rybalchenko et al. 1994).

The hydraulic properties of the aquifer complexes vary considerably, but the total transmissivity of the Palaeogene aquifer complex is in the range 1000–4500 $m^2 day^{-1}$. The transmissivity of the Cretaceous aquifer complex is somewhat lower at 60–80 $m^2 day^{-1}$. The transmissivity of the argillaceous horizons totals an estimated $1.2 \times 10^{-4} m^2 day^{-1}$. The aquifer horizons of the Cretaceous and Palaeogene complexes contain confined groundwaters of calcium bicarbonate type, with a total mineralization of 300–400 $mg l^{-1}$.

The main recharge of groundwater to the aquifer complexes is derived from infiltration of precipitation falling on the outcrops of the aquifer units in the east and south–east parts of the region and, to a lesser degree, from vertical infiltration over the entire area underlain by the aquifer complexes. That the groundwater flow in the aquifer complexes is governed largely by the local infiltration and topography is testified by piezometric heads decreasing with depth in interfluve areas, and by piezometric heads increasing with depth in the river valleys, such as the Tom. This indicates that the River Tom is the main local recipient of groundwaters in the Meso-Cenozoic cover, and also that there is a theoretical possibility that migration of groundwater could occur upwards to the River Tom through aquitard layers from the lower aquifer horizons. The regional groundwater flow has a predominantly westerly and northwesterly direction and is characterized by estimated filtration velocities of 3–5 m year^{-1}, with a hydraulic gradient in the range 1×10^{-3} to 1×10^{-4}.

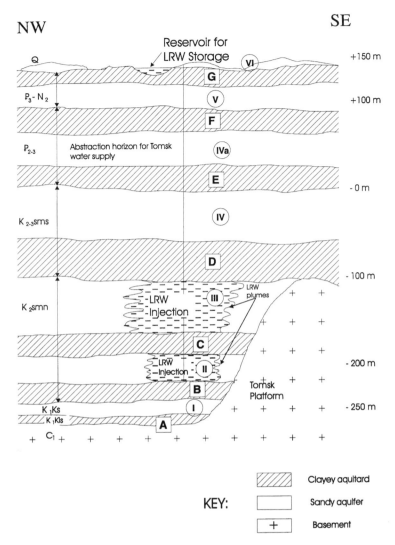

Fig. 3. Schematic cross-section of aquifer system near Seversk, showing location of injection and abstraction horizons. (Modified after Tomsk Oblast' State Committee for Ecology 1994.)

The use of groundwater for water abstraction and waste disposal

The most important sites are the Tomsk groundwater abstraction system and the zones of injection of LRW near Seversk.

The Tomsk groundwater abstraction system

The Tomsk groundwater field, within the Ob–Tom triangle (Fig. 2), is the main source of water-supply for the city of Tomsk. The abstrac-

tion system is also one of the largest in Russia, abstracting groundwater from Palaeogene sediments of horizon IVa. It has been operational since December 1973, the abstraction rate increasing from $30\,000$–$50\,000\,m^3\,day^{-1}$ in 1974 to $230\,000$–$240\,000\,m^3\,day^{-1}$ in 1995. The design of the system incorporates two lines of water wells (184 wells total), located along the western bank of the Tom from SE to NW (sections 1 and 2) and across the Ob–Tom triangle in its northern part (section 3). The total length of the abstraction system exceeds 55 km. The depth of the water wells varies from 80 to 170 m, with an average abstraction per well of 2500–$3000\,m^3\,day^{-1}$.

Intensive abstraction has perturbed the natural groundwater regime and has drawn down groundwater levels in both the Palaeogene complex and in the overlying Neogene–Quaternary deposits. The piezometric levels and chemical quality of the groundwater are observed via a network of observational wells, established in all the horizons: Neogene–Quaternary, Palaeogene, Cretaceous and Palaeozoic aquifer horizons. The number of observational wells is large and their regular, long-term observation has permitted an objective assessment of the state of the aquifer and the influence of the groundwater abstraction.

The non-uniform distribution of abstraction from the wellfield, which is particularly intensive on the southern section of the field, has resulted in a decrease in piezometric levels in the operational horizon of some 10–15 m, and in the overlying horizon by 5–10 m. The net result of abstraction is that an ellipsoidal 'cone' of depression has been generated along the line of water wells. The width of the cone is up to 10–20 km in the southern part of wellfield and 5–8 km in the north (Shvartsev et al. 1990). Unfortunately, no reliable information is accessible on the true extent of the cone of depression on the eastern bank of the Tom, due to the fact that the territory of the the SCC at Seversk is closed for civilian authorities. LRW injection occurs about 14 km east of the Tomsk abstraction well-line.

A theoretical analysis of the size and configuration of the cone of depression within the bounds of the Ob–Tom triangle suggests that, in the east, it may already be impinging on the site of the LRW injection. There may thus be an overlap between the cone of depression of the abstraction line and the doming of piezometric levels caused by LRW injection. In order to further study the effects of abstraction on piezometric levels, a non-steady state numerical groundwater flow model of the Ob–Tom triangle was created, named CAM. CAM is a two-aquifer model, incorporating the Palaeogene (IVa) and Neogene–Quaternary (V) aquifers, and enables the study of the interaction of these two aquifers under various external and internal boundary conditions. The model permits the non-steady-state solution of problems and is able to assist with the hydrodynamic optimization of the operation of the components of the Tomsk water abstraction wellfield.

LRW injection polygons

During the first years of SCC's activity, LRW was stored in surface reservoirs, pulp-storages and natural ponds. Due to the considerable quantities of waste disposed of in this fashion, aerosol pollution of land adjoining surface storage reservoirs has occurred. It is also possible that some of the LRW has infiltrated into the ground through the base of these reservoirs and ponds. By the mid-1950s the significant potential danger of this situation was recognized and a decision concerning deep geological disposal was taken. Geological exploratory investigation in 1963 of one particular site, known as platform 18, led to the experimental injection of technogenic medium-level radioactive waste from the surface storage reservoirs, via a system (or polygon, as it is referred to in Russia) of injection wells into aquifer horizon II at a depth of 300–325 m.

In subsequent years two full-scale polygons for waste injection entered into operation: since 1967 at platform 18 for disposal of low-level, non-technogenic radioactive waste into horizons II (depth of injection 350–400 m) and III (depth of injection 280–300 m) and, since 1978, at a site to the southwest of platform 18 (platform 18a) for the disposal of technogenic high- and medium-level wastes into horizon II. Until 1982, wastes were deposited in the boreholes under their own hydraulic head, but since that date wastes have been actively injected under pressure.

It should be noted that argillaceous horizon D of 25–85 m thickness is the main aquitard isolating the LRW-containing horizons from the overlying aquifer horizons, namely IV (the buffer aquifer) and the Oligocene aquifer IVa (exploited by the Tomsk wellfield). It is believed (Rybalchenko et al. 1994) that all the aquifer horizons in the stratigraphic section are hydraulically connected via sandy windows through the argillaceous horizons, with the exception of aquitard D, which has not been shown to contain such windows in the course of studies and exploratory works.

Low-level non-technological wastes (i.e. wastes not produced by reprocessing of plutonium, see p. 24 in Rybalchenko et al. 1994) are injected as typically alkaline solutions with pH ranging from 6 to 10.5, mineralization up to 20 g l^{-1} and total radioactivity activity in the order 100–10 000 Bq l^{-1}. The solutions contain nitrate and sulphate ions, calcium and magnesium carbonates and detergents. The main radionuclides present include strontium-90, caesium-137, ruthenium-106, caesium-144 and plutonium-239. Tritium is present at low concentrations.

Medium-level technological (i.e. resulting from reprocessing of plutonium) waste solutions have mineralizations up to 300 g l^{-1} and are mildly alkaline, containing sodium salts (nitrates, acetates, carbonates and sulphates), and dissolved silica. The specific radioactivity is in the order 10^8–10^9 Bq l^{-1}, and the radionuclide content is similar to that in low-level waste.

High-level technological wastes are injected with acidic solutions (Solodov 1998) of pH 2–3 and have total mineralization up to 200 g l^{-1}. The macrocomponents of these solutions are sodium nitrate and soluble complexes containing heavy metals (iron, chromium, manganese and nickel) which are the products of corrosion. The specific radioactivity of the wastes is of the order of 10^9–10^{11} Bq l^{-1}, which is sufficient to cause heating of the strata close to the injection wells. Experimental data suggest that the temperature in the injection horizon may reach 130–160 °C (Rybalchenko et al. 1994) compared with the the natural groundwater temperature of 7–10 °C.

The injection wells used for LRW injection have a special multicolumn or 'telescopic' design. The management of the distribution of waste injection and monitoring of the geological environment are performed via a network of control-observational wells established within all the water-bearing horizons. There are in excess of 250 such observation wells. The injection of LRW, in terms of its distribution in space and time, is managed on the basis of the volume and composition of the waste, total quantities of injected substances and radionuclides, and operational parameters of the injection wells. Monitoring of the geological environment comprises the observation of the distribution of waste components within the aquifer(s), the piezometry of the injection and adjacent horizons, the temperature, the radionuclide content and the chemical composition of the groundwater. This monitoring network is run by the authorities of the closed city of Seversk, although data from a selection of these wells is released to form part of the Tomsk Groundwater Monitoring Network (GWMN; see below and Fig. 2a). The pressures of injection are typically of the order of 1.0–2.5 MPa, while the average volumes of liquid injected vary from 300 up to 1000 m^3 day^{-1} for medium- and high-level waste and about 5500 m^3 day^{-1} for low-level waste.

Injection of waste is performed sporadically, at typical intervals of several months. The total volume of injected waste in the years up to 1993 is believed to exceed 35 million m^3 (Rikhvanov 1994). As a result of this waste injection, a dome has developed in the piezometric surface, the groundwater pressures exceeding their natural values by 10–15% in the central part of the injection area (Rybalchenko et al. 1994).

The radius of piezometric influence of injection in horizon III, according to our estimates, may be as much as 5–8 km and may impinge upon the River Tom.

According to data collected in 1994 by the geological service of the SCC from horizons II and III, the injected contaminants are located within the bounds of ellipsoidal 'plumes' of area 1.6 km^2 around platform 18, and 1.4 km^2 around platform 18a. The degree of pollution of groundwater within these plumes is rather non-uniform. With increasing distance from the injection wells, the degree of both radioactive and other contamination falls off sharply such that at a distance of 100–200 m it does not exceed the limits established by the state standard for potable waters and norms of radiation safety. Nevertheless, in several cases, the penetration of ammonium ions, sulphates, beta- and gamma-activity into the buffer aquifer horizon (IV) has been observed in observation well data (Rybalchenko et al. 1994). This may testify to the unreliability of aquitard D as a containment horizon or, conceivably, to the migration of waste materials up poorly sealed annuli around the casing of observation and injection wells.

Prognostic numerical modelling of the distribution of LRW, performed by independent experts, has suggested that from the time of eventual cessation of LRW injection, the centre of contamination will migrate beyond the borders of the area covered by SCC's permit to discharge waste under the effect of natural groundwater flow within 500–570 years for many components of the waste, and within 660–730 years for the less mobile plutonium. Modelling thus suggests that groundwater may become contaminated beyond the bounds of the SCC area within a period of less than 1000 years. If this is the case, the centre of pollution will reach as far as the edge of the Seversk drinking water abstraction wells, both those currently in operation and planned, which are located on the eastern bank of the Tom at a distance of 1–2 km from the river.

The results of the prognostic modelling must be regarded as preliminary due to the unavailability of the proven hydrogeological information from SCC, which would enable a more realistic simulation of contaminant migration from the injection polygons.

The principles of the Tomsk groundwater monitoring network

The analysis of hydrodynamic and hydrogeochemical conditions in the present complex ecological situation requires a monitoring system that is able to observe and rapidly react to changes in groundwater and environmental conditions. The main objective of a groundwater monitoring network (GWMN) in the Tomsk area is to collect information on ground and surface waters suitable for observing changes in the geohydrosphere related to the injection of LRW into Cretaceous aquifer horizons and the simultaneous abstraction of potable water from Palaeogene aquifers by the Tomsk and Seversk wellfields. The current status of the GWMN is shown in Fig. 2(a) (Seversk) and Fig. 2(b). The GWMN is not yet regarded as complete, however, and further development of the network is planned.

The basic principles for establishing the GWMN are as follows:

(i)   a unified methodology for supervising the monitoring of piezometric levels and groundwater quality, co-ordinated with the needs of the users of the aquifer systems;

(ii) the use of automated information systems (AIS) for storage, analysis and processing of primary and manipulated data from the GWMN;

(iii) the visualization of primary, computationally manipulated and prognostic data by means of geographical informational systems (GIS) complemented by specific software;

(iv) that the GWMN should provide the data necessary for making administrative decisions to prevent or minimize any damage to the geological environment during injection of LRW or groundwater abstraction.

The monitoring system is based on the acquisition of authentic data from a specifically designed network of monitoring wells on the state of the geological environment and changes brought about by anthropogenic activity. The network is run by various organizations (Fig. 2b) and in the past was subject to a significant variety of monitoring techniques and practices. It was thus necessary to develop a co-ordinated programme for co-operative monitoring to provide a unified regime of observations. This co-ordinated programme was developed by Tomsk Geomonitoring and was agreed upon by all aquifer users possessing a state license for groundwater abstraction or for disposal of waste into water-bearing horizons.

The GWMN consists of a network of observation wells situated throughout the Ob–Tom triangle, in association with the Tomsk groundwater wellfield, the Seversk groundwater abstractions and the polygons of injection of LRW. Water levels and groundwater quality are monitored for each member of the aquifer sequence (though not necessarily at every monitoring point), and the sampling and monitoring are regulated by uniform rules. The GWMN is managed by Tomsk Geomonitoring, who collect and collate the observational data. The monitoring of the plume of LRW within the boundary of the injection polygon area is, however, performed under a special programme by the geological service of the SCC. This SCC monitoring also covers the entire sequence of aquifer horizons.

All retrospective data as well as those data collected by GWMN are stored in specialized databases, with data tied to the geographic location of the observation points. The database also includes information on the observation infrastructure itself and other non-time-variant data, including the design features of the wells, the geological and hydrogeological structure of sections, the results of experimental infiltration and migration studies, geophysical studies, hydraulic testing, etc. Observations on water level and quality are stored in the form of time series data. The database structure also allows the effective control of the integrity of the information, the specification of various kinds of output documents, and data sampling and grouping for statistical and mathematical processing. The database also serves as the basis for CAM modelling.

The graphical representation of the GWMN data is based on a fully functional GIS, implementing direct access to AIS and CAM data. It allows the presentation of various types of information in the form of, for example, maps, diagrams, charts, geological and hydrogeological sections.

A system of variably scaled CAM numerical models will eventually allow the simulation of a wide range of prognostic problems of groundwater flow and contaminant migration. These models incorporate a maximum of detail regarding the geological and hydrogeological structure of the object being modelled. The main benefit of the CAM system design is an ability to create large-scale models of several separate objects of research, and then to embed them into the fundamental regional model. The objects of research may be polygons of injection of LRW, groundwater abstractions, or surficial sources of pollution. The results of prognostic and simulation modelling serve as initial data for administrative decision-making to prevent or reduce hydrogeological damage, to optimize the operational conditions of injection and/or abstraction, or to optimize the location of monitoring points or frequency of monitoring.

Conclusions

The pollution of underground waters due to the disposal of liquid radioactive waste by injection may have important ecological consequences. First of all, it has the potential to degrade the potable properties of groundwater and hence threatens the conditions necessary for a sustainable groundwater-based water supply. Beside radioactive contamination, a significant danger is also posed by nitrates, sulphates and heavy metals present in the LRW solutions. Many of the substances in the LRW may have health-related consequences and may affect the blood formation system and genetic material. One other consequence of the pollution of groundwater is the reduction in the oxygen content of the water which, in itself, may affect the aquifer's ability to attenuate other contaminants.

Thus, given these potential dangers, it has been found necessary to create a complex groundwater monitoring network to observe the condition of groundwater and the natural environment in the zone of influence of the LRW injection polygons. This system has the ability to provide early warning of dangerous levels of groundwater contamination and may assist in providing information to design a containment programme for such an incident. It is of course regarded as necessary to continue such monitoring after the termination of active LRW injection as the contaminants represent a potential long-term danger to groundwater quality and hence also to other components of the natural environment.

In summary, the implementation of the observational principles described above will allow the establishment of a geoenvironmental monitoring system which can be used to reliably manage the geological situation in a region where a potential conflict exists between LRW injection and the large-scale abstraction of groundwater.

References

BRADLEY, D. J. & JENQUIN, U. P. 1995. Radioactive inventory and sources for contamination of the Kara Sea. *In*: STRAND, P. & COOKE, A. (eds) *Proceedings of the International Conference on Environmental Radioactivity in the Arctic*, Oslo, Norway, 21–25 August 1995. Statens Stralevern (Norwegian Radiation Protection Authority), Østerås, Norway, 51–56.

EDWARDS, R. 1996. Hot ice could contaminate fish. *New Scientist*, **2019**, 7.

NATO 1995. *Cross border environmental problems emanating from defence-related installations and activities. Final Report, Volume 1: Radioactive contamination, phase I: 1993–95*. NATO/CCMS (Committee on the Challenges of Modern Society)/NACC, Report No. 204.

RIKHVANOV, L. P. 1994. *Sostoyaniye okruzhayuschey sredy i zdorovya naseleniya v zone vliyaniya Sibirskogo khimicheskogo kombinata. Analitichesky obzor nauchno issledovatelskikh otchetov [State of the environment and people's health in the area of influence of the Siberian Chemical Combine]*. Tomsk Polytechnic University, Tomsk [in Russian].

RYBALCHENKO, A. I., PIMENOV, M. K., KOSTIN, P. P. *et al.* 1994. *Glubinnoye zakhoronenie zhidkikh radioaktivnykh otkhodov [Deep disposal of liquid radioactive wastes]*. IzdAT Publishing House, Moscow [in Russian].

SHVARTSEV, S. L., RASSKAZOV, N. M. & MAKUSHIN, U. V. 1990. Problemy ratsialnogo ispolzovaniya i okhrany podzemnykh vod Ob–Tomskogo mezhdurechya [Problems of rational use and protection of groundwater in the Ob'–Tom' triangle]. *Chelovek i Voda. Tezisy dokladov k nauchno-prakticheskoy konferentsii Vodnye resursy Tomskoy Oblasti, ikh ratsialnoye ispolzovaniye i okhrana [Humans and Water, Proceedings of the Scientific/Applied Conference on Water Resources of Tomsk oblast': their rational use and protection]*. Tomsk, 71–74 [in Russian].

SOLODOV, I. N. 1998. The retardation and attenuation of liquid radioactive wastes due to the geochemical properties of the zone of injection. *This volume*.

——, VELICHKIN, V. I., ZOTOV, A. V., KOCHKIN, B. T., DROZHKO, E. G., GLAGOLEV, A. V. & SKOKOV, A. N. 1994. *Distribution and geochemistry of contaminated subsurface waters in fissured volcanogenic bedrocks of the Lake Karachai area, Chelyabinsk, southern Urals*. Lawrence Berkeley Laboratory, University of California (Earth Sciences Division) Report No. **LBL-36780/UC-603**.

STRAND, P. & COOKE, A. (eds) 1995. *Proceedings of the International Conference on Environmental Radioactivity in the Arctic*, Oslo, Norway, 21–25 August 1995. Statens Stralevern (Norwegian Radiation Protection Authority), Østerås, Norway.

—— & HOLM, E. (eds.) 1993 *Proceedings of the International Conference on Environmental Radioactivity in the Arctic and Antarctic*, Svanhøvd Environmental Centre, Kirkenes, Norway, 23—27 August 1993. Statens Stralevern (Norwegian Radiation Protection Authority), Østerås, Norway.

TOMSK OBLAST' STATE COMMITTEE FOR ECOLOGY 1994. *Sostoyanie okruzhayushei sredi i zdorove naseleniya v zone vliyaniya Sibirskogo Chimicheskogo Kombinata [The State of the Environment and the Health of the Population in the Zone of Influence of the Siberian Chemical Combinate]*. Ministry of Environmental Protection and Natural Resources/Tomsk Oblast' State Committee for Ecology and Natural Resources, Tomsk 1994 [in Russian].

—— 1995. *OBZOR: ekologicheskoe sostoyanie, ispolzovanie prirodnich resursov, ochrana okruzhajushei sredi. Tomskoi oblasti v 1994 godu [OBZOR: Ecological Condition, Use of Natural Resources and Environmental Protection: Tomsk Oblast' 1994].* Tomsk Oblast' State Committee for Ecology and Natural Resources, Tomsk 1995 [in Russian].

Editors' note

For those interested in reading more about the Siberian plutonium reprocessing plants and associated waste disposal issues, information can be accessed on the Internet at http://www.bellona.no/e/russia/sibir/index.htm.

The retardation and attenuation of liquid radioactive wastes due to the geochemical properties of the zone of injection

IGOR N. SOLODOV

Department of Rare Metals Geology and Radiogeoecology,
Institute of Geology of Ore Deposits, Petrography, Mineralogy and Geochemistry,
Russian Academy of Sciences, Staromonetnii per. 35, Moscow 109017, Russia

Abstract: Natural geochemical factors are of significant importance for the design of underground waste disposal systems and their subsequent operation. Analysis of geochemical similarities between liquid radioactive waste (LRW) disposal sites and sulphuric acid uranium-leaching mines in sandy aquifer horizons, has proved to be a productive approach, permitting predictive understanding of the hydrogeochemical processes which might develop in aquifer horizons subject to the injection of LRW. The approach allows the attenuative geochemical properties of terrigenous aquifers to be estimated. It is demonstrated that a number of geochemical processes (neutralization, reduction, sorption, precipitation, pore occlusion and radiolysis) in the injection zone of such aquifers may lead to attenuation, degradation and solid-phase immobilization of contaminants associated with LRW.

This paper discusses a small proportion of the ongoing work at the nuclear facility of Tomsk-7 in Siberia to study the geochemical and hydrogeological conditions surrounding the disposal of liquid radioactive waste (LRW) in sandy aquiferous strata via injection at depths of several hundred metres. Current studies include an intense network of observation wells around the injection site, numerical hydraulic and geochemical modelling, sampling and geophysical logging of boreholes in the contaminated zone. This paper focusses on one aspect of the work, the redox and acid-base conditions which lead to the existence of 'geochemical barriers' to contaminant migration. The study draws both on real data from the Tomsk-7 site, modelling works and, to a large extent, on analogy with the processes recognized to occur at uranium leach-mining sites. The paper does not set out to exhaustively assess all the attenuation mechanisms (e.g. adsorption, radioactive decay, ion exchange, complexation) taking place in connection with LRW disposal; nor does the paper focus on specific radionuclide contaminants but on changes taking place in major ion chemistry.

A short history of site selection procedures for the disposal of LRW from Russian radiochemical enterprises

Until the end of the 1950s, a huge volume of LRW was accumulated at the locations of Russian radiochemical enterprises producing radioactive materials for nuclear weapons

(Table 1). The resulting problems threatened the viable continuation of such industrial activities and prompted the initiation of research on technologies for underground disposal of LRW (Rybalchenko *et al.* 1994). Special prospecting studies aimed at selection of sites suitable for LRW disposal were carried out in the vicinity of the radiochemical enterprises, namely (Fig. 1):

- the Siberian Chemical Industrial Complex (SCIC), at Seversk (or Tomsk-7), near Tomsk city;
- the Mining–Chemical Industrial Complex in Krasnoyarsk district (MCIC);
- the Nuclear Production Association 'Mayak' in Chelyabinsk district (NPA 'Mayak').

The purported environmental impacts of these industrial plants are discussed by Tomsk Oblast' State Committee for Ecology (1994, 1995), Ilyinskikh *et al.* (1995), Zuev & Shvartsev (1995) and Edwards (1996).

During site selection, the following factors were taken into account (Rybalchenko *et al.* 1994):

- the acceptable distance of a disposal site from a LRW source (less than 20 km);
- the presence within the geological column of stratified terrigenous sedimentary rocks with significant porosity and matrix permeability;
- the presence within the geological column of aquifer horizons with suitable hydraulic and geochemical properties for LRW injection;
- the presence of low-permeability confining/buffer horizons isolating the injection zone from the surface;

SOLODOV, I. N. 1998. The retardation and attenuation of liquid radioactive wastes due to the geochemical properties of the zone of injection. *In*: MATHER, J., BANKS, D., DUMPLETON, S. & FERMOR, M. (eds) *Groundwater Contaminants and their Migration*. Geological Society, London, Special Publications, **128**, 265–280.

Table 1. *Yearly quantities of highly radioactive waste produced at the Siberian Chemical Combinate, Tomsk, and quantities of radionuclides emitted to the atmosphere (after Tomsk Oblast' State Committee for Ecology 1994)*

Nuclide	Ci per year
Atmospheric emissions	
H-3	160
C-14	17
Kr-85	330 000
I-131	2
Xe-131m	190
Xe-133	540
High-level waste produced	
Ce-144	5.8×10^7
Zr-65	4.3×10^7
Nb-95m	4.3×10^7
Ru-106	7.8×10^6
Rh-106m	7.8×10^6
Pr-144m	2.7×10^6
Cs-137	2.7×10^6
Sr-90	2.5×10^6
Y-90m	2.5×10^6
Tc-99	3.3×10^2
Am (total)	1.6×10^2
Am-241	1.6×10^2
Cm (total)	9.7×10^1
I-129	8.0×10^{-1}
Am-242m	8.5×10^{-2}
Am-243	1.8×10^{-2}
Cm-242	2.0×10^{-3}

- a relatively slow rate of groundwater exchange with surface water.

The three radiochemical enterprises are located at the peripheral zone of the West Siberian epi-Palaeozoic platform. According to Rybalchenko *et al.* (1994), Grabovnikov (1993), Pinneker (1991), Rogovskaya (1991) and others, the terrigenous Mesozoic–Cenozoic rocks comprising the platform's sedimentary cover form a huge asymmetrical depression with relatively gently dipping western and steeper eastern flanks. This structure forms, from the hydrogeologist's viewpoint, the West Siberian artesian megabasin. The megabasin contains, in vertical section, two main aquifer systems of permeable rocks. These are separated by regionally extensive clayey deposits of Oligocene–Turonian age, which are absent only at the eastern and southeastern flanks of the basin. The upper aquifer system of the basin undergoes direct groundwater exchange (i.e. recharge and discharge) with the surface water and is thus unsuitable for LRW disposal. Most favourable for LRW disposal are the aquifer horizons in the lower system in the central part of the

basin. The terrigenous strata of the lower aquifer system are subject to only an extremely slow rate of water exchange due to their depths ranging from 400 m to 3000 m. The regional discharge area is at the northern margin of the megabasin, which forms a hydraulically open hydrogeological structure. According to the criteria for site selection, this northern zone is thus unsuitable for LRW disposal, as long-term isolation of injected wastes cannot be guaranteed.

From the standpoint of LRW underground injection, NPA 'Mayak' in the Chelyabinsk region is located in the least favourable conditions. It lies on the eastern slope of the Ural mountain chain, underlain by fractured basement rocks. The western limit of sedimentary cover lies some 40 km east of the plant's territory.

At Tomsk, the SCIC disposal sites lie in the vicinity of aquifers which contain water of drinking quality. In fact, injection takes place in Cretaceous strata at depths of some 300–400 m, whereas groundwater is abstracted for public supply from Palaeogene aquifers on the opposite (western) side of the River Tom at depths of between 100 and 200 m (Lgotin & Makushin 1998). Some discharge of radionuclides also takes place to surface water, via a small tributary of the River Tom, passing through the territory of the closed area of Tomsk-7.

The MCIC disposal sites in Krasnoyarsk region are located near a large fault which currently appears to act as a hydraulic barrier. Its future, on the thousand-year timescale, is less certain. Nevertheless, the SCIC and MCIC disposal sites have continued to be assessed as ecologically safe during their *c.* 30 year history, largely due to the presence of favourable local geological factors, and also due to the fact that the process of LRW injection is well managed and controlled.

Potential sites for disposal of LRW from NPA 'Mayak', Chelyabinsk

Specialist site investigations were carried out during the 1960s and 1970s along the western flanks of the West Siberian artesian basin, within its marginal hydrochemical zone. These revealed approximately N–S trending, tectonically complex, synclinal graben structures, infilled with Triassic–Jurassic terrigenous sediments. Hydraulic testing within boreholes in these deposits revealed only a low specific capacity (Grabovnikov 1993). The inadequate hydrogeological characteristics and the considerable distance (*c.* 150 km) from the LRW

Fig. 1. Outline map of southern Siberia and the central Asian republics, showing the approximate locations (as stars) of the main Siberian nuclear facilities (near Tomsk, Krasnoyarsk and Chelyabinsk) and the Kysyl-Kum mining area in Uzbekistan.

source led experts to conclude that the site was unsuitable for LRW disposal by injection.

The exclusion of this LRW disposal option and the subsequent adoption of a management strategy of interim storage of LRW in natural and human-made lakes and reservoirs resulted in radioactive contamination of the surface environment and groundwater. The scale of this contamination is without parallel in Russia and almost certainly in the rest of the world (Drozhko *et al.* 1993; Solodov *et al.* 1994*a*).

Currently, research is under way to examine more thoroughly the option of LRW disposal within geological structures formed by ancient river channels at the western flank of the megabasin. These structures have previously been intensively mapped during prospecting for uranium deposits (Lisitsyn *et al.* 1994). Nevertheless, it should be noted that the final solution for LRW at Chelyabinsk is likely to be LRW solidification for subsequent underground disposal in a borehole repository within the NPA 'Mayak' territory (i.e. within 10–20 km of the source).

There are two main reasons why this solution is regarded as the most favourable:

- the lack of safe means for long-distance transport of high-level LRW;
- the negative public reaction to the perceived hazard of radioactive contamination from liquid wastes.

Roll-front uranium ore deposition and sulphuric acid leach-mining of uranium ores as analogues for LRW injection disposal systems

Lisitsyn *et al.* (1994) have noted that an analogy exists between the conditions of formation of exogenic–epigenetic roll-front uranium deposits (Galloway 1978) and LRW injection disposal systems. This analogy implies that the geological factors (i.e. a mineralogically and hydrochemically controlled oxidation/reduction front) resulting in the precipitation of ore elements and their subsequent long-term preservation have much in common with the factors which would ensure safe LRW injection disposal.

In addition to this natural analogue, the author of this paper has investigated the subsurface processes of leaching or 'solution mining' of uranium ore from exogenic uranium deposits. These studies indicate that in such systems, subsurface processes result in strong retardation of migrating technogenic contaminants (i.e. produced by technological processes), as well as petrogenic and ore elements, including radionuclides. These processes, typically using sulphate leaching agents for the uranium ore, are similar to the processes observed in LRW injection disposal systems where the injected solution is a radioactive sodium nitrate brine. Thus we can treat underground leaching systems as anthropogenic analogues of LRW injection disposal systems. This conclusion is supported by the data obtained from an underground leach-mining site at the Bukinaii stratiform infiltration-type uranium deposit in southeastern Kysyl-Kum, Uzbekistan.

Anthropogenic redistribution of elements in the Bukinaii uranium deposit, Southeastern Kysyl-Kum, Uzbekistan

The identification of mechanisms promoting contaminant retardation at geochemical barriers in underground leaching systems is of obvious significance for the prediction of mechanisms and scale of attenuation of LRW at injection disposal sites. The Bukinaii uranium deposit is located on the western periclinal flank of the South Nuratinsky mountain belt, on the slope of the Karakatinsky artesian basin. Fractured and folded basement rocks of Palaeozoic age outcrop in the core of the mountain belt. In cross-section, the sedimentary cover comprises inter-layered sands and clays of Turonian age, overlain by Palaeogene chemogenic clayey sediments and reddish, continental sandy and clayey Neogene sediments. Among the Turonian deposits are two sandstone aquifer horizons. The uranium ore bodies are located in the lower part of the upper horizon, with a thickness of 23–27 m (Fig. 2).

Groundwater flow in the vicinity of the ore is from east to west at a rate of around $1 \, \text{m year}^{-1}$, the hydraulic gradient being around 0.002–0.005. The ore-bearing sand sequence is lithochemically subdivided into four layers. The first three layers account for 75–100% of the sand's total thickness and are ore-bearing. The

curved, ellipsoidal main ore body is formed at the edge of the oxidation zone within the most permeable sands of layer 2 (counting from the bottom). The sandy layer contains only a very low silt/clay fraction and hydraulic conductivities may reach $K_f = 8.6 \, \text{m day}^{-1}$. The limbs of the ore body encroach upwards and downwards by the flow of the ore-forming solutions into layers 1 and 3, which have a higher silt/clay content and K_f of 1.0 and $3.4 \, \text{m day}^{-1}$, respectively. In the lower layer, which is only present in the southern part of the deposit, the silt/clay fraction can reach 7%, with K_f values of $2.5 \, \text{m day}^{-1}$. Typically, the coefficient of hydraulic conductivity anisotropy for these alluvial sands is close to $K_f(x)/K_f(z) = 2$.

The ore-bearing sands can be classified as aluminosilicic mesomict quartz–feldspar sands with a kaolinite–hydromica intergranular matrix. Quartz and feldspars are present in the sands in all fractions from pelitic to coarse-grained. In the sandy fraction (grain size $> 0.1 \, \text{mm}$), quartz and plagioclase are present in almost equal quantities in all layers (34–40% of the total rock mass). The quantity of potassium feldspar increases from 4% to 10% from the bottom to the top of the sequence. Mica minerals are present in the sands in quantities less than 5% and titanates less than 1.8%. The main authigenic minerals are calcite (up to 2%) and pyrite (0.2–1.8%). Accessory minerals include rare grains of garnet, epidote, tourmaline, sphene, apatite, monazite and zircon.

Ore mineralization is dominantly represented by nasturan (or uraninite, UO_2) with a lesser quantity of U in sorbed forms on titanates. The redox potential in the oxidation zone is typically $c. +150 \, \text{mV}$; in the ore zone it varies from $+100$ to $-160 \, \text{mV}$, while in the reduced greyish sands beyond it can become as low as $-330 \, \text{mV}$.

The subsurface leach-mining was carried out over $c.$ 20 years and the formation of the sulphate brine lens was observed throughout this period and for six years after cessation of mining activities (Solodov et al. 1994b). The lens was formed by the injection into the ore-bearing aquifer of solutions comprising $10–25 \, \text{g} \, \text{l}^{-1} \, H_2SO_4$. The composition of the lens was investigated by analysis of pore waters extracted from core samples from the observation boreholes. By the end of the production period, the areal extent of the sulphate brine lens had only migrated beyond the limit of the ore body by an insignificant amount (less than 30–60 m). In cross-section, sulphate solutions occupied the entire thickness of the permeable sands (Fig. 3).

During the mining process, the natural groundwaters (of calcium–sodium sulphate–chloride

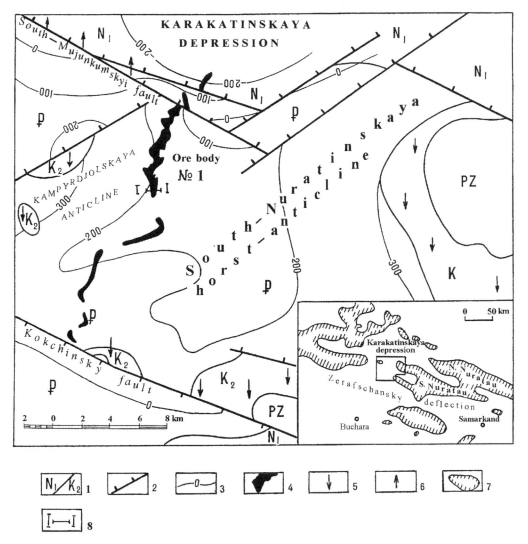

Fig. 2. Structural–geological sketch of the region of Bukinaii uranium deposit, Uzbekistan. 1, Palaeozoic folded basement (PZ), Upper Cretaceous sandy–clayey deposits (K2), Palaeogene carbonate and clayey rocks (P), Miocene sandy clayey rocks (N1); 2, faults (tick on downthrown side); 3, relief of top surface of Upper Cretaceous; 4, ore body No. 1, projection onto the earth surface; 5, 6, recharge (5) and discharge (6) areas for the ore-bearing aquifer; 7, Palaeozoic basement on the regional scheme; 8, section for which detailed data on technogenic geochemical barriers was obtained.

type with pH 7.4 and total salinity of $1.7\,g\,l^{-1}$) were replaced in the vicinity of the ore by acid (pH 1–2) sulphate brines. The resultant salinity in the main volume of the lens varied from 33 to $45\,g\,l^{-1}$, resulting in densities of 1.030–$1.045\,g\,cm^{-3}$. The brine solutions had the following per cent-equivalent cationic (Mg, 32; Al, 32; Fe, 18; NH_4^+, 5; Ca, 4.5; Na, 4; K, 1; Mn, 0.7) and anionic (SO_4^{2-}, 96; NO_3^-, 2.6; Cl^-, 1.4; PO_4^{3-}, 0.6) compositions and could be categorized as weakly

acidic brines of ferroaluminium–magnesium sulphate type. At the frontiers of the lens, where pH values typically vary from 1 to 2.5, maximum concentrations of elements are greater than those outlined above, and are specified in Table 2. There is potential for migration of these elements from the lens frontier out into the sandy aquifer.

The Cl^- concentration in solutions at the lens frontier is equivalent to its concentration in natural underground water of $300\,mg\,l^{-1}$. The

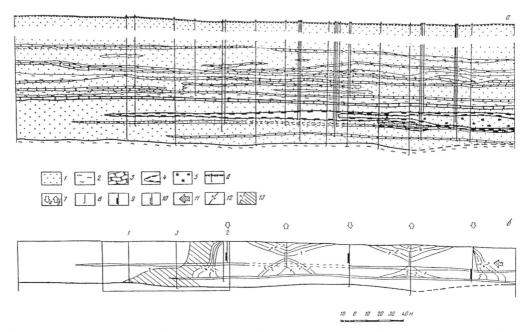

Fig. 3. Cross-sections for the Bukinaii deposit: (**a**) geological structure and ore distribution; (**b**) lens of residual sulphate solutions. On the left of section (b), the location of accumulation of transported solutes at the geochemical barrier is marked by a frame. 1, greyish sands; 2, clays and siltstones; 3, sandstone with carbonate cement; 4, ore body contour; 5, oxidation zone with limonite-containing yellow sand; 6, borehole; 7, zones of technogenic solutions: (a) injection and (b) discharge; 8–10, boreholes: observational (8), injection (9), discharge (10); 11, direction of the natural groundwater flow; 12, pH value isolines; 13, area occupied by technogenic solutions and position of pH 7 isoline.

total salinity at the lens frontier reaches $65\,\mathrm{g\,l^{-1}}$, resulting in densities of $1.055–1.065\,\mathrm{g\,cm^{-3}}$.

The initial mineral composition of quartz–feldspar sand with kaolinite–hydromica intergranular matrix is not essentially altered by leaching with sulphate solutions, despite the significant leaching of certain elements from the host sediment material. This observation indicates essentially incongruent dissolution of the host rock minerals.

As the residual solutions migrate along the aquifer with the regional groundwater flow and come into contact with unaltered rock, several geochemical barriers successively develop. The main barriers are pH and redox barriers.

pH barriers

The pH geochemical barrier develops first and is caused by the neutralization of acidic sulphate solutions by calcite. After calcite dissolution is complete (or after secondary precipitates of gypsum on calcite surfaces preclude further reaction), the main neutralizing reactions occur with the aluminosilicate minerals.

Laboratory-scale reactor simulation

The neutralization by calcite of synthetic acidic sulphate solutions, with macrocomponent composition characteristic of *in situ* brines, was investigated in a series of laboratory-scale experiments. Portions of mineralized solution were neutralized in a closed reactor by the addition of calcite (of grain size 0.1–1 mm) in quantities that simulated the range of calcite content in aquifer sands. The equilibrium macrocomponent composition of the solution was then analysed.

The solid precipitate resulting from calcite addition was subjected to in-reactor ageing for three months, after which time the wet, *in situ*, precipitate volume was measured. Experimental results of calcite neutralization of acidic sulphate solution indicated that the initial volume of solid phases increases during this ageing period. With a calcite addition equivalent to 0.21% of rock mass, the volume of solid phases increases by 0.4 vol. % during ageing, while with a calcite addition equivalent to 4.03% of rock mass, the volume increase goes up to 20.2 vol. % (Table 3). This volume increase has significant implications for clogging of the aquifer porosity.

Table 2. *Concentrations of selected components in groundwater at the periphery of the sulphate brine lens, Bukinaii uranium deposit, Uzbekistan*

	Concentration (mg l^{-1})
Petrogenic/ore components	
Fe^{II}	3300
Al	2860
Mg	2400
Fe^{III}	1500
Ca	730
Na	650
PO_4^{3-}	580
K	570
SiO_2	560
Mn	100
U	86
Zn	17.5
Ni	12.3
Ce	8.35
Sr	8.2
Rb	8.0
Li	6.2
Y	5.67
Co	5.6
Nd	4.85
Cr	4.05
Ti	3.44
La	3.15
V	2.65
Ba	2.05
Pb	1.13
Cu	1.07
Technogenic components	
SO_4^{2-}	46 600
NO_3^-	750
NH_4^+	533
Conservative components	
Cl^-	300

Previous investigations (Rafalsky 1978; Solodov *et al.* 1993, 1994*b*) have shown that at the *in situ* neutralizing barrier (in the direction of pH change from 1 to 7.4), the following mineral phases are sequentially formed as precipitates (Fig. 4):

- amorphous silica (its precipitation begins immediately after the addition of minor quantities of calcite and ends at pH 4);
- an association of gypsum, hydrogoethite and hematite (α-Fe_2O_3);
- an association of alunite ($KAl_3(OH)_6(SO_4)_2$), amorphous ferric hydroxide, jarosite ($KFe_3^{III}(OH)_6(SO_4)_2$) and gibbsite ($Al(OH)_3$), which precipitate almost simultaneously;
- an association of rhodochrosite ($MnCO_3$), siderite ($FeCO_3$), aragonite and newberyite ($MgHPO_4.3H_2O$).

Precipitation of ammonium and potassium alum ($KAl(SO_4)_2.12H_2O$) has not been found to take place in such conditions. The formation of minerals of the halotrichite group ($Fe^{II}Al_2(SO_4)_4.22H_2O$) has not been clearly demonstrated although is regarded as probable to explain the volume increases in precipitates during ageing, which cannot be explained by the detected mineral phases.

Observation boreholes and in situ *core sampling*

During the course of investigations of the Bukinaii site, core samples were extracted from observation boreholes drilled into the lens frontier six years after the cessation of sulphuric acid injection into the aquifer. By examination of the cores, the following sequential reaction phases were demonstrated:

(i) initial rock sulphatization (SO_4^{2-} as secondary solid phases reaches 2.8 mass % of whole rock);
(ii) carbonatization (carbonate as CO_2 in secondary phases up to 0.3 mass %);
(iii) phosphatization (P as P_2O_5 as secondary phases increases to 0.08 mass %).

The following parameters present in secondary phases (in mass % of whole rock) were also detected in core samples from the lens frontier: SiO_2 0.009, Ca 0.13, Al 0.12, Fe^{III} 0.04, Fe^{II} 0.22, Mg 0.16 (Fig. 5). These quantities were calculated by comparing the total content of these parameters in affected cores with those in background, unreacted cores. These estimates were also confirmed by assessing progressive loss of these parameters from solution to secondary solid phases.

Observation boreholes confirm that as the technogenic leaching solutions move across the neutralization barrier, they undergo attenuation with respect to Fe, Al, Be, Cr and Th, due to basic hydrolysis and precipitation. Precipitation of Ca, Ba and Ra sulphates occurs and is spatially related to technogenic gypsumization. Ca and Sr are also partially removed from solution as carbonates. As a result of sulphate precipitation, the sulphate concentration in groundwater decreases from 13 to 1.5 g l^{-1} across the transition zone.

Pore space occlusion

The precipitation of the various solid phases results in a decrease in porosity of the sandy

Table 3. *Neutralization of mineralized sulphate solution by calcium carbonate*

Solution (mmol l⁻¹)	Converted to rock mass %	Converted to rock vol. %*	Volume of precipitate at the end of experiment vol. %	pH	HCO_3^-	SO_4^{2-}	Ca	SiO_2	Al	Fe II	Fe III	Mn	U
Initial solution†				0.92	0.0	280	11.6	3.11	30.8	19.3	4.19	1.0	0.028
4.87	0.05	0.02	0.0	1.03	0.0	274	15.3	2.40	30.8	19.3	4.19	1.0	0.028
14.7	0.15	0.05	0.0	1.06	0.0	267	14.0	2.93	30.8	18.7	4.80	1.0	0.028
20.9	0.21	0.08	0.5	1.09	0.0	**263**‡	**14.0**	2.43	30.4	18.7	4.80	1.0	0.028
41.9	0.42	0.15	4.2	1.18	0.0	242	13.0	2.10	30.4	18.9	4.60	1.0	0.028
59.8	0.60	0.21	7.3	1.25	0.0	219	13.0	2.50	29.8	19.2	4.60	1.0	0.028
81.7	0.82	0.29	9.0	1.43	0.0	197	12.8	2.66	30.0	18.7	4.80	1.0	0.028
122	1.22	0.44	12.9	2.20	0.0	152	11.8	2.90	**28.8**	18.9	**4.05**	1.0	**0.026**
139	1.39	0.50	15.3	2.95	0.0	131	12.0	**2.96**	27.2	19.8	3.29	1.0	0.023
159	1.59	0.57	17.5	3.60	0.0	115	11.6	1.20	16.8	19.8	0.30	1.0	0.002
199	1.99	0.71	18.7	3.97	0.2	68.5	11.6	0.32	2.52	18.8	0.05	0.98	0.002
211	2.11	0.75	18.6	4.63	1.2	60.8	11.6	0.14	0.05	**18.3**	0.011	1.0	0.002
256	2.56	0.91	19.1	4.97	7.8	63.8	11.6	0.12	0.019	17.8	0.009	1.0	0.002
301	3.01	1.08	19.9	6.10	8.2	58.6	12.2	0.12	0.019	14.2	0.009	**0.84**	0.002
403	4.03	1.44	20.5	7.52	7.8	51.9	12.3	0.24	0.019	13.9	0.009	0.72	0.002

Constituent concentration in solution at the end of the experiment (mmol l⁻¹)

* A density value of 2.71 g cm⁻³ (from Naumov *et al.* 1971) was used in the conversion of calcite (mmol l⁻¹) to volumetric per cent content.

† Variations in Cl, K, Na and Mg content were not recorded during the course of the experiment. Concentrations of these elements in the initial solution were 8.46, 3.32, 11.3 and 46.5 mmol l⁻¹, respectively.

‡ Bold type marks the commencement of component transfer from solution to solid phase.

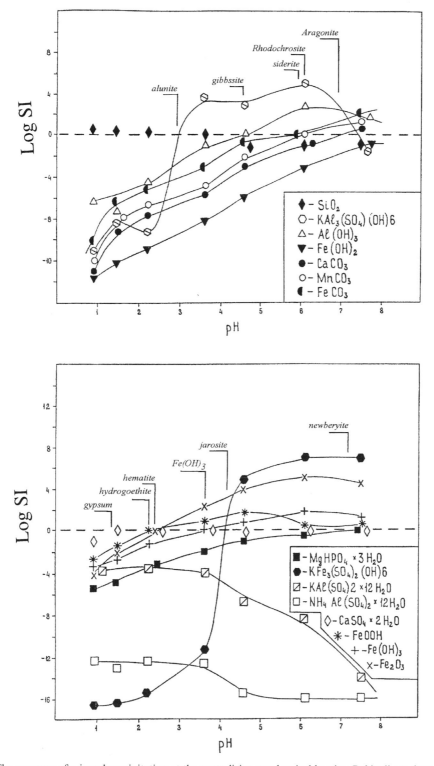

Fig. 4. The sequence of mineral precipitation at the neutralizing geochemical barrier, Bukinaii ore deposit.

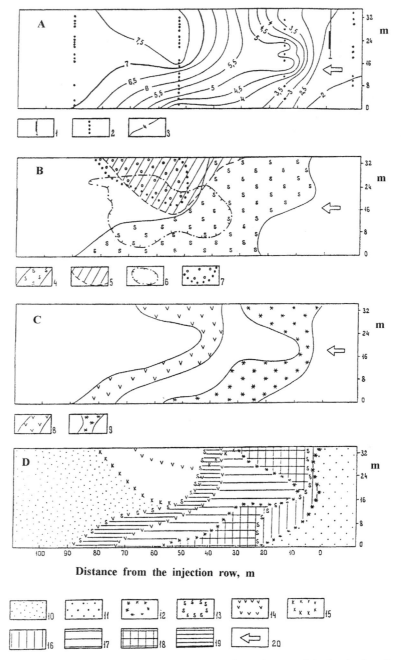

Fig. 5. Section of sulphate brine lens frontier (**a**) and geochemical zonality at the neutralizing barrier (**b, c, d**), according to data obtained six years after the cessation of mining production. (**a**) pH distribution in pore solutions: 1, injection well filter; 2, points of control borehole core sampling; 3, pH isolines. (**b**) 4, sulphatization; 5, carbonatization; 6, phosphatization of sandy deposits; 7, zone of denitrifying bacterial activity and of suspected chemical and biochemical gas generation. (**c**) Accumulation of Fe^{II} (8) and Fe^{III} (9) in the host rock. (**d**) Generalized material distribution at the neutralizing geochemical barrier between leached (10) and unaltered (11) aquifer sands, and of technogenic substances (in the solid phase): 12, Fe^{III}; 13, sulphates; 14, Fe^{II}; 15, carbonates; 16, zone of co-precipitation of Fe^{III} and Al; 17, zone of co-precipitation of Fe^{III}, Al, Ca, SO_4^{2-}, P, U, Si; 18, zone of most intense pore space occlusion by mineral precipitates; 19, zone of co-precipitation of Fe^{II}, $CaSO_4$, $CaCO_3$; 20, direction of groundwater flow.

deposits. The extent of pore space sealing depends on the neutralization capacity of the specific rock lithology and the relative proportion of micropores with a dimension of less than $3\,\mu m$, which will not transmit many colloid particles (Solodov et al. 1994b).

It was established via borehole investigation that the maximum pore space occlusion by solid phases develops in the thin, fine-grained sands of the upper layer, having the lowest permeability of the ore-bearing horizons ($<1\,m\,day^{-1}$) and neutralization capacity (about 3% mass), as well as the highest proportion of pore spaces $<3\,\mu m$ (83% of total porosity). In the upper layer, hydraulic conductivity is reduced by one to two orders of magnitude, from $c.\,10°\,m\,day^{-1}$ to 10^{-1} to $10^{-2}\,m\,day^{-1}$. Permeability inhibition only occurs to a limited extent in the medium- and coarse-grained sands of the second layer. In the other ore-bearing layers the process of pore space occlusion is developed to an intermediate extent.

Redox barrier

A free gaseous phase is observed below the top boundary of the ore-bearing layer where the sands possess their maximum carbonate content. The CO_2 and N_2 gases detected in the investigated system are also formed as a result of microbiological processes promoted by natural subsurface microflora.

It should also be noted that in natural aquifer systems, denitrifying bacteria may be active both in zones of active and depressed water exchange. Fedorov & Sergeeva (1956) suggest that this is due to the bacterial communities being able to utilize oxygen in two forms. In aerobic conditions they utilize molecular oxygen, while in anaerobic conditions they utilize nitrate species. The bacterial nitrate requirement increases by an order of magnitude in anaerobic conditions as compared with conditions when both oxygen and nitrate are present in the water. For this reason, the transition from aerobic to anaerobic conditions is favourable for nitrate reduction to molecular nitrogen. Thus, the redox barrier in the ore-bearing horizon acts as a biogeochemical barrier for the transport of nitrates.

The redox transition zone (or barrier) also attenuates sulphates and some heavy metals. Sulphate-reducing bacteria, active at neutral pH and at redox potential less than $+150\,mV$, occur on the distal side of the neutralization barrier, reducing sulphate to sulphide and resulting in the precipitation of chalcophile elements.

The sharp change from aerobic to anaerobic conditions, with a decrease of redox potential from $+450\,mV$ to as low as $-330\,mV$ develops over a short distance (of between 62 and 100 m). This redox front coincides with the frontier of the sulphate brine lens and also with the pH transition from $c.\,1$–2 to 7.4. The redox front corresponds to a reducing geochemical barrier at which precipitation of Se, As, U, Mo and V and a number of other elements, whose solubility depends on oxidation state, occurs. Zones of sulphate reduction and denitrification typically have a discrete spatial relationship to this redox barrier; in these zones precipitation of the chalcophile sulphides, and decreases in NH_4^+ (from 40 to $2.5\,mg\,l^{-1}$) and NO_3^- (from 85 to $5\,mg\,l^{-1}$) concentrations, respectively, are observed.

Combined effects of pH, redox and hydraulic barriers

The combined influence of the processes of formation of solid and gaseous phases, by neutralization and redox reactions, which approximately coincide spatially in the upper layer, leads to a marked attenuation of the most hazardous macro- and microcomponents. Pore-sealing at the neutralization barrier by mineral precipitates leads to the development of a hydraulic barrier.

The neutralization and redox barriers developed in the aquifer are mobile in space, as the lens is displaced by regional groundwater flow. However, the rate of barrier movement is less than that of the groundwater flow. Therefore, one would expect the extent of the sulphate brine lens to be slowly 'squeezed' under the effects of neutralization until it eventually completely disappears. The quantity of accumulated solid phase precipitate at the geochemical barriers will gradually increase and, consequently, as a result of occlusion of pore space, the flow rate and rate of attenuation of hazardous substances will decelerate. It has been calculated that, assuming an initial lens of width 400 m, with the neutralization and redox barriers being displaced by regional groundwater flow, the complete attenuation of the brine lens would take some 80 years and the resulting zone containing solid phase precipitates would have a width half that of the initial lens ($c.\,200\,m$).

Hypothetical attenuation of LRW components at the injection disposal site at the Siberian Chemical Complex, Tomsk Oblast'

It has already been observed that there is an analogy between the attenuation of hazardous components at neutralization, redox and hydraulic barriers which develop in the vicinity of underground uranium leaching operations and processes which might be expected in the subsurface at LRW injection locations, such as that of the Siberian Chemical Industrial Complex in Tomsk Oblast'. The analogy is not perfect: sulphate is the dominating anion at the uranium mining site, while nitrate is dominant in the LRW stream (although sulphate becomes increasingly important at intermediate distances from the injection wells as nitrate is broken down and aquifer sulphide is oxidized). According to Grabovnikov (1993) and Rybalchenko et al. (1994), the injection horizon at the Tomsk site (conventionally referred to as aquifer horizon II; see Figure 3 in Lgotin & Makushin 1998) is composed of continental greyish sandy deposits of the Senonian stage of the Upper Cretaceous. It is separated from the silty-clayey schists of the Palaeozoic basement by c. 140 m of underlying sediments which include low-permeability kaolinitic clays, weathering crusts of Triassic–Jurassic age, multi-coloured Gotteriv-Barremian clays, c. 18–48 m of Aptian sands (aquifer horizon I) and Albian kaolinitic clays (20-35 m thick).

Overlying the injection horizon (aquifer II), five further water-bearing horizons (III, IV, IVa, V, VI) are present; they are separated by low-permeability, dominantly clayey layers. The lateral continuity of these clayey layers is the subject of some debate, as is the hydraulic significance of geophysically detected lineaments traversing the area (Tomsk Oblast' Committee for Ecology 1994). The total thickness of the sedimentary rock sequence, overlying aquifer II, is about 350 m (Fig. 6). Although high level wastes are injected into aquifer horizon II, low- and medium-level wastes are additionally injected into horizons II and III. Investigation of the geochemical and hydrogeological conditions prevailing in the injection horizons is currently being carried out, involving several organizations. The techniques used in these investigations, including descriptions of sampling techniques and geochemical modelling codes, are discussed by Solodov et al. (1993, 1994a, b).

The injection sites are located at the south-eastern part of the Obsky artesian basin. The discharge area of the lower water-bearing complex, which includes horizons I and II, is at the confluence of the Ob' and Tom' rivers. The natural flow of groundwater in the region of the disposal sites is dominantly WSW, in the direction to the Tom', with a flow rate of 3–5 m year^{-1}. The hydraulic gradient is estimated as 0.0009. In the area enclosed by the Ob' and Tom' rivers,

south of their confluence, groundwater for public supply is abstracted from Palaeogene aquifers of the upper horizons (dominantly horizon IVa), from long lines of abstraction wells parallel to the rivers, on the inner side of the 'V' formed by the confluence (see Figure 2b in Lgotin & Makushin 1998).

Aquifer horizon II comprises greyish quartz–feldspar medium- to fine-grained sands with varying clay content. The percentage composition is typically as follows:

- quartz: 70–80%,
- feldspars (orthoclase, microcline, plagioclases): 0–15%,
- micas (muscovite, phlogopite, vermiculite) and hydromicas: 3–5%,
- clay minerals of the kaolinite (3–5%) and montmorillonite (8–10%) groups,
- carbonates and sulphides are present in minor quantities (<1%).

Organic substances are also present in the aquifer horizon. The total thickness of the horizon is 30–50 m, of which c. 13–30 m is formed of silty or clayey layers. As the standard sand porosity is 35%, the effective porosity of the entire horizon is 5–14%. The hydraulic conductivity is estimated as 0.7–0.9 m day^{-1}, while the hydraulic head at the top of the horizon is 300–320 m (relative to the top of the horizon). It is evident from this information, that horizon II is not hydrogeologically dissimilar to the ore-bearing layer I of the Bukinaii uranium deposit.

High-level LRW is filtered prior to injection in order to separate out colloid particles. The LRW is then doped with complexation agents to prevent the precipitation of hydrolysable elements in the near-well-screen zone of the injection boreholes. An optimum concentration of heat-generating radionuclides is maintained by dilution of the solution, and the acidity is maintained at pH 2.

The injected solutions are basically sodium nitrate brines with a salinity up to 220 g l^{-1}. The solutions contain macroquantities (from hundreds of milligrams up to tens of grams per litre) of nitrates, acetates, oxalates, sulphates, silicic acid, detergents, sodium, iron, aluminum, chromium, manganese, nickel and other components.

Four boreholes are used at the site for LRW injection. Injection was performed 1–3 times per year as quanta of about 1–2 m^3, either by free inflow or under an excess pressure of 0.5–0.6 atmospheres. Solutions of nitric acid (pH 1–2) were injected prior to LRW injection in order to create an acid subsurface environment and to prevent hydroxide precipitation. Nitric

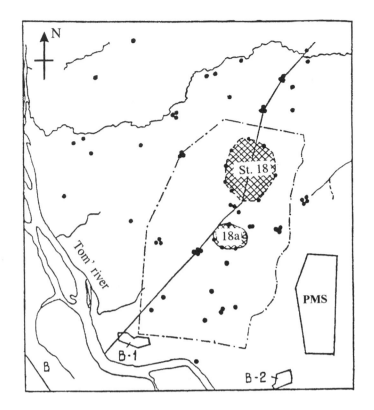

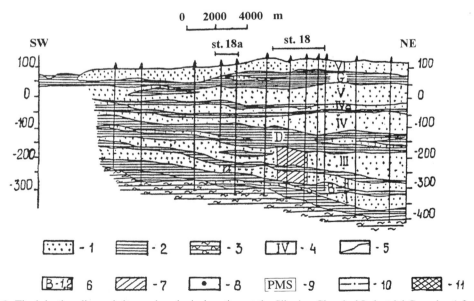

Fig. 6. The injection disposal sites and geological section at the Siberian Chemical Industrial Complex (after Rybalchenko *et al.* 1994). Station 18a is the site of disposal of high- and medium-level LRW. 1, permeable rocks; 2, low-permeability rocks; 3, low-permeability rocks of the Palaeozoic basement; 4, conventional aquifer index; 5, line of cross-section; 6, underground water abstraction from aquifers overlying the LRW injection aquifer; 7, portion of the injection aquifer occupied by disposed LRW; 8, observation wells; 9, Tomsk petrochemical plant; 10, Tomsk-7 official boundary; 11, disposal sites.

acid is also injected post-LRW-injection in order to push the 'slug' of LRW away from the well-screen of the borehole. As a result of this process, a lens of technogenic solutions, with a diameter of 1.1–1.5 km, has formed in the injection horizon (as of 1993). The large volume of this lens is due to the injection of large quantities of non-radioactive acidic solutions to promote the migration of waste away from the wells, rather than to the small volumes of injected high-level radioactive waste itself. This lens can be subdivided into three geochemical zones. Laboratory studies and geochemical modelling can shed light on the types of process likely to be occurring in each of the zones. The conceptual model indicated by such studies is described in the following paragraphs.

The technogenic solutions in the central zone of the lens, within a few tens of metres of the injection boreholes, have a temperature in the range of 80–150 °C, are also highly acidic (pH 1–2), and lead to intense radiolysis of the water, dissolved components and host rock minerals. The combined effect of these aggressive conditions is the intense decomposition of rock minerals and the transition of matter into the dissolved state. Computer-based physico-chemical modelling indicates that the solubility of quartz (6 mg l^{-1} at 15 °C) increases to c. 26–115 mg l^{-1} at a temperature of 80–150 °C (although these estimates must remain approximate, as modelling of brines would be preferable using a pitzer-type approach for which high-temperature thermodynamic data are unavailable). The highly acid conditions of the central zone will also cause intense kaolinization of feldspars, micas and hydromicas, as well as the dissolution of montmorillonite. The high radioactivity of the technogenic solutions leads to the destruction of crystal lattices of certain rock-forming minerals, significantly increasing their solubility. Nitrates and organic compounds incorporated in the technogenic solutions are radiolytically, and also microbiologically, destroyed, with the generation of N_2, CO_2, O_2, H_2 and CH_4. The spontaneous emission of these gases is documented in the injection wells. The radiogenic oxygen produced by the radiolytic breakdown of nitrate creates somewhat oxidizing conditions in the aquifer, leading to the oxidation of sulphide minerals in the aquifer matrix and the increasing dominance of sulphate in the groundwater.

The boundary between the central and the intermediate zone is marked by a relatively sharp decrease in temperature. In the intermediate zone, a partial neutralization of nitric acid occurs at a relatively moderate temperature, although the groundwater environment remains acid with a pH of around 2 to 4. This zone acts as a transitional zone for substances leached from the central zone and transported by the injection solutions. However, it would be expected that significant precipitation of many dissolved substances takes place in the intermediate zone due to the temperature decrease and to sorption. In this zone, the temperature and salinity of the groundwater return to ranges which can be more readily tackled by standard geochemical modelling techniques, not requiring a pitzer approach.

A marginal zone occurs at the periphery of the nitrate brine lens. Here, neutralization of the acid is completed, and the solutions' pH values increase from 4 to 7–7.5. A neutralizing geochemical barrier develops, beyond which it is likely that a reducing redox barrier has developed. Dissolved radionuclides and other stable components are transferred to the solid phase at these barriers. One can also expect the development of a hydraulic barrier with a sharp decrease of hydraulic conductivity, due to porosity clogging with solid precipitates. The exact nature of solid-phase precipitates at Tomsk-7 is still under investigation, although some similarity with those observed at Bukinaii might be expected. At such distances from the injection wells, hydrogen carbonate is typically the dominant anion.

The transferral of radionuclides from the dissolved to the solid phase is likely to occur in all zones. Radionuclide sorption by kaolinite is believed to be the prevailing process in the central zone, whereas, in the intermediate and marginal zones, co-precipitation and sorption on the surface of newly formed minerals are likely to be the dominating radionuclide attenuation reactions.

Thus, 'solid-phase immobilization' of LRW will take place at the periphery of the brine lens due to the accumulation of its components at the neutralizing and redox barriers. In the case of the development of a hydraulic barrier, these effects will be amplified by a hydraulic containment effect. A further important factor for attenuation of stable hazardous pollutants (nitrates, acetates, oxalates, etc.) is their radiolytic destruction.

The combined effect of these geochemical and hydraulic barriers is expected to be the significant attenuation of the majority of contaminant parameters. The actual results of wider monitoring of the contaminant plume from the injection wells are discussed in more detail by Lgotin & Makushin (1998). The author has consciously not discussed the behaviour of specific radionuclides at the redox and neutralization barriers. It is believed, however, that the barriers would be effective at attenuating many of the most toxic radionuclides; for example, plutonium and uranium exhibit drastically reduced solubility under reducing and alkaline conditions. It is, however, recognized that the geochemical barriers would be less effective for more soluble nuclides such as those of iodine or technetium (VII).

Conclusion

The author's long-term investigations of the subsurface geochemistry of uranium leaching mines, based on exogenous uranium deposits in sandstone aquifers, demonstrates that residual sulphate solutions in the ore-bearing horizons, after the cessation of mining operations, are subject to geochemical processes resulting in the retardation and precipitation of both contaminants derived from the ore and host rock as well as from the injected leachants. These processes have much in common with those which develop in the sandy aquifers typically used for disposal of high-level sodium-nitrate-based LRW. This conclusion is supported by a comparison between the geological and hydrogeological situations present in the stratiform infiltration-type uranium deposit at Bukinaii (Kysyl-Kum, Uzbekistan), mined by sulphate-based underground leaching, and the underground disposal site for high-level LRW at the Siberian Chemical Industrial Complex (Tomsk Oblast', Russia).

At Bukinaii, the injection of sulphuric acid leachants has resulted in intense redistribution of the geochemical components of the ore-bearing aquifer, and has had a major impact on the geological structure of the deposit, and its hydrological, lithological and geochemical characteristics. Long-term observation and specialized experimental studies have allowed the characterization of the composition of the sulphate-brine lens resulting from the injection of acid into the ore-bearing quartzofeldspathic clayey sands. Observations have demonstrated the neutralization of the sulphuric acid concomitant with the displacement of residual sulphuric solutions into unaltered parts of the ore-bearing horizon by the regional groundwater flow. As a result of this process, neutralizing, redox and hydraulic barriers develop at the frontier of the contaminated lens. Technogenic solutions are subject to significant attenuation as they encounter these barriers, manifested by the precipitation of solid phases. These precipitates occlude pore space and cause a sharp decrease in permeability. It is estimated that the sulphate solution lens, with its initial width of 400 m, will be almost entirely attenuated within 80 years, resulting in a volume of aquifer containing solid phase precipitates of diameter $c. 200$ m. Research is currently progressing on the crucial problem of the degree to which such secondary phases may be resolubilized and eventually remobilized in the future.

Based on the analogy between the attenuation processes affecting technogenic solutions at geochemical barriers at the sites of uranium leaching and sites of LRW injection disposal, one can make a plausible prediction for radionuclide migration and attenuation at the Siberian Chemical Industrial Complex injection site, near Tomsk. It is regarded as highly likely that neutralizing and redox chemical barriers would develop at the frontier of the LRW-contaminated lens. These barriers would contribute to radionuclide immobilization in the solid phase, resulting in significant contaminant attenuation. One would also predict the development of a hydraulic barrier at the zone of most intense precipitation of solid phases. Furthermore, radiolytic destruction of non-radioactive contaminants (nitrates, oxalate, acetate, etc.) will contribute to contaminant attenuation.

Thus, the reliability and safety of LRW injection disposal at Russian sites (with suitable characteristics for implementation of this option) is supported not only by an analysis of regional and local hydrogeological conditions (Rybalchenko et al. 1994), but also by analyses of geochemical processes providing for retardation and attenuation of radionuclide migration. Ultimately, one might predict that such aquifers are, to a large extent, 'self-cleaning' with respect to injected solutions of radionuclides and other contaminants, due to their incorporation into solid phases. As at Bukinaii, the long-term stability of the solid phases and their potential for remobilization is clearly an area where considerable research effort needs to be expended.

The analysis of processes at LRW injection sites is largely based on analogy with uranium mining activities. Confirmatory field investigations at LRW sites would lend added support to the analysis presented in this paper.

The author expresses his gratitude to the Director of the Department of Rare Metals Geology and Radiogeoecology (IGEM RAC), V. I. Velichkin, for his constant interest in the investigations described in this paper and for his support. Discussions and encouragement from D. Banks are gratefully acknowledged.

References

DROZHKO, E. G., SHARALAPOV, V. I. & POSOKOV, A. K. 1993. History, contamination and monitoring of water bodies at the P/A 'Mayak'. *In*: Ahlstroem, P. E. *et al.* (eds) *Proceedings of 1993 International Conference on Nuclear Waste Management and Environmental Remediation, Vol. 2: High-level Radioactive Waste and Spent Fuel Management.* The American Society of Mechanical Engineers, New York.

EDWARDS, R. 1996. Hot ice could contaminate fish. *New Scientist*, **2019**, 7.

FEDOROV, M. V. & SERGEEVA, R. V. 1956. Cyanide influence on nitrate reduction by denitrifying bacteria. *Proceedings USSR Academy of Sciences (Doklady Akademii Nauk SSSR)*, **108**(6), 1182 [in Russian].

GALLOWAY, W. E. 1978. Uranium mineralization in a coastal-plain fluvial aquifer system: Catahoula Formation, Texas. *Economic Geology*, **73**, 1655–1676.

GRABOVNIKOV, V. A. (ed.) 1993. *Hydrogeological Investigations for Justification of the Underground Injection Disposal of Industrial Wastes.* Nedra, Moscow [in Russian].

ILYINSKIKH, N. N., ADAM, A. M., NOVITSKII, V. V., ILYINSKIKH, E. N., ILYIN, S. Y. & KUDRIAVTSEV, D. P. 1995. *Mutagenic Consequences of Radiating Pollution of Siberia.* Siberian Medical University, Tomsk [in Russian with English summaries].

LGOTIN, V. & MAKUSHIN, U. 1998. Groundwater monitoring to assess the influence of injection of liquid radioactive waste on the Tomsk public groundwater supply, Western Siberia, Russia. *This volume.*

LISITSYN, A. K., MARKOV, S. N. & POPONINA, G. Yu. 1994. The Dalmatovskoe deposit at the Trans-Ural region as an example of a geological site suitable for safe disposal of radioactive wastes. *Geology of Ore Deposits*, **35**(4), 360–368 [in Russian].

NAUMOV, G. B., RYDJENKO, B. N. & KHODAKOVSKY, I. L. 1971. *Handbook of Thermodynamic Species.* Atomizdat, Moscow [in Russian].

PINNEKER, E. V. (ed.) 1991. *Resources of Fresh and Low-Salinity Underground Waters in the Southern Part of the West-Siberian Artesian Basin.* Nedra, Moscow [in Russian].

RAFALSKY, R. P. 1978. Chemistry of metal leaching. *Atomnaya Energiya (Moscow)*, **44**(3), 1–35 [in Russian].

ROGOVSKAYA, N. V. 1991. *Regularities in the Organization of the Underground Hydrosphere on the Platform Regions.* Nauka, Moscow [in Russian].

RYBALCHENKO, A. I., PIMENOV, M. K., KOSTIN, P. P. *et al.* 1994. *Deep Disposal of Liquid Radioactive Wastes.* IzdAT, Moscow.

SOLODOV, I. N., KIREEV, A. M., ZELENOVA, O. I. & UMRIKIN, V. A. 1993. Technogenic oxidation changes in reduced uranium-bearing sandy deposits of marine genesis. *Lithology and Mineral Resources*, **28**(6), 84–96 [in Russian with English translation in 1994].

——, VELICHKIN, V. I., ZOTOV, A. V., KOCHKIN, B. T., DROZHKO, E. G., GLAGOLEV, A. V. & SKOKOV, A. N. 1994a. *Distribution and geochemistry of contaminated subsurface waters in fissured volcanogenic bedrocks of the Lake Karachai area, Chelyabinsk, southern Urals.* Lawrence Berkeley Laboratory, University of California (Earth Sciences Division) Report No. LBL-36780/UC-603.

——, SHUGINA, G. A. & ZELENOVA, O. I. 1994b. Man-made (technogeneous) geochemical barriers in the mineralized horizons of hydrogene uranium deposits. *Geochemistry International*, **31**(10), 110–127.

TOMSK OBLAST' STATE COMMITTEE FOR ECOLOGY 1994. *The state of the environment and the health of the population in the zone of influence of the Siberian Chemical Combinate.* Ministry of Environmental Protection and Natural Resources/Tomsk Oblast' State Committee for Ecology and Natural Resources, Tomsk.

—— 1995. *OBZOR: Ecological condition, use of natural resources and environmental protection: Tomsk Oblast' 1994.* Tomsk Oblast' State Committee for Ecology and Natural Resources, Tomsk.

ZUEV, V. A. & SHVARTSEV, S. L. 1995. Ecological-hydrogeological evaluation of radiation–chemical pollution of the Naumovskaya zone (Tomsk area). *Abstracts of the International Conference on Fundamental and Applied Problems of Environmental Protection*, 12–16 September 1995, Tomsk State University, Vol. **4**, 80–82.

Section 7: Groundwater pollution by exotic organics: acid tars, pesticides and phenols

Tars are generally fairly refractory and consist largely of hydrocarbons of high molecular weight which are poorly soluble, poorly biodegradable and of low volatility. Their behaviour over long periods in the subsurface is not, however, especially well-understood. The situation is even more complicated when the tars in question are acidic tars from re-refining processes, mixed with concentrated sulphuric acid and disposed of in conjunction with foundry wastes and oil-saturated bentonites. In this case the potential arises for the migration of both inorganic and organic contaminants and the mobilisation of metals from the aquifer matrix in a low pH environment. This is the situation described by **Banks** *et al.* Against the background of this complex potential contaminant source, they examine a limited selection of chemical parameters to identify potential groundwater flow pathways in the complex, mined Coal Measures aquifer system of northeastern England. They also attempt to illustrate how a relatively limited amount of contaminant data, coupled with actual and assumed knowledge of flow pathways, can be utilised in the selection procedure for a remedial option.

Gore & Campbell introduce us to a site in the West Midlands of England where phenolic wastes had previously been dumped in a quarry in low-permeability marls. The wastes were happy where they were and had no plans to go anywhere ... until it was decided to build a large road embankment across the site. It was feared that the overloading caused by the embankment would squeeze contaminated pore waters from the waste, which might then overflow into minor surficial aquifers and surface waters. The authors describe how this situation was monitored and alleviated during the embankment construction operation. An unusual, but interesting problem ...

The final two papers of the book concern another contaminant type which is a cause of major concern, namely pesticides. **Chilton** and colleagues introduce us to this topic and discuss the behaviour of pesticides in the groundwater environment and also touch upon the climatic controls on pesticide occurrence and behaviour. The paper by **Sweeney** *et al.* describes every hydrogeologist's nightmare (or possibly dream scenario if you work in the private consultancy sector!). The site in question, a quarry, was licensed for the disposal of pesticide wastes largely derived from a major, but nameless, British agrochemicals concern. The quarry is unfortunately in a karstified Jurassic limestone and in the near vicinity of several major groundwater pumping stations. The authors describe possible management techniques for the situation, maybe in the future involving the hydraulic manipulation of the aquifer unit to minimise the impact on the existing potential receptors.

This is the final chapter of the book and hence the final editorial intrusion. As you worked your way through the case studies, you will have become increasingly aware that contaminant hydrogeology is still a very young science. Just as considerable leaps forward have been achieved in recent years in understanding groundwater pollution, we are nevertheless a long way from achieving a sufficient understanding and definition of the problems confronting us to allow their resolution at a practical level.

The longevity of groundwater pollution and increasing pressure on resources are raising the stakes. It is clear that many exciting challenges lie ahead for the hydrogeologist, not least of which will be the fundamental issue of quantifying the main controls on contaminant behaviour. We also need to balance the real needs for groundwater protection and clean-up with objective criteria defining the significance of likely impacts within the framework of emerging risk-based methodologies. Treatment of uncertainty will always remain an important aspect of groundwater science, and probabilistic methods are likely to be of prime importance for understanding groundwater systems, just as they are in helping us to define cleanup goals.

The experience and understanding gained in recent years from case studies (including those found in these pages) and ongoing monitoring provide us with the critical results of the system response to contaminant inputs. Greater successes can be predicted for the future as this feedback loop becomes more firmly established. We hope that this book becomes the first of a series documenting this progression, and that you may be inspired to contribute to the next volume.

We hope you have found the book enjoyable and informative. We've looked at MTBE, solvents and pesticides, and are currently taking bets on the next 'hot' groundwater pollutant topic to come into fashion. Our money is on artificial oestrogens or possibly viral contaminants, and the use of DNA as a pollutant tracer. You will have to wait for the next Special Publication on 'Groundwater Contaminants and their Migration' to find out.

Contaminant migration from disposal of acid tar wastes in fractured Coal Measures strata, southern Derbyshire

D. BANKS[1,2], N. L. NESBIT[3], T. FIRTH[1,4] & S. POWER[5,6]

[1] *Scott Wilson Kirkpatrick, Bayheath House, Rose Hill West, Chesterfield,*
Derbyshire S40 1JF, UK.
[2] *Present address: Norges Geologiske Undersøkelse, Postboks 3006-Lade,*
N7002 Trondheim, Norway.
[3] *Aspinwall & Company, Walford Manor, Baschurch, Shrewsbury, Shropshire SY4 2HH, UK.*
[4] *Present address: Rendel Geotechnics, 61 Southwark Street, Southwark,*
London SE1 1SA, UK.
[5] *Section for Engineering Geology, Department of Geology, Royal School of Mines,*
Imperial College of Science, Technology and Medicine, Prince Consort Road,
London SW7 2BP, UK.
[6] *Present address: Mott Connell Ltd, 12th Floor, Sun Hung Kai Centre, 30 Harbour Road,*
Wanchai, Hong Kong

Abstract. In excess of 60 000 m^3 of acidic tar wastes from a lubricating oil factory have been disposed in old clay pits at a site on Coal Measures strata near Belper, southern Derbyshire, UK. A site investigation was carried out to assess the nature and volume of the waste, the extent of contaminant migration and possible remediation options. The wastes have migrated as free phase, gravity-driven flows of semi-fluid tar. These tar flows have led to contamination of on-site ponds by hydrocarbon emulsions and dissolved phase contaminants. Limited groundwater contamination by dissolved phase contaminants has also occurred. At present, no impact on off-site surface water receptors has been detected and no risk to any water abstraction has been identified. The site investigation has confirmed the exceedingly complex nature of the Coal Measures sequence. The Coal Measures exhibit some aspects of a multi-layer aquifer system, dominant fracture flow mechanisms, strong anisotropy governed by fracture orientation and a heterogeneous and discontinuous distribution of groundwater heads. Indications of strong vertical head gradients, possibly related to under-drainage by mine workings, were detected.

Introduction

The study site is located near the town of Belper in southern Derbyshire and comprises a number of old clay pits in Lower Coal Measures strata which have been used for the disposal of acid tars – a waste product from the re-refining of used lubricating oils. Derbyshire County Council (DCC) commissioned an investigation with the following objectives:

- define the area and volume of these pits;
- define the extent to which they have contaminated the surrounding soils and groundwater;
- assess the potential for migration of contamination off-site;
- recommend alternatives for remediation of soil and groundwater.

This information was required as part of a proposal to redevelop the site. DCC commissioned Aspinwall & Company and a team of associated consultants to carry out the investigation. Scott Wilson Kirkpatrick was employed to manage the site investigation and assist in interpretation. The site investigation was executed by Wimpey Environmental Limited, while a geophysical survey was carried out by the British Geological Survey. Funding for the project has been provided by English Partnerships (formerly the Department of the Environment) through a Derelict Land Grant.

Objective of this paper

The objective of this paper is not to provide an in-depth case analysis of all the data resulting from the study of the tar pits. In particular, the paper does not discuss details of chemical analyses of contaminated soil, geotechnical properties of the site, nor the detailed hydrochemistry of the groundwater at the site. Indeed, due to

BANKS, D., NESBIT, N. L., FIRTH, T. & POWER, S. 1998. Contaminant migration from disposal of acid tar wastes in fractured Coal Measures strata, southern Derbyshire. *In*: MATHER, J., BANKS, D., DUMPLETON, S. & FERMOR, M. (eds) *Groundwater Contaminants and their Migration*. Geological Society, London, Special Publications, **128**, 283–311.

the fact that there was no evidence of significant off-site migration of contaminants in ground- or surface-water, the intensity of hydrochemical investigation was not as great as would have been desirable from a research-oriented study. Nevertheless, the study has provided useful information on the hydrogeology and, in particular, the contaminant flow pathways within the Coal Measures aquifer system. It is this aspect on which the paper focuses, although it also aims to place such information in the context of the site as a whole, and the possible remedial options for clean-up. To support this objective the paper draws upon the distribution of a limited range of inorganic chemical parameters. The sampling of organic parameters was not intense enough to allow an elucidation of groundwater flow characteristics to be based on the distribution of such parameters.

Origin of acid tar wastes

Acid tar wastes arise from several different industrial refining processes, including benzole refining, oil re-refining and white oil production (Singleton 1987). The literature on the environmental impact of disposal of acid tar wastes is not large, but studies of varying extent have been carried out in the UK (Singleton 1987) at Ravenfield Quarry, South Yorkshire (Smith & Bromley 1982); Whitehall Road Tips, West Yorkshire; Hoole Bank, Cheshire; Llwyneinion Brickworks, near Wrexham, Clwyd, and internationally at Dubova, Czechoslovakia and Incukalns, Latvia (Latvian Geological Survey, pers. comm.). Extensive investigations have also been carried out at Sydney Tar Ponds in Canada, although the wastes here are mainly derived from coking processes in the steel industry (Environment Canada 1987–1990, 1989).

The acid tar wastes at the study site were historically produced by a local lubricating oil manufacturer, using a multi-stage re-refining process to remove contaminants from used oils. This process included washing the oils with concentrated sulphuric acid (Singleton 1987). The tar pits are also reported to contain oil-saturated clays (Fuller's Earth) and heavy-metal-containing foundry sand wastes.

The wastes produced had very a low pH (often <1) with up to 30% H_2SO_4. Class compositional analysis of organic components indicated the tar to consist of around 67% polar, tarry type resins of very high molecular weight, but with a significant (29%) component of saturated, high distillate hydrocarbon, possibly derived from weathering of the tar. Simulated distillation of the solvent extract from the tar indicated it to be predominantly composed of components with a higher boiling point than C_{23} *n*-alkane.

Additionally, free, acidic oils are encountered on the site. These may be associated with excessive use of sulphuric acid in the re-refining process, but may also be derived from weathering of the tars. The oil typically consists of 87% saturated hydrocarbon with boiling points in the C_{16}–C_{23} range, 7% aromatics and 6% polar resin.

The Fuller's Earth materials indicate the cyclohexane-extractable material adsorbed in the Fuller's Earth to consist of 85% saturated hydrocarbons, 7% polar resins and 7% aromatics. The hydrocarbons are dominantly in the C_{16}–C_{23} range. Analytical results are detailed in Table 1.

The site

Location

The study site (Fig. 1) is split into two distinct areas, the larger to the east (approximately 14 ha), and the smaller to the west (approximately 6 ha) of a major N–S dual-carriageway trunk road.

The northern part of the area west of the trunk road is dominated by an exposed sandstone bedding plane slope, dipping approximately 18° E. It is likely that this was exposed by the excavation of the overlying clay (weathered mudstone and shale) down to the sandstone. This sandstone is referred to as the Lower Sandstone. The southern part of this western area is covered by rough pasture and falls gently in a southerly to southwesterly direction.

The area east of the trunk road contains the main areas of acid tar. Part of the eastern boundary of the site is marked by a small stream. To the north of the site are open fields in use for crops and grazing. To the south of the site are fields and residential housing. The majority of the site to the east of the trunk road has a very uneven topography, but overall there is a fall from west to east. Three water-filled lakes or 'lagoons' (named the 'southern', 'middle' and 'northern' respectively) and some smaller ponds, occur east of the trunk road. The lagoons were created by excavation of clay. Between the 'middle' lagoon and the trunk road, sandstone bedding planes are exposed, dipping to the east at a shallow angle, again believed to have been exposed by removal of the overlying clay.

In total, 13 acid tar disposal pits are known to have existed; those to the west of the trunk road

Table 1. *Analytical results for soils and wastes found on the study site. All units (except pH, CEM, CV and EC are in milligrams per kilogram*

Type	Supposed 'clean' soils		Bentonite from pits		White acid material Borehole S2	Contaminated mudstone Trial pit TP32 at 1.1 m	Free tar product	
	Trial pit P9 at 2.1 m	Trial pit TP3 at 1.8 m	Trial pit TP11 at 2.2 m	Trial pit TP17 at 0.8 m			Trial pit TP49 at 0.0 m	Borehole S6 at 3.6 m
pH	7.6	6.8	6.0	5.7	1.4	5.8	0.8	3.9
CEM %	0.08	0.03	19. 6		2.69	2.37	12.36	1.77
EC				63.3			45 000	
Fe	37 200		3470	33	28 800			
Acid sol. SO_4^{2-}	994	697	56 600	35 500	221 000	2030	130 000	36 400
Cl^-				63.3		<12.5	71.6	20.3
Al					16 500			
Mn					537			
As	27		8		8			
Cd	<1		<1		<1			
Cr	15		8		31			
Cu	46		54		72			
Hg	<2		<2		<2			
Ni	33		<2		24			
Pb	69		46		96			
Se	<0.5		<0.5					
Zn	70		42		697			
Co					12			
Mo					<2			
V					54			
Phenol	0.6	<0.5	9.6	27.5	1.6	3.4	4.2	19.1
PAH	26	<10	442	636	267	179	196	1210
Min. oil				259 000		18 400	174 000	
Aromatics				22 040			22 400	
NSO				19 930			407 000	
TEM				302 000			605 000	
CV				15.98				
PCBs						2		
Cyanide (total)				<10	<10	<10	<10	<10
Sulphide				29	<10	<10	<10	<10
Ammonia				9	105	28	16	21
Thiocyanate				<5			9	<5

CEM = cyclohexane extractable matter; Phenol, total monohydric phenols; PAH, polynuclear aromatic hydrocarbons (total); Min. oil, total mineral oil; NSO, heterocyclic polar resins (including nitrogen, sulphur and oxygen-containing species); TEM, toluene extractable matter; CV, calorific value ($MJ kg^{-1}$); PCB, polychlorinated biphenyls; EC, electrical conductivity ($\mu S cm^{-1}$).

are referred to as tar pits A–H, and those to the east as tar pits 1–4 and 1A (see Fig. 1).

History of the site

This area of Derbyshire has been mined, principally for coal and clay, for at least several hundred years. Indeed, old maps of the site show at least three vertical shafts, for the circulation of air and the removal of coal and spoil (Fig. 1).

The coal seams that, according to British Coal, have been worked in the immediate vicinity of the site are summarized in Table 2 and shown in Fig. 2. There is a possibility that unrecorded workings may exist below the site, particularly at shallow depth. The Mickley Thin Seam outcrops at the eastern edge of the site and recorded workings on this seam are at least *c.* 50 m farther east.

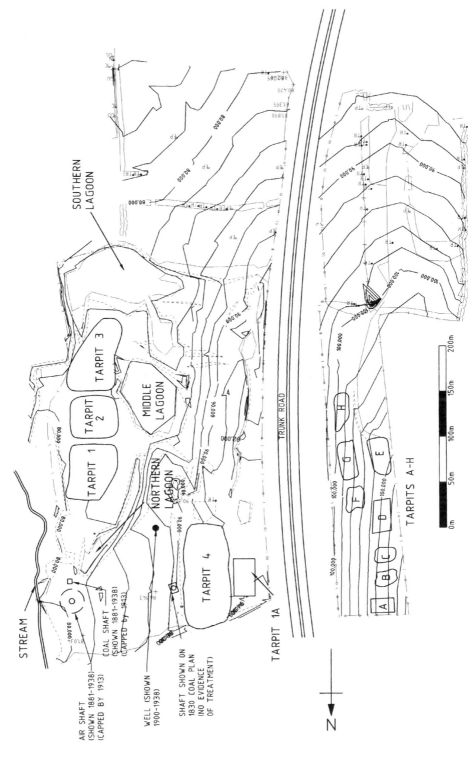

Fig. 1. Overview plan of the study site. Contours in mOD.

Table 2. *Worked coal seams at the study site*

Seam	Approximate depth (m)	Type of workings	Period of working
Mickley Thin (or Upper Brampton)	0	Pillar and stall	Up to mid-1950s (east of the site)
Kilburn	75	Uncertain	1830–1935 Opencast immediately west of site
Alton	235	Longwall	1959–1965
Belperlawn	265	Longwall	1967

DCC granted planning permission for various opencast pipe- and brick-clay workings from the materials overlying the Morley Muck coal in April 1961. The clays were worked down dip using dragline techniques, leaving sequences of N–S striking high walls. Permission to continue clay working after 1980 was refused in order to allow reinstatement of the site to commence and to prevent damage to the existing tar pits (see below) by ongoing extraction works.

Acid tar disposal and subsequent activities

Belper Rural District Council granted planning permission on 11 August 1972 to tip tar wastes and ash in quarried clay-pit voids on the site. It is known that other materials, including oil-saturated Fuller's Earth and foundry sand, were additionally deposited with the tar. Attempts to stiffen, stabilize and/or neutralize

the tar have been made by addition of lime and ashes, but these appear to have been unsuccessful.

East of the trunk road, acid tar disposal continued up to August 1977 when the County Surveyor refused a licence to continue tipping under the Control of Pollution Act. The tar appears to have remained mobile and deposits of tar are currently visible at numerous localities around the site as isolated up-wellings and as thin flows up to 40 m in length and 10–15 m in width (Fig. 3a,b). In some locations the tar appears to have flowed through soils, and even bedrock fractures, to well up several metres away from the point of deposition. 'Glacier'-like tongues of tar can be seen entering the middle lagoon, releasing brownish emulsion-like contamination into the water. Tar up-wellings also appear to contaminate the southern lagoon where a leachate containing floating non-aqueous phase, emulsion phase and (presumably) dissolved

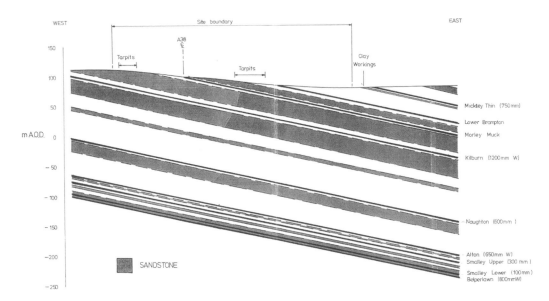

Fig. 2. Geological cross-section through the study site, based on available geological maps. The horizontal scale is equal to the vertical scale.

(a)   (b)

Fig. 3. Photographs showing (**a**) tongue of tar welling up from near tar pit 2 and entering the middle lagoon (width of flow is around 60 cm) and (**b**) flow of tar along a fracture in Coal Measures strata.

phase contaminants can be seen seeping out of the soil at the side of the pond.

To the west of the trunk road, eight pits were used for tipping between 1972 and 1976. In 1976, however, the road cutting was widened to create a deeper cut for the road. This led to instability of the west side of the cutting and, consequently, these eight tar pits were disturbed. A programme of remediation was carried out: the tar from pits B–H seems to have been removed from the site, and some of the tar in pit A was probably redeposited in pit 1A, east of the trunk road, while part was left *in situ*. To the west of the trunk road and at the northern-most end of the site, a few tar seeps are visible in the exposed rocks. The tar appears to have moved laterally from tar pit A along fractures in the sandstone (Fig. 3b).

The tar appears to be generally solid during the winter, but, in the course of site visits during April and May 1994, it was noted that the tar was becoming more fluid and mobile and beginning to flow, indicating that contami-nation may be mobilized during the warmer months of the year.

Surface water and water abstractions

The northern part of the eastern boundary of the site is marked by a small stream, which is a tributary of the Bottle Brook. The Bottle Brook itself feeds into the River Derwent just north of Derby. The Bottle Brook is classified by the NRA (1991) as a Class 2 (fair) watercourse, i.e. it is suitable for potable supply after advanced treatment, it supports coarse fisheries, and it may have moderate amenity value.

No licensed or unlicensed ground or surface water abstractions are known to exist in the immediate vicinity of the site which might be considered to be at risk from contaminants migrating from the study site.

Site investigation

The site investigation was carried out between January and February 1994. A number of inves-tigative techniques were used to evaluate the ground conditions and the extent of the contam-ination. The works comprised the following:

(c)

(d)

Fig. 3. Photographs showing (**c**) excavation of tar pit 4 and (**d**) drilling of an over-water borehole from a floating pontoon on the middle lagoon.

- a topographical survey,
- a geophysical survey by the British Geological Survey,
- 44 static cone penetration tests (CPTs),
- 13 cable percussive boreholes,
- 34 rotary drill-holes,
- 72 machine-dug trial pits/trenches,
- one over-water borehole (middle lagoon),
- six over-water dynamic cone penetration tests (middle lagoon),
- a depth sounding survey of the north and south lagoons.

The rotary cored boreholes were located so as to obtain a definition of the stratigraphy of the site. Sixteen of the rotary boreholes (and

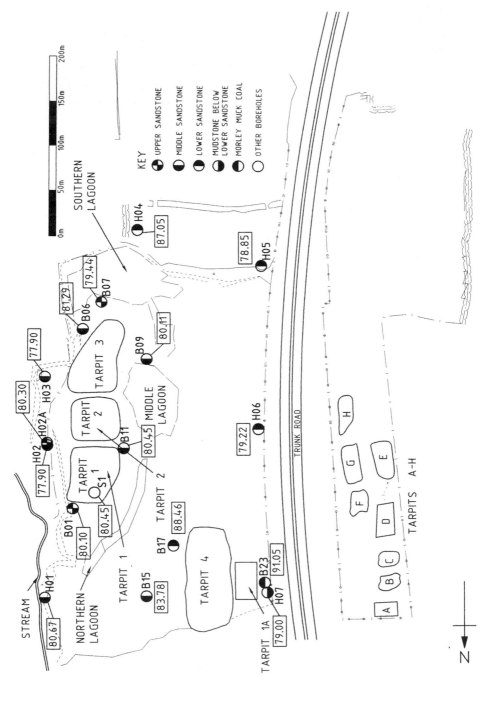

Fig. 4. Site plan showing locations of hydrogeological boreholes and water levels (mOD) on 26 April 1994.

one percussion borehole) were installed with monitoring wells to enable monitoring of hydrogeological conditions and collection of groundwater samples (Fig. 4; Table 3).

Most of the trial pits (Fig. 5a) were designed in order to locate and define the outlines and volumes of the tar pits (Table 4), and to allow the collection of soil samples. However, 20 of the pits were excavated in areas outside the tar pits to confirm the site geology.

A percussion borehole was located in the centre of each tar pit to define its composition and total depth, and other boreholes were located around the edge of each tar pit to ascertain any migration of tar or tar leachate through the surrounding strata (Fig. 5b).

A geophysical survey was carried out by the British Geological Survey (Busby *et al.* 1994) in order to ascertain if electrical and electromagnetic geophysical techniques (electromagnetic induction, DC vertical electrical resistivity sounding and very low frequency electromagnetic induction (VLF)) could be of assistance in elucidating the thickness and nature of surface backfill over tar as well as the depth and extent of the tar pits. The overall conclusion of the survey was that the methods employed were useful for defining the boundaries of the tar pits and in obtaining a first estimate of the depth profile, but that further calibration information from boreholes was necessary to interpret an exact depth and stratification profile.

Soil and water sampling methods

Soil samples were taken either directly from the bucket of the excavator machine (in the case of trial pits) or the shell of the cable percussion rig (in the case of boreholes) using a stainless steel trowel. Care was taken to remove visible dirt from the trowel between samples and, if necessary, the trowel was washed with a mild detergent solution to minimize cross-contamination. Soil samples were stored in labelled soil containers at $<4\,°C$, prior to despatch to the laboratory, with the appropriate chain of custody documentation.

Water samples were taken from selected trial pits and surface water features using a stainless steel bucket, which was washed with mild detergent solution and rinsed with water between samples. Groundwater samples were taken from boreholes following purging and development using a small stainless steel electrical submersible pump. Groundwater was flushed from the wells until field measurements of pH and electrical conductivity (EC) were constant, at which stage samples of groundwater were taken into a combination of plastic and glass containers with preservatives appropriate to the parameters of concern. The samples were transported to the laboratory under the same conditions as for the soil samples. pH and EC were measured both in the field and the laboratory.

Sampling of soils and groundwater was undertaken in accordance with Aspinwall & Company's in-house procedures which are based on current best-practice guidance as laid down, for example, by Scottish Enterprise (1994) and CIRIA (1995).

The chosen analytical technique for metals analysis was a laboratory-documented in-house method based upon inductively coupled plasma optical emission spectroscopy (ICP-OES). The analyses refer to filtered water samples and acid-digested dried soil samples. Sulphate analyses were performed by documented in-house methods based upon British Standard BS 1016 Part 11: 1977. Chloride was determined by spectrophotometric techniques.

Geology

The site lies on the Lower Coal Measures (Upper Carboniferous), which consist chiefly of mudstone, siltstone, sandstone, seat earth and coal in repetitive cyclothems (Gibson *et al.* 1908; Frost & Smart 1979). The strata dip at approximately 1 in 4 (15°) towards the east and are summarized in a cross-section in Fig. 6.

There are several productive coal seams in the vicinity of the site, as detailed in Table 2. Other 'non-productive' seams in this area, which may have been subject to small-scale undocumented working include the Lower Brampton, Morley Muck, Naughton (Norton), Smalley Upper and Smalley Lower seams.

Three significant bands of sandstone outcrop within the site boundary, one (the Lower Sandstone) to the west of the trunk road and two (the Middle and Upper Sandstones) to the east. Geological maps (1:10560 Sheet SK34NE) indicate that a possible southwest to northeast trending fault, with down-throw to the southeast, occurs at the northeast corner of the site. This fault may pass under the southern end of tar pit No. 4.

Over a substantial part of the site, siltstone and sandstone outcrops are seen at ground level. A discontinuity survey was carried out (Power 1994; Richard Moore, Scott Wilson Kirkpatrick, unpublished data), the results of which are presented as a stereonet in Fig. 7. Two main joint sets exist in addition to the sub-horizontal

Table 3. *Characteristics of hydrogeological monitoring boreholes*

Monitoring borehole	Ground level (mOD)	Filter pack (depth) (m)	Filter pack (mOD)	Piezo. tip (depth) (m)	Piezo tip (mOD)	Aquifer*	Water level 26April 1994 (depth) (m bgl)	Water level 26April 1994 (depth) (mOD)
H1	80.45	31.5–41.35	48.95–39.10	32.5–38.5	47.95–41.95	S_M	+0.22 (art.)†	80.67
H2	79.5	30.0–40.0	49.5–39.5	31.0–37.0	48.5–42.5	S_M	1.60	77.90
H2A	79.5	11.5–16.5	68.0–63.0	12.43–15.43	67.07–64.07	S_U	+0.80 (art.)	80.30
H3	82.70	25.0–40.0	57.7–42.7	32.5–38.5	50.2–44.2	S_M	4.80	77.90
H4	81.05	34.5–40.0	46.55–41.05	36.0–39.0	45.05–42.05	S_L	+6 (art.)	87.05
H5	91.30	21.0–30.0	70.3–61.3	22.5–28.5	68.8–62.8	S_L	12.45	78.85
H6	92.40	31.0–40.0	61.4–52.4	32.5–38.5	59.9–53.9	S_L	13.18	79.22
H7	94.20	30.0–41.25	64.2–52.95	34.5–40.5	59.7–53.7	Mudstone under S_L	15.20	79.00
B1	80.60	4.78–25.20	75.82–55.4	5.28–24.28	75.32–56.32	S_U	0.50	80.10
B6	83.45	16.0–25.0	67.45–58.45	17.5–23.5	65.95–59.95	S_M	2.16	81.29
B7	80.95	6.5–12.0	74.45–68.95	8.5–11.5	72.45–69.45	S_U	1.51	79.44
B9	83.05	1.0–12.0	82.05–71.05	8.5–11.5	74.55–71.55	S_M	2.94	80.11
B11	83.20	17.0–25.0	66.2–58.2	18.0–24.0	65.2–59.2	S_M	5.53	77.67
B15	85.35	8.5–13.48	76.85–71.87	9.5–12.5	75.85–72.85	S_M	1.57	83.78
B17	87.75	17.0–25.0	70.75–62.75	18.0–24.0	69.75–63.75	S_M	+0.71 (art.)	88.46
B23	94.00	7.0–12.0	87.0–82.0	8.0–11.0	86.0–83.0	S_L Morley Muck	2.95	91.05
S1	82.75	2.0–9.1	80.75–73.65	5.6–8.6	77.15–74.15	Made ground, tar pit 1	2.30	80.45

* S_U, S_M and S_L, Upper, Middle and Lower Sandstones respectively.
† art. = artesian

bedding-parallel joint set, and these discontinuities occur with reasonable regularity across the whole site. The two main non-bedding parallel joint directions are c. 160° and 60° (i.e. NNW–SSE and WSW–ENE). The NNW–SSE set gives the impression of being the most well developed, most continuous and with the largest aperture at outcrop.

The importance of bedding-plane parallel fractures at depth in the bedrock is unknown. The openness and frequency of such fractures typically decreases with depth in many fractured formations, due to increasing lithostatic pressure.

The made ground

Large areas of the study site are covered by fill materials or 'made ground'. It is evident from the logs of the exploratory holes and the descriptions of the types of fill that none of the fill is 'engineered', and it has not been emplaced in layers and compacted. The investigation showed that the fill is highly variable in terms of its distribution, composition and geotechnical properties, probably reflecting a random sequence of tipping across the site. The fill consisted of free tar and oil product, tarry silts, foundry sands, mining and quarry spoil, ash, clinker, Fuller's Earth, topsoils, compacted bricks and compacted gravels. Natural, dominantly silty-clayey superficial materials were also encountered in some areas.

Conditions in lagoons

One over-water borehole (OW1) and six dynamic penetration tests (DPTs) were carried out in the bed of the middle lagoon. The depth of the lagoon bed level was approximately 1.50 m below water level. The following typical sequence was interpreted from the borehole and DPTs:

- 0–2 m below the lagoon bed: intermixed Fuller's Earth and tarry silt materials
- 2–3.5 m below the lagoon bed: mixed tarry silt and mining spoil materials
- >3.5 m below the lagoon bed: mining spoil materials with occasional pockets of tarry/oily water.

Hydrogeology

Coal measures hydrogeology

The Coal Measures strata are regarded as being a 'minor aquifer' by the National Rivers Authority.

These aquifers 'seldom produce large quantities of water for abstractions, but are important both for local supplies and in supplying baseflow for rivers. In certain local circumstances, minor aquifers can be highly vulnerable to pollution' (NRA 1992).

The Coal Measures strata consist of repetitive cyclothems of shales, sandstones and coal. It is generally considered that the sandstone strata are significantly more permeable than the shales. The sequence typically has low porosity, and groundwater flow in undisturbed Coal Measures is dominated by fracture flow (Cripps et al. 1993). Examination of outcrops indicates that, in the rather more massive sandstones, fractures are typically sparser, but of greater aperture than in the mudstones and shales, resulting in higher hydraulic conductivities.

Many hydrogeologists regard the Coal Measures as a series of separate sandstone aquifers separated by low-permeability shale and mudstone aquitards (Smith et al. 1967; Price 1996). The current authors feel that the contrast in permeability between the shales and sandstones at this particular site is insufficient to support this simple model and the strata must thus be regarded as a partially discontinuous, heterogeneous and anisotropic fractured aquifer system; studies such as the current one suggest that fracture permeability is not negligible even in the mudstones. What is certain is that a two-dimensional conceptual model of head distribution and groundwater flow is inadequate. In a study of a waste disposal site located on Scottish Coal Measures, Harrison et al. (1981) note that 'Correlation of water levels between boreholes is not valid unless they intersect the same stratigraphic level. The concept of a regional water-table is not tenable, hydraulic gradients being determined by essentially local factors.' In addition, in a fractured system, groundwater flow may not be exactly in the direction of the regional head gradient; water and contaminant transport will be constrained by fracture discontinuities.

Mining voids as flow pathways The exploitation of coal has occurred in the UK since Roman times (Bromehead 1948) but it was not until the early 17th century that the mining industry expanded rapidly. Most of the Coal Measures of the UK have been actively mined for coal, often at several different horizons, as is the case at the study site. Longwall mining techniques result in collapse of the mined voids behind the working face, leading to an elevated transmissivity along the 'goaf'-filled mined seam and enhanced fracturing above the seam due to collapse. Pillar and stall workings may be more

(a)

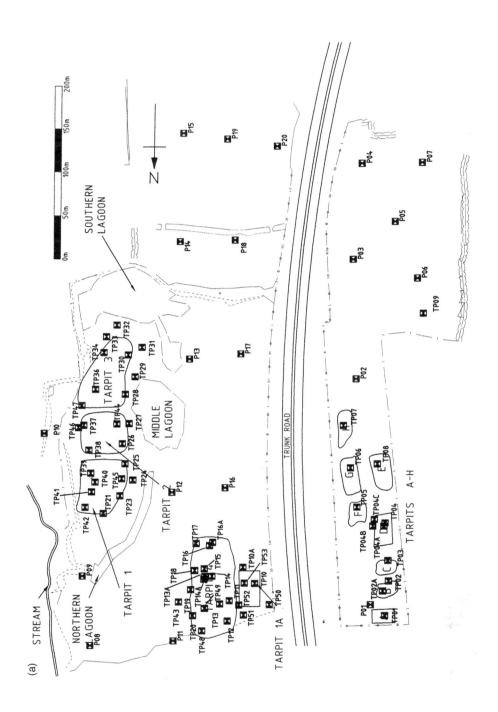

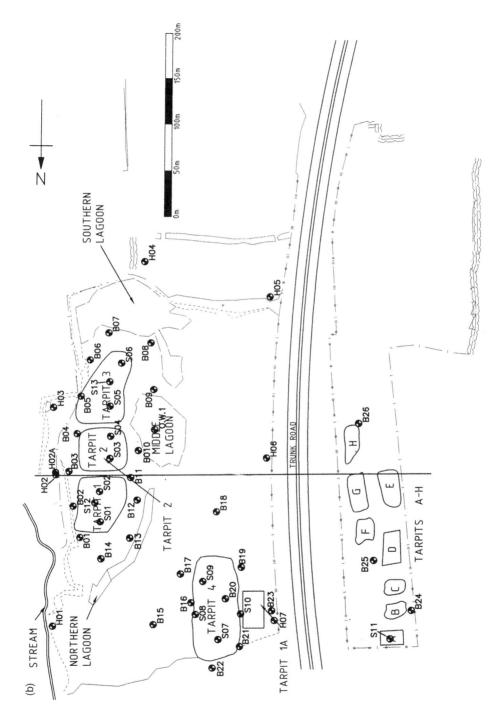

Fig. 5. Site plans showing locations of (**a**) trial pits and (**b**) boreholes. The line shows the section represented in Fig. 6.

Table 4. *Approximate tar pit volumes*

Pit ID	Estimated volume (m^3)		Rockhead material
	Tar pit (made ground)	Additional contaminated rock	
1	17 800[*]	5 000	Mudstone
2	12 800[*]	4 000	Mudstone
3	16 000[*]	7 000	
4N	3 300[*]	1 700	Siltstone
4S	9 000	3 000	Mudstone
1A	2 900	0	Mudstone
A	1 400 (400)[†]	160	Mudstone

Notes:

[*] Figures based on BGS survey (Busby *et al.* 1994).
[†] One-metre layer of tarry silt and oily bentonite, giving *c.* 400 m^3 contaminated fill.
Additional contaminated rock is estimated, on the basis of observations during drilling, at *c.* 2 m thickness below the tar pits.
Total volume of contaminated fill = 62 200 m^3; estimated volume of underlying contaminated rock = 21 100 m^3.

resistant to collapse and can remain open as extremely transmissive groundwater flow pathways. In addition, abandoned shafts may also form preferential pathways for groundwater movement and may increase vertical connectivity

between different geological strata. Thus, the mined Coal Measures form an 'anthropogenic' aquifer where groundwater flow pathways may be dominated by mined cavities and collapsed seams, which may stretch over great distances.

Mined voids can act as preferential pathways through which contaminated groundwater may potentially travel at great speeds and with little attenuation. Typical velocities from the Forest of Dean coalfield (Aldous *et al.* 1986, Aldous & Smart 1988) in Gloucestershire, proven by tracer tests, are summarized in Table 5. Even in the 3.6 km tracer test described in the table, the tracer breakthrough was abrupt and short-lived, indicating conduit flow with little longitudinal dispersion. In a separate study, disposal of liquid waste into abandoned coal mines in Lanarkshire, Scotland, in the early 1970s (Henton 1974) resulted in pollutant transport to a mine portal (1 km distant) at a mean velocity of 0.015 m s^{-1}.

As regards the study site, the area in the immediate vicinity has been subject to intensive mining activity in the past (Griffin 1971). Collieries extracting coal from deep workings in the vicinity have included Kilburn, Denby Hall, Denby Drury Low, Salterwood and Ryefield.

It is likely that workings below the site were related (or, at least, hydraulically interconnected) with these collieries. All of these collieries are now abandoned and are not actively pumped.

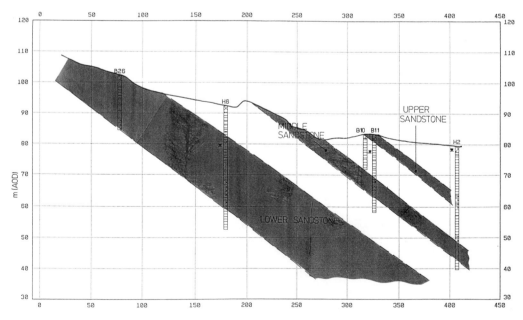

Fig. 6. Geological cross-section through the study site, based on site investigation data, from borehole B26 to H2. The line of section is shown on Fig. 5(b) and runs from west (left) to east (right). The vertical scale is in mOD (i.e. metres above sea level), and the horizontal scale is in metres.

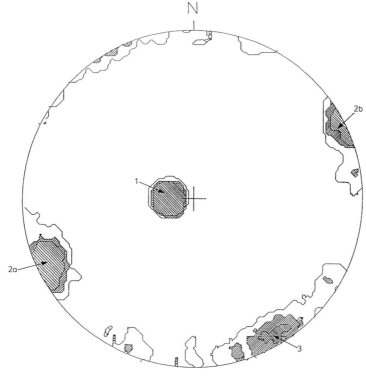

TOTAL NUMBER OF POLES = 118

	3 POLES = 2·54%
	6 POLES = 5·08%
	9 POLES = 7·62%

NO.	DISCONTINUITY	APPROX. DIP	DIP DIRECTION
1	BEDDING PLANE	16°	090°
2a	JOINT PLANE	90°	062°
2b	JOINT PLANE	90°	062°
3	JOINT PLANE	86°	325°

Fig. 7. Stereographic net showing poles to mapped fracture planes.

However, it is possible to regard the coal workings of East Derbyshire and Nottinghamshire as forming, to a large extent, a hydraulically continuous, anthropogenic, multi-layer, rapid flow aquifer system (Dumpleton & Glover 1995). There will also be hydraulic interconnection between the various worked seams via shafts and other vertical connections.

Currently, minewater levels in the Nottinghamshire/Derbyshire coalfield are managed on

Table 5: *Summary of tracer tests in abandoned coal mines of the Forest of Dean (after Aldous & Smart 1988)*

Saturation conditions	Flow conditions	Distance of travel (km)	Dye recovery (%)	Travel velocity	
				m day^{-1}	m s^{-1}
Unsaturated	Low	0.5	108	1 840	0.021
	High	0.5	63	160 000	1.85
Saturated		3.6	1	460	0.0053

Table 6. *Estimates of apparent hydraulic conductivity in overflowing monitoring boreholes*

Borehole	Q $(\text{m}^3\,\text{s}^{-1})$	s (m above well top)	L (m)	T_a^* $(\text{m}^2\,\text{s}^{-1})$	K_a^* $(\text{m}\,\text{s}^{-1})$	$K_a^\dagger$ $(\text{m}\,\text{s}^{-1})$
H1	6.3×10^{-5}	0.22	9.85	3×10^{-5}	3×10^{-4}	3×10^{-5}
H2A	4.6×10^{-5}	0.65	5.0	8×10^{-5}	2×10^{-5}	1×10^{-5}
H4	4.2×10^{-4}	6	5.5	8×10^{-5}	1×10^{-5}	1×10^{-5}
B17	3.2×10^{-5}	0.67	8.0	5×10^{-5}	7×10^{-6}	5×10^{-6}

T_a, apparent transmissivity, K_a, apparent hydraulic conductivity, L, length of borehole section open to formation, s, head above ground level, Q, flow rate under artesian pressure at ground level. *Banks (1992). †Somerville (1986).

a regional scale by pumping at strategically located shafts (Awbery 1988; Lemon 1991; Banks *et al.* 1997), the nearest of which are detailed in Table 9.

Morton and Blackwell lie some 14 km NNE of the study area and pump from a higher stratigraphic level than the seams below the tar pits. Woodside lies some 4 km to the SE and pumps from (among other horizons) workings in the Kilburn seam. Woodside Pumping Station is believed to drain water from over 20 abandoned pits, including the group around the study site. Pumped water is discharged into the Nut Brook. British Coal are of the opinion that water from workings below the study site would drain down-dip towards Woodside rather than to Blackwell or Morton (A. Barnes, British Coal, pers. comm.).

In addition, a major opencast site, Kirk Opencast, is being operated some 2–3 km ENE of the study site. The site currently extends to the Deep Hard seam at some 40 m depth, where inflows from old workings were encountered. The site is dewatered by pumping to the Bottle Brook. The opencast has historically extended as deep as 88 m, to the Piper seam, where again, water flowed in from old workings (P. Reeves, NRA & P. Day, British Coal Opencast, pers. comm.). Both the Deep Hard and Piper seams lie stratigraphically above the highest seams below the study site, but dewatering at Kirk may draw on water from the workings below the tar pits via vertical connections (shafts) between stratigraphically separate seams.

Fractures as flow pathways It is generally agreed (e.g. Williams & Aitkenhead 1991; Cripps *et al.* 1993) that, in unmined Coal Measures strata, fractures are the dominant pathways through which groundwater flow occurs. The bulk hydraulic conductivity of the rock depends upon the characteristics of the fractures, including the spacing, aperture, interconnection and

degree of infilling. These are closely related to the lithology, bedding, weathering/alteration state and structural setting of these rocks. The strong and brittle sandstones are generally more permeable than the more ductile mudstones as fractures in the former tend to be open and interconnected throughout the mass of the rock compared to the finer-grained members in which they are shorter, narrower and commonly blocked by degradation products.

It is possible, by measurement of fracture properties (such as orientation, spacing, aperture, roughness and infilling), to estimate the bulk hydraulic conductivity of the rock in certain directions (Hoek & Bray 1981). Alternatively, Cripps *et al.* (1993) have carried out empirical measurements on the hydraulic conductivity of the Coal Measures in Sheffield by means of *in situ* tests in boreholes drilled for a new sewer. A total of 18 tests were carried out, largely in mudstone or interbedded mudstone and siltstone and the mean hydraulic conductivity recorded was $1.31 \times 10^{-5}\,\text{m}\,\text{s}^{-1}$ (range 4.2×10^{-7} to $1.5 \times 10^{-5}\,\text{m}\,\text{s}^{-1}$). The hydraulic conductivity data were related to a number of rock quality parameters and it was concluded that, for the mudstone/siltstone lithology, the bulk hydraulic conductivity increases as fracturing intensity increases.

Matrix hydraulic conductivity Intergranular flow in Coal Measures strata is usually subordinate to flow along fractures (an exception being in the upper part of a sequence where sandstone may have disaggregated to yield a weathered sandy deposit). A study of the matrix hydraulic conductivity of sandstone from the Lower Coal Measures has been carried out by Younger (1992) by studying thin sections and by carrying out permeameter tests on samples from Northumberland. Results fell in the range 10^{-8}–$10^{-5}\,\text{m}\,\text{s}^{-1}$ for hydraulic conductivity and 3–22% for porosity.

Hydrogeology of the study site

Hydraulic conductivity estimates A considerable number of water strikes during drilling occurred in the mudstone. Water was also encountered in the sandstone members and at coal seams. This pattern suggests that the mudstone contains fractured zones where hydraulic conductivities reach values comparable with those in the sandstones.

Hydraulic conductivity was estimated by measurement of the specific capacity of the four boreholes which were artesian, by allowing the flow to stabilize and then measuring flow rate (Q), under a measured artesian head (s). The methods of Banks (1992) and CIRIA (Somerville 1986) were then used to estimate apparent transmissivity and hydraulic conductivity of the aquifer sections corresponding to the well filter length. The results are given in Table 6, although the values of hydraulic conductivity derived must only be regarded as 'order of magnitude' estimates, as a number of assumptions are implicit in the analytical methods, possibly the most important being that of two-dimensional radial inflow. The estimates indicate a hydraulic conductivity of around $10^{-5}\,\mathrm{m\,s^{-1}}$, which is typical of a rather well-fractured mudstone or moderately-fractured sandstone. The K_a values also refer only to the rock within the immediate vicinity of the well, rather than giving a 'large scale' value for K_a.

Falling head tests were carried out in non-artesian monitoring wells. Hydraulic conductivity was estimated using solutions given in BS 5930 (British Standards Institution 1981) and Hvorslev (1951), and it was found that the results from the two techniques were generally similar. Rock Quality Designation (RQD – a measure of core recovery expressed as a percentage of drilling length; see British Standards Institution 1981) and Fracture Index (FI – a measure of fracture intensity, the number of clearly identifiable fractures per metre of core; see Geotechnical Engineering Office of Hong Kong 1988) were calculated from cores for the sections of borehole corresponding to the well filter. Figure 8 displays the results of conductivity testing as follows:

- conductivity versus average depth of the well screen (Fig. 8a);
- conductivity versus per cent sandstone over the well screen (Fig. 8b);
- conductivity versus average RQD over the well screen (Fig. 8c);
- conductivity versus average FI over the well screen (Fig. 8d).

Several interesting observations can be made from this figure. Overall the hydraulic conductivities recorded fall within the range of results for matrix conductivity of sandstones given by Younger (1992) and also the bulk conductivities of siltstone and mudstone given by Cripps *et al.* (1993) (see above). As can be seen from Fig. 8(a), the hydraulic conductivity appears to increase with depth below ground level. As fracturing is more intense near the surface, Figs 8(c) and 8(d) appear to suggest that hydraulic conductivity increases with increasing fracture spacing. This appears to contradict the conclusions of Cripps *et al.* (1993) and may be due to the fact that fractures close to the surface may be clogged by weathering products. It may also reflect the fact that Cripps *et al.* investigated a single lithological group (mud- and siltstones), whereas in this study the sequence also contained sandstones. It should be remembered that fracture transmissivity is proportional to the cube of fracture aperture for plane parallel fractures and thus a few, widely spaced but open fractures (in sandstone) may be more transmissive than many densely spaced fractures of low aperture (in mudstone, for example). Hence, care should be taken when trying to estimate hydraulic conductivity from parameters such as RQD and FI, especially in the type of near-surface investigation common at contaminated sites.

Water-table elevation The groundwater head distribution at the study site is found to be complex and three-dimensional, where terms such as confined and unconfined have little meaning. In such a situation, water level readings in deep piezometers represent water pressures at depth in the aquifer and not necessarily the level of the water-table. The water-table is, for practical purposes, defined as the interface between saturated and unsaturated strata. The elevation of the water-table is best approximated from groundwater readings in shallow piezometers and trial pits, where hydrogeological conditions approximate to unconfined.

On this basis, the water-table at the study site appears to follow a pattern typical for relatively low-permeability rocks (Thorne & Gascoyne 1993). It seems to be a subdued reflection of the surface topography lying at only 1–2 m below ground surface. The general fall in water-table is from west to east, providing a lateral head gradient driving shallow groundwater through the superficial materials generally towards the stream. However, groundwater flow and contaminant transport in the deeper fractured aquifer will be governed by the orientations of

Table 7. *Groundwater quality recorded from samples from drilled boreholes, following development pumping*

	Background	B1	B6	B7	B9	B11	B15	B17	B23	H1	H2	H2a	H3	H4	H5	H6	H7
pH	5.8–7.3	6.4	6.5	4.2	6.4	6.7	5.8	6.7	6.9	7.1	5.8	4.8	5.9	5.7	5.6	6.6	6.7
EC (μS cm^{-1})	200–530	906	1740	1610	2600	732	1790	1580	1630	702	2710	5400	2260	3000	2760	1530	720
Ca (mg l^{-1})	14–60	128		133				232		77			324	651			94
Mg	5–13	81		114				101		41			184	312			51
Fe	<0.01–0.12	0.06		209				0.23		<0.01			1.37	12			0.04
SO$_4^{2-}$	26–70	334	977	1360	1850	153	1160	369	425	90	1940	5040	1420	2520	2040	572	94
Cl$^-$	14–45	24	30	12	88	25	34	235	225	27	42	105	34	89	37	74	38
As (μg l^{-1})		<40		<40				<40									
Cd	<1–5	10		10				10		<10			20	20			10
Cr		70		150				90		<20			20	160			60
Cu	<2–4	10		10				10		<10			80	20			10
Ni	<16–17	70		210				70		<20			80	120			60
Pb	<1–5	80		140				90		<30			<30	100			80
Zn	<7–80	30		3030				<10		<10			320	20			10
TOC (mg l^{-1})		8.3	51.9	29.7	90.7	4.6	13.3	6.8	8.7	11	16.7	178	18.7	52.8	57.4	12.9	10.8
COD		<5	170	300	345	<5	35	10	10	5	45	150	60	130	180	25	10
BOD		<3	304	<3	6.2	<3	4.2	<3	<3	<3	<3	<3	<3	<3	<3	<3	<3
Phenol (mg l^{-1})		<0.1	<0.1	0.2	<0.1	<0.1	<0.1	<0.1	<0.1	<0.1	<0.1	<0.1	<0.1	<0.1	<0.1	<0.1	<0.1
Min. Oil		<1	6	7	4	<1	1	<1	<1	<1	<1	1	<1	<1	<1	<1	<1
Amm-N (mg l^{-1})		1	2.2	1.6	2.1	1	202	1.6	1.2	0.4	0.3	2.2	0.6	0.4	1.1	1.6	0.9

Note: Ranges of background concentrations are taken from data on shallow Coal Measures groundwaters in areas free of point-source contamination in Derbyshire and Yorkshire by Banks (1997). pH and EC for the samples from the study site are laboratory determinations.

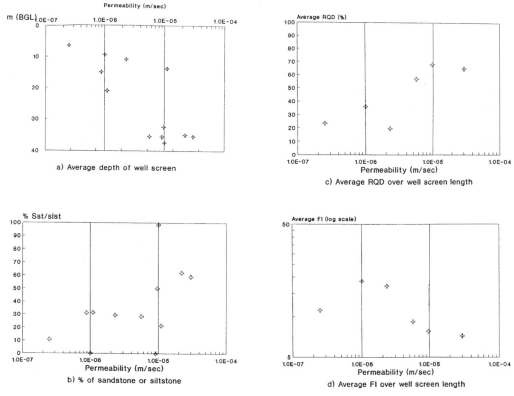

Fig. 8. (**a**) Hydraulic conductivity versus average depth of the well screen (in metres below ground level). (**b**) Hydraulic conductivity versus per cent sandstone/siltstone over the well screen (for non-cored holes, this is estimated from drilling cuttings). (**c**) Hydraulic conductivity versus average RQD (British Standards Institution 1981) over the well screen (as RQD increases, the fracture intensity decreases). No points are plotted for holes drilled by open-hole (uncored) techniques. (**d**) Hydraulic conductivity vs average FI (Geotechnical Engineering Office of Hong Kong 1988) over the well screen (as FI increases, the fracture intensity increases). No points are plotted for holes drilled by open hole (uncored) techniques. (All after Power 1994.)

the most extensive fracture sets of the greatest aperture (Bradbury & Muldoon 1994). At the study site, these are probably the bedding-plane parallel (i.e. sub-horizontal) set and a prominent sub-vertical NNW to SSE fracture set. The former will not introduce azimuthal anisotropy in the flow, but the latter fracture set may tend to induce a component of flow to the SSE.

Groundwater head distribution Coal Measures strata form areas of significant and variable topography. Even under natural conditions, therefore, vertical hydraulic head gradients are important; upward head gradients (and artesian boreholes) being found in valleys and downward gradients in interfluves (Toth 1963). Indeed, at the study site, boreholes H1, H2A, H4 and B17 were observed to be overflowing at the surface.

This pattern is further complicated by the existence of mine workings. If the mine system is pumped or freely drained, this may result in regional lowering of water-tables (Anon 1993; Younger & Sherwood 1993) and/or high downward vertical groundwater head gradients (and thus potential for downward groundwater flow) in the overlying strata, which will remain saturated if the amount of water percolating into the mine is matched by the recharge to the system from above. This latter situation was encountered by Harrison *et al.* (1981), causing migration of contaminants from a landfill downwards towards underlying abandoned coal workings. A number of intermediate permeable horizons were also discovered in the strata, through which lateral movement could occur. In the computer model of Harrison *et al.* (1992), the head distribution in a fractured

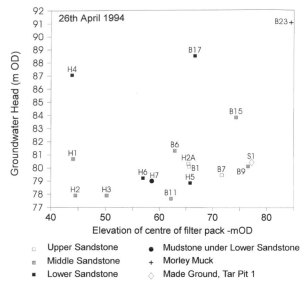

Fig. 9. Groundwater head in boreholes at study site, plotted against elevation of filter pack (both in metres above sea level).

clayey aquitard subject to downward flow into an underlying aquifer was simulated. The main conclusion reached from this simulation was that 'because of the inherent difficulties associated with identifying [the most influential] fractures in cores ... and because such fractures can produce a highly erratic head distribution in the aquitard, there will be much uncertainty in predicting contaminant transport in such sparsely fractured aquitards.'

Cessation of pumping or drainage may result in rising groundwater levels, resaturation of overlying strata and re-establishment of a drainage pattern dominated by more natural lateral head gradients.

Groundwater levels, as measured in boreholes on 26 April 1994, are plotted in Fig. 4. There appear to be strong vertical groundwater gradients at the site, with a general decrease in groundwater head with depth. Apparent downward vertical head gradients between the pairs of piezometers H2 and H2A, and H7 and B23 (as of 26 April 1994) are 0.114 and 0.47 m m^{-1} respectively. Plots of piezometric heads against elevation of filter pack (Fig. 9), tentatively suggest a general decline in head with decreasing elevation. Filter packs are completed at a variety of stratigraphical levels, the majority of packs traversing both mudstone and silt/sandstone horizons. The exception to this trend is borehole H4, a somewhat anomalous borehole in terms of stratigraphy recorded,

and very anomalous in terms of its very high artesian head (some 6 m). No satisfactory explanation has been found for this borehole's behaviour, except possibly that it would appear to indicate a discontinuity in the facies or fracture system of the aquifer; conceivably even a fault.

In summary, it is believed that vertical head gradients and possibly also vertical (downward) flow are dominant at depth over lateral head gradients. A probable explanation for this is the existence of coal workings at depth beneath the site, which may be connected to pumping mine systems. The overall effect of these is to 'drain' the site from below.

However, the existence of artesian boreholes (H1, H2A and H4) indicates that upwards head gradients are also of importance locally, particularly, but not exclusively (e.g. borehole B17), in the vicinity of the stream, a typical discharge zone, implying that the stream is influent in character. It is thus not possible to draw a simple hydrogeological plan or cross-section through the site, illustrating contours on piezometric head. That is, the aquifer cannot be adequately conceptualized as a uniform porous medium; it is a complex, discontinuous system of fractures interacting with a porous aquifer horizons. The sandstone horizons may also impart a multi-layer character to the aquifer, although the existing data are too sparse to provide clear evidence of this.

Hydrochemistry and groundwater contamination

A selection of hydrochemical parameters measured in groundwater in the monitoring wells and in the lakes, stream and trial pits, are shown in Tables 7 and 8. In Table 7, values are compared with concentrations recorded in ostensibly uncontaminated shallow Coal Measures groundwaters in Derbyshire and Yorkshire by Banks (1997). In most boreholes, concentrations of sulphate, calcium, magnesium and heavy metals are generally elevated with respect to background, although only in B7 and H2a is the pH significantly depressed, suggesting that acid hydrolysis of aquifer materials (carbonates, hydroxides and possibly silicates) is buffering the acidity derived from the tar pits. Sulphate and electrical conductivity (EC) appear to be among the best indicators of contamination, although the tar pits also appear to be a source of dissolved chloride. The boreholes exhibiting the highest sulphate and EC generally also show concentrations of heavy metals above the background, although the extent to which these are derived from the waste or from dissolution of aquifer components remains unclear. Concentrations of phenols and dissolved mineral oil are generally low, confirming the rather inert nature of the organic component of the tars. Detailed analyses of the dissolved organic components have not been carried out as yet.

The distribution of hydrochemical parameters in groundwater often provides useful information on the groundwater flow regime. The following three parameters in groundwater from the monitoring boreholes have been plotted on maps of the site (Fig. 10):

- electrical conductivity (EC);
- sulphate (SO_4^{2-});
- pH.

Contamination from the tar pits has the effect of lowering the groundwater pH and increasing concentrations of chloride, sulphate and conductivity. Maps of the three parameters all reveal a similar picture. Boreholes H1 and H7, at the north end of the site, both yield neutral pH levels and low SO_4^{2-} and EC (and Cl^-), despite the fact that H1 lies down-gradient of tar pits 1A and 4. In contrast, boreholes H4 and H5, well to the south of the tar pits, appear to be significantly contaminated, with pH 5.6–5.7; EC 2760–3000 $\mu S\,cm^{-1}$, SO_4^{2-} 2040–2520 $mg\,l^{-1}$ and Cl^- 37–89 $mg\,l^{-1}$. No contaminant sources lie immediately up-gradient of H4 and H5. This implies that groundwater flow is *not* parallel to the prevailing hydraulic gradient (Bradbury &

Muldoon 1994), but is probably controlled by fracture pathways. These induce a north to south component of groundwater flow in the Coal Measures strata, probably controlled by the main NNW–SSE fracture set. This does not, however, imply that contaminant transport will only take place along that NNW–SSE fracture set. The intersecting fracture sets (both sub-vertical and sub-horizontal) will effectively impart a high degree of macro-scale dispersion of contaminant transport, with significant components in both southerly (e.g. contamination in H4 and H5, controlled by the NNW–SSE fractures) and easterly (contamination in H2 and H2A, controlled by the prevailing head gradient acting on bedding plane-parallel and other sub-vertical fracture sets) directions.

On the basis of the parameters mentioned above, borehole H2A yields the most contaminated water, with 105 $mg\,l^{-1}$ Cl^-, 5040 $mg\,l^{-1}$ SO_4^{2-}, 5400 $\mu S\,cm^{-1}$ EC, and pH 4.8. Although it cannot be assumed that H1 and H7 are uncontaminated, it appears that their groundwater approaches 'background' concentrations more nearly than that of the other boreholes.

It is noteworthy that the very shallow borehole (B23) yields considerably higher chloride, sulphate and EC values than the deeper, adjacent H7. This could be interpreted as an indication of a shallow westerly component of groundwater flow and contaminant transport from tar pits 4 and 1A, down towards the road cutting. This implies that the road cutting may affect the drainage direction of the water-table, at least locally.

Conceptual model of groundwater flow

During drilling, no significant difference was noted in the frequency of groundwater strikes between the sandstones and the mudstones, although higher hydraulic conductivities were associated with sand/siltstone horizons as opposed to mudstones (Fig. 8b). This picture only offers partial support to the generally held view that Coal Measures sandstones are significantly more permeable than mudstones (Smith *et al.* 1967). In the absence of conclusive evidence, the Coal Measures series at the study site is regarded as a single, anisotropic, heterogeneous and possibly partially discontinuous aquifer.

Groundwater flow in the aquifer system appears to occur in different directions at different aquifer levels. It should, however, be stressed that the data on which the conceptual model is based, are too sparse to adequately define the complex three-dimensional nature of the flow

Table 8. *Water quality in samples from surface water features, selected trial pits and borehole S5 (centred in one of the tar pits)*

Type	Stream along eastern boundary		Lake water		Waters from trial pits and boreholes in tar pits				Reference groundwater
	Stream (upstream site)	Stream (downstream site)	Middle lagoon	South lagoon	Trial pit 42	Trial pit 27	Trial pit 34	Borehole S5	Borehole H1
pH	7.6	7.9	3.0	7.1	7.1	6.0	6.6	0.9	7.1
EC ($\mu S\,cm^{-1}$)		668	1320	339	1380	2000	1560	97 000	702
Ca ($mg\,l^{-1}$)			83	33			126		77
Mg			53	24			115		41
Na			26	9			66		
K			3	4			15		
Fe			25.4	0.19			0.3		<0.01
SO_4^{2-}			754	96	179	1270	124	183 000	90
As ($\mu g/l^{-1}$)			<40	<40			<40		<10
Cd			10	<10			<10		<20
Cr			70	<20			50		<10
Cu			130	<10			<10		<20
Ni			270	<20			30		<30
Pb			50	<30			30		<10
Zn			3900	20			<10		<10
TOC ($mg\,l^{-1}$)	9.4	8.3	9.5	7.8	27	280	29.5	2960	11
COD					115	1150			5
BOD					4.5	23.6			<3
Phenol ($mg\,l^{-1}$)	<0.1	<0.1	<0.1	<0.1	<0.1	0.4	<0.1	0.2	<0.1
Min. Oil	<1	<1	2	4	19	44	152	38	<1
Amm-N ($mg\,l^{-1}$)					2	1.2	0.3		0.4

Note: See Fig. 5 for locations. pH and EC are laboratory determinations.

Table 9. *Pumping shafts from coal workings in the vicinity of the study site*

Pumping shaft	Quantity pumped (1987) in m³ year⁻¹ *	Location	Depth
Blackwell A-Winning	1 550 000	Immediately S of Tibshelf, at approx. Grid Ref SK 43 58	To Black Shale horizon
Morton	1 450 000	Immediately W of Tibshelf, at approx. Grid Ref SK 41 60	
Woodside	5 640 000	West of Ilkeston, near Stanley and Mapperley, at approx. Grid Ref SK 40 44	To Kilburn Seam

* Lemon (1991).

system. The main points of the conceptual model are as follows:

- In the saturated part of the fill and weathered overburden materials, which, although heterogeneous, may be considered to be more isotropic than the underlying Coal Measures, groundwater flow is likely to follow the general gradient of the water-table, i.e. eastward, over most of the site, towards the small stream.
- In the Coal Measures proper, a lateral component of head gradient is likely to occur, also directed from west to east. Extant fracture directions will, however, constrain the direction in which groundwater can flow. The dominant NNW–SSE fracture set appears to induce a SSE component to the flow of groundwater and associated contaminants. An indication that this is occurring may be deduced from the distribution of hydrochemical parameters.
- Vertical gradients also appear to be of great importance deeper in the Coal Measures, with groundwater flow presumably being drawn down to old coal workings, being pumped from distant pumping stations.
- Locally, upwards hydraulic gradients are indicated by several artesian boreholes. These may reflect remnants of natural vertical head gradients, i.e. groundwater discharge to stream valleys. The combination of upward and downward vertical head gradients is not necessarily contradictory, but offers evidence for the discontinuous nature of the fractured Coal Measures aquifer.

Contaminant fate in the hydrosphere

Dissolved contaminants

There are four routes by which dissolved contaminants could potentially reach watercourses via the hydrosphere at the study site:

(1) *Via surface run-off.* During normal weather conditions such as were encountered during the study period, no run-off of contaminated surface water from the site appears to occur. This observation is supported by the fact that tar-pit related contamination was not found in the stream at the east of the site, either during this investigation or during previous monitoring by the NRA (P. Reeves, NRA, pers. comm.). This does not preclude the possibility that, during extreme rainfall or snow-melt events, run-off of contaminated surface water from the site to the stream might occur.

(2) *Via shallow groundwater in superficial materials.* The lack of hydraulic conductivity measurements in the fill precludes any quantitative assessment of this possibility although observations of water seepages into trial pits indicate low values of hydraulic conductivity and hence slow groundwater flow rate. The fact that contamination has not been found in the stream adjacent to the site suggests that contaminant concentrations of a magnitude which would be unacceptable to the NRA (i.e. infringing Water Quality Standards) do not reach the stream via this pathway.

(3) *Via lateral groundwater flow in the Coal Measures.* Hydrochemical data and the observed fracture pattern suggest that off-site migration of contaminants is occurring to a limited extent in a generally southeasterly and, to some extent, easterly direction. It is possible that some of the groundwater flowing through the site via this mechanism eventually discharges to the stream further south of the site.

(4) *Downward flow to underlying coal workings.* No determination of vertical hydraulic conductivity has been carried out. On reaching any coal workings, any contaminants would be subject to a large factor of dilution before being discharged at Woodside pumping station, bearing in mind that Woodside receives water from over 20 abandoned collieries, each with its own complex of workings draining extensive areas.

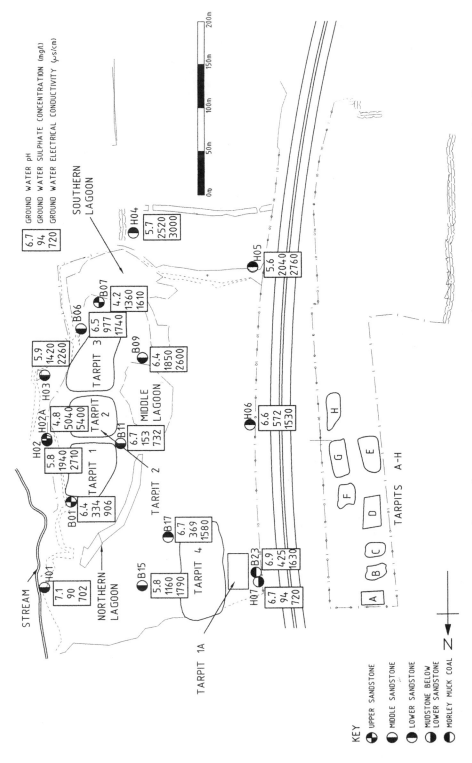

Fig. 10. Site plan showing pH, concentration of SO_4^{2-} (mg l^{-1}) and electrical conductivity (μS cm^{-1}) in groundwater sampled from monitoring wells.

Comparing the area of the contaminated site (c. 70 000 m^2) with the potential catchment area underdrained by Woodside (say, 5 × 5 km as a conservative estimate) a dilution factor of the order of *at least* 10^2 to 10^3 may be expected. (The yearly pumping rate of Woodside of 5 × 10^6 m^3 year^{-1}, coupled with a typical groundwater recharge rate, is consistent with a catchment area of at least that order of magnitude.) A more disturbing scenario would be that of opencasting of coal (with concomitant pumping) adjacent to the study site. This scenario is currently the subject of a planning debate.

Light non-aqueous phase liquids (LNAPLs) It is known that LNAPLs are generated by the tar pits as they can be observed as films on the surface of the southern and middle lagoons and also in groundwater samples taken from boreholes B7 and S1. The fact that no seepage of LNAPLs has been observed in the stream may be indicative of their limited migration, although very few data exist to draw definitive conclusions on the presence of LNAPLs within the aquifer. Most of the hydrogeological boreholes are screened at some depth below the water-table and would not be expected to detect LNAPL phases.

Dense non-aqueous phase liquids (DNAPLs) The tar itself can be regarded as a DNAPL, as it tends to sink on reaching the lagoons, but it appears to be too viscous to migrate downwards to any great extent in the aquifer, only being encountered within the Coal Measures in very few near-surface fractures. A further, largely unexplored possibility, is that of the generation of less viscous DNAPLs by the weathering of the tar.

Future development of groundwater conditions at the site

There is some evidence to suggest that the study site is partially 'underdrained' by deep coal workings. At present, these are pumped at some distance from the study site by British Coal. Many mines have, however, been closed in the recent past and such dewatering cannot be guaranteed in the future. If pumping ceases, groundwater heads would be expected to recover in the coal workings and the 'under-drainage' effect would cease. This would lead to an increase in groundwater heads in deep boreholes and, possibly, also a small rise in the water-table elevation. This would also imply a decrease in downward groundwater flow and a concomitant increase in lateral groundwater flux from the site towards the stream.

Consideration of a potential rise in groundwater heads and in lateral groundwater flux in the future must be taken into account in the assessment of remedial options for the site.

Remediation options

The fractured, anisotropic and exceptionally complex nature of groundwater flow at the study site renders any attempt at groundwater clean-up extremely expensive and with a low chance of success. This observation, coupled with the current lack of any measurable impact on off-site surface water receptors and the lack of water abstractions in the vicinity which might be considered to be at risk argue strongly against an attempt to remediate the quality of the groundwater in the Coal Measures aquifer system. Nevertheless, the fact that the waste is measurably contaminating groundwater below the site and that the groundwater, with possible future cessation of mine dewatering, may undergo changes in flow regime, render the removal of the contaminant source (i.e. the tar pits) highly desirable. Other factors in favour of some form of remedial action include the following:

- odour;
- risk to users of the site and wildlife from direct contact with the wastes;
- highly visible contamination of soils and surface water lagoons;
- political pressure to clean up what is perceived to be a highly undesirable site.

The excavation of the contaminated materials is seen as being desirable. Simply capping the tar waste to reduce infiltration of rainfall and direct contact with the waste would be cheap and would lead to a reduction in risk. It would, however, not fully address the issue of groundwater contamination and would be unlikely to contain the tar, given its ability to flow under loading and emerge at unpredictable points. The possibility of off-site disposal by landfill is regarded as unrealistic, due to risks associated with transport and the difficulty of finding a landfill operator willing to take the waste. Several methods (Sims 1990) of on-site and off-site waste treatment were assessed by Aspinwall & Co., including those detailed in Table 10.

Incineration was considered the most attractive treatment option on a combination of financial and environmental grounds, and this was compared with other isolation/encapsulation options as detailed in Table 11.

Table 10. *Summary of assessment of potential remedial options at the study site*

Treatment	Assessment
Ex situ biological treatment (biofarming)	Tars too refractory and acidic
In situ biological treatment (bioremediation)	Tars too refractory and acidic
In situ aqueous, solvent or steam extraction	Low tar solubility, high viscosity and low matrix permeability render techniques unrealistic
Cement stabilization/solidification	Treatment shown to have poor success rate in soils with high organic content as organics impede cement set
Vitrification	Energy demand and cost excessive. Presence of combustible material precludes use of technique
Incineration	Applicable to acid tars subject to control of acidity pre-treatment
Vapour extraction	Low volatility of tars precludes use
Soil flushing	Low solubility of tars precludes use

The site was subject to a semi-quantitative risk assessment to identify critical risk pathways and those which would be affected by the various remediation options. The options were subject to cost–benefit appraisal, with an outcome as summarized in Table 12. The final selection of remediation option is currently under discussion.

Discussion: drawbacks of the study

As in most similar, single-phase studies, the site investigation exhibited some limitations, many of which could only be identified in retrospect. The study achieved its main objectives (see 'Introduction'), but the following limitations imposed some uncertainties in data interpretation:

- the construction of boreholes for monitoring specific aquifer horizons (at depth) was unsuitable for monitoring of LNAPL (and DNAPL) phases;
- lack of more sophisticated hydraulic testing;
- in retrospect, geophysical logging (to determine stratigraphy and flowing fracture locations) would probably have been more cost-effective than intensive coring;
- lack of component-specific analysis of organic components (soon becomes highly expensive);
- the monitoring borehole array, although giving an adequate two-dimensional coverage, was shown to be inadequate to cover a highly three-dimensional aquifer system such as the Coal Measures;
- lack of repeat sampling.

Conclusion

The industrial revolution of the 17th and 18th centuries resulted in a massive explosion of urban population and industry dominantly centred on the coalfields of the UK. The growth of manufacturing industries drove the demand for large quantities of coal to provide power to the urban centres. Three centuries later a large portion of the once-extensive active coal industry has been closed with an associated decline in the heavy industry. A legacy of this industrial activity nevertheless remains today in the form of large areas of land and groundwater (Dowse & Selly 1975) contaminated by a wide variety of potentially hazardous materials.

The mechanisms of flow, contaminant transport and evolution of groundwater chemistry are therefore crucial to successfully tackling the remediation of contaminated sites on the British coalfields. The acid tar pits site in southern Derbyshire illustrates the urgent need for future research into this task; the Coal Measures strata of the UK are among the most complex, interesting and yet under-researched aquifer systems of the UK, exhibiting the following properties:

- Some aspects of multi-aquifer character; sandstone horizons alternating with mudstones and shales. The hydraulic roles of these units appear to be much more complex than the simplistic model of sandstone aquifers and mudstone/shale aquitards often presented; the mudstones having the potential to contain significantly transmissive discontinuities.
- The strata are dominated by fracture flow mechanisms, although intergranular porosity is likely to represent a significant sink for diffusion of contaminants. The hydraulic conductivity of the sequence cannot be simply estimated from semi-quantitative rock-quality description factors. Lithology must also be taken into account; poorly fractured sandstones being potentially more conductive than highly fractured mudstones.

Table 11. *Detailed assessment of selected incineration and encapsulation remediation options*

Remedial option	Advantages	Disadvantages
1. Excavate and place in an on-site containment cell	• Waste totally contained. Minimal environmental contamination. • Free tar is encapsulated and prevented from migration • Cheaper than remedial treatment of waste • Reduces odour problem • Chemical/physical stabilization of wastes could be attempted prior to emplacement • Greater flexibility for maximum utilization of the site in future development	• Leachate produced will be contained and concentrated with waste remaining intact indefinitely • Exact quantity of waste uncertain prior to excavation • Extraction may destabilise waste • Problems with groundwater management during extraction • Requires long-term maintenance and management • Long-term efficacy/performance of gas vents and geomembranes unknown
2. *In situ* capping	• Aesthetically preferable to 'do-nothing' • Leachate generation reduced • If excavation is minimized, little risk of contaminant mobilization or spillage during construction • Risk of direct exposure to users of site reduced • Odour problems will be reduced • Cheaper than total containment	• Continued leaching of contaminants to groundwater • Area of tar pits sterilized for future development • Waste is still vulnerable to future changes in environmental conditions (e.g. cessation of mine pumping) • Full containment of tar is not achieved, and tar could still conceivably migrate from the tar pits • Requires long-term maintenance and management • Long-term efficacy of gas vents unknown • Long-term performance of geomembranes unknown
3. Excavate and dispose at an off-site landfill		• Considered unrealistic due to difficulty in finding a landfill prepared to accept the waste.
4. 'Do-nothing' option	• Low cost • No significant current water pollution risk identified with *in situ* waste • No possibility of further mobilization of contamination due to remedial works	• Aesthetically and 'politically' problematic • Continued odour problems • Risk associated with continued public access to site • Site unsuitable for redevelopment • Continued opportunity for flow of tar • Possibility of increased lateral run-off of dissolved contaminants caused by future cessation of mine pumping • Future legislation could impose one of the more costly options at a future date. • Requires long-term maintenance of site fences, etc., and monitoring of waste
• 5. On-site treatment of heavily contaminated wastes by fluidised bed incineration; on-site encapsulation of moderately contaminated materials	• Waste totally contained • Free tar destroyed • Greater flexibility for maximum utilization of site in future development • Leachate in cell less strong than for Option 1	• As Option 1 • Requires HMIP approval
• 6. Off-site treatment of heavily contaminated wastes by high temperature incineration; off-site landfill of moderately contaminated materials	• Creates a clean site, free from development constraints • Provides a complete solution to environmental contamination • Protects all sensitive targets, on- and off-site • Does not require long-term site management	• High cost • Need to neutralize tars before incineration • Exact quantities of waste uncertain prior to excavation • Problems with groundwater management during excavation • Significant traffic on local roads
• 7. *In situ* treatment of waste materials		• Unviable: no treatment methods identified which are suitable for waste and soil-types involved.

Table 12. *Summary of cost:benefit analysis of remedial options.*

Option no.	Description	Permanence	Risk reduction	Cost estimate (£M)
1	On-site encapsulation, *ex situ*	Medium term	Medium/high	>1.8
2	On-site capping, *in situ*	Medium term	Medium	1.0
4	Do nothing (and monitor)	Short term	Short term	0.1
5	On-site treatment and encapsulation	Long term	High	3.3–4.5
6	Off-site treatment and landfill	Long term	Almost total	24.8–42.3

- Mine workings, shafts and collapsed strata above mine workings represent rapid transport pathways for water and contaminants
- Vertical groundwater head gradients are highly important, being induced both by topographic effects and under-drainage by coal workings.
- The aquifer system is strongly heterogeneous, anisotropic and discontinuous in nature.

The combination of these factors implies that the costs of an adequate hydrogeological investigation of contamination in Coal Measures aquifers are extremely high, requiring a three-dimensional array of piezometers and monitoring wells. It also renders the quantitative prediction of contaminant transport and the remediation of contaminated groundwater in such aquifers an exceptionally complex task.

The authors wish to acknowledge the contribution made to the study by D. Cragg and R. Moore of Scott Wilson Kirkpatrick; H. Mallett, S. Pollard and P. Crowcroft of Aspinwall; and P. Hossain of Camp Dresser & McKee. We also thank the following for the information and useful discussions they have contributed to the study: S. Pearson, P. Clark, A. Wood, C. Beech of Derbyshire County Council; T. Barnes, M. Allen, B. Brown and P. Day, of the former British Coal; and P. Reeves of the National Rivers Authority. G. Reeves of Newcastle University and M. de Freitas of Imperial College were of great assistance to S. Power during his MSc thesis on the study site. The Natural Environment Research Council (NERC) are thanked for funding S. Power's MSc. Finally, we thank Derbyshire County Council for permission to publish the results of this study.

Postscript

The environmental and coal industries of Britain have undergone significant metamorphosis since the study described here. The role of the National Rivers Authority has been taken over by the new Environment Agency. British Coal is now largely defunct, its regulatory function having been taken over by the Coal Authority and its operational function by the private sector, dominated in the East Midlands by the firm RJB Mining.

References

ALDOUS, P. J. & SMART, P. L. 1988. Tracing groundwater movement in abandoned coal mined aquifers using fluorescent dyes. *Ground Water*, **26**, 172–178.

——, —— & BLACK, J. A. 1986. Groundwater management problems of abandoned coal mined aquifers: a case study of the Forest of Dean. *Quarterly Journal of Engineering Geology*, **19**, 375–388.

ANON 1993. River in jeopardy as Durham coal mines close. *ENDS Report*, **223**, 11–12.

AWBERY, H. G. 1988. The protection of the Nottinghamshire coalfield by the Bentinck Colliery mine water concentration scheme. *International Journal of Mine Water*, **7**(1), 9–24.

BANKS, D. 1992. Estimation of apparent transmissivity from capacity testing of boreholes in bedrock aquifers. *Applied Hydrogeology*, **4/92**, 5–19.

—— 1997. Hydrogeochemistry of Millstone Grit and Coal Measures groundwaters, south Yorkshire and north Derbyshire, UK. *Quarterly Journal of Engineering Geology*, **30**, 237–256.

——, BURKE, S. P. & GRAY, C. G. 1997. Hydrogeochemistry of coal mine drainage and other ferriginous waters in north Derbyshire and south Yorkshire, UK. *Quarterly Journal of Engineering Geology*, **30**, 257–280.

BRADBURY, K. R. & MULDOON, M. 1994. Effects of fracture density and anisotropy on well head protection area delineation in fractured rock aquifers. *Applied Hydrogeology*, **3/94**, 17–23.

BRITISH STANDARDS INSTITUTION 1981. BS 5930: *Code of Practice for Site Investigation*. BSI, London.

BROMEHEAD, C. E. N. 1948. Practical geology in ancient Britain: Part 1, The metals. *Proceedings of the Geologists' Association*, **59**, 65–76.

BUSBY, J. P., RAINES, M. G., EVANS, A. D. & OGILVY, R. D. 1994. *Tar pits; geophysical survey and interpretation*. British Geological Survey Technical Report WN/94/1C.

CIRIA 1995. *Remedial Treatment for Contaminated Land. Vol. III: Site Investigation and Assessment*. CIRIA (Construction Industry Research and Information Association) Special Publication No. 103, London, UK.

CRIPPS, J. C., DEAVES, A. P., BELL, F. G. & CULSHAW, M. G. 1993. The Don Valley Intercepting Sewer Scheme, Sheffield, England: An investigation of flow into underground workings. *Bulletin of the Association of Engineering Geology*, **15**(4), 409–425.

DOWSE, L. H. & SELLY, K. H. 1975. Groundwater pollution in an industrialized part of the Trent Basin. *Water Pollution Control*, **74**, 526–541.

DUMPLETON, S. & GLOVER, B. W. 1995. *The impact of colliery closure on water resources, with particular regard to NRA Severn–Trent Region*. British Geological Survey, Technical Report Series WD/95.

ENVIRONMENT CANADA 1987–1990. *Sydney tar ponds clean-update*. Newsletter June 1987–March 1988; April 1988–March 1989; April 1989–March 1990. Sydney Tar Ponds Clean-up Project, Canada.

—— 1989. PAHs in the tar ponds: the sources, dangers and destruction of polynuclear aromatic hydrocarbons. *Sydney tar ponds clean-up factsheet.* Sydney Tar Ponds Clean-up Project, Ministry of Supply and Services, Canada.

FROST, D. V. & SMART, H. G. O. 1979. *Geology of the Country around Derby*. Geological Survey Memoir for 1 : 50,000 geological sheet 125. IGS, London.

GEOTECHNICAL ENGINEERING OFFICE OF HONG KONG 1988. *Geoguide 3: Guide to Soil and Rock Descriptions*. Hong Kong Government.

GIBSON, W., POCOCK, T. I., WEDD, C. B. & SHERLOCK, R. L. 1908. *The Geology of the Southern Part of the Derbyshire and Nottinghamshire Coalfield*. Memoir of the Geological Survey of Great Britain.

GRIFFIN, A. R. 1971. *Mining in the East Midlands, 1550–1947*. Frank Cass & Co.

HARRISON, I. B., PARKER, A. & WILLIAMS, G. M. 1981. *Investigation of the landfill at Eastfield Quarry, Fauldhouse, West Lothian, Scotland*. IGS Report 81/13. IGS, London.

HARRISON, B., SUDICKY, E. A. & CHERRY, J. A. 1992. Numerical analysis of solute migration through fractured clayey deposits into underlying aquifers. *Water Resources Research*, **28**, 515–526.

HENTON, M. P. 1974. Hydrogeological problems associated with waste disposal into abandoned coal workings. *Water Services*, **78**(944), 349–353.

HOEK, E. & BRAY, J. W. 1981. *Rock Slope Engineering*. 3rd edition. Institute of Mining and Metallurgy, London.

HVORSLEV, M. J. 1951. *Time Lag and Soil Permeability in Ground Water Observations*. US Army Corps of Engineers Waterway Experimentation Station, Bulletin 36.

LEMON, R. 1991. Pumping and disposal of deep strata minewater. *Mining Technology*, March, 69–76.

NRA 1991. *The Quality of Rivers, Canals and Estuaries in England and Wales*. Water Quality Series No. 4, National Rivers Authority.

—— 1992. *Policy and Practice for the Protection of Groundwater*. National Rivers Authority.

POWER, S. 1994. *Contaminant Transport in the British Coal Measures Strata*. MSc thesis, Department of Geology, Royal School of Mines, Imperial College of Science, Technology and Medicine, London.

PRICE, M. 1996. *Introducing Groundwater*. Chapman & Hall, London.

SCOTTISH ENTERPRISE 1994. *How to Investigate Contaminated Land*. Scottish Enterprise, Glasgow.

SIMS, R. C. 1990. Soil remediation techniques at uncontrolled hazardous waste sites: a critical review. *Journal of the Air and Waste Management Association*, **40**, 704–732.

SINGLETON, R. G. 1987. *An Evaluation of the Treatment and Disposal Options for Acid Tars*. MSc Thesis, Imperial College, London.

SMITH, E. G., RHYS, G. H. & EDEN, R. A. 1967. *Geology of the Country around Chesterfield, Matlock and Mansfield (Memoir of the Geological Survey of Great Britain, Explanation of One-Inch Geological Sheet 112, New Series)*. Institution of Geological Sciences, HMSO, London.

SMITH, E. T. & BROMLEY, J. 1982. The British Experience – Clean Up. *In: Hazardous Waste Disposal*. Ann Arbor, 192–193.

SOMERVILLE, S. H. 1986. *Control of Groundwater for Temporary Works*. CIRIA (Construction Industry Research and Information Association), Report 113.

THORNE, G. A. & GASCOYNE, M. 1993. Groundwater recharge and discharge characteristics in granitic terranes of the Canadian Shield. *In: Proceedings of the XXIVth Congress of the International Association of Hydrogeologists*, 28 June–2 July 1993, Oslo, Norway, 368–374.

TOTH, J. A. 1963. A theoretical analysis of groundwater flow in small drainage basins. *Journal of Geophysical Research*, **68**(16), 4795–4811.

WILLIAMS, G. M. & AITKENHEAD, N. 1991. Lessons from Loscoe: the uncontrolled migration of landfill gas. *Quarterly Journal of Engineering Geology*, **24**, 191–207.

YOUNGER, P. L. 1992. The hydrogeological use of thin sections: inexpensive estimates of groundwater flow and transport parameters. *Quarterly Journal of Engineering Geology*, **25**, 159–164.

—— & SHERWOOD, J. M. 1993. The cost of decommissioning a coalfield; potential environmental problems in county Durham. *Mineral Planning*, **57**, 26–29.

Great Bridge Marl Pit: a case study in the prevention of contaminant migration

BARRY C. GORE & IAN M. CAMPBELL

Scott Wilson Kirkpatrick & Co Ltd, Bayheath House, Rose Hill West, Chesterfield, Derbyshire S40 1JF, UK

Abstract: Great Bridge Marl Pit is a former brick pit located in the West Midlands. The pit is infilled by an assortment of materials including a wide range of toxic and aggressive substances (most significantly phenolic liquids). Phase 2 of the Black Country Spine Road (opened 1995) crosses the pit on a raised embankment.

Prior to the construction of the road, the site was derelict and contaminated groundwater within the pit was thought to present only a minor risk to the adjacent River Tame and virtually no risk to local groundwater abstractions. Mobile contaminants within the pit were contained by a combination of the geometry of the pit, the location of the bulk of the contaminative materials within the base of the excavation and the low permeability of the base of the excavation.

However, impact of road construction over the site, in particular the preconsolidation of the pit infill by surcharging, was considered to have the potential to mobilize significant volumes of contaminated groundwater from within the pit and therefore to present a considerable risk of pollution of the River Tame.

To minimize the potential for contaminant migration, groundwater flow within the Marl Pit and in its immediate vicinity was controlled during the surcharging period by the installation of a groundwater abstraction system. Contaminated water abstracted during the surcharging was disposed of to trunk sewer and treated at the local water treatment works. The efficiency of the groundwater abstraction system was assessed by monitoring groundwater pressures and concentrations of contaminants in a number of strategically located boreholes around the margins of the site.

During the surcharging period the groundwater control system successfully controlled excess groundwater pressures generated beneath the surcharge and there was no evidence of an increase in contaminant migration from the pit; in fact a considerable reduction in the volume of contaminated groundwater present within the pit is likely to have occurred.

The area of the West Midlands of the UK, known as the Black Country, has a long history of industrial activity, the decline of which has left a legacy of large areas of derelict land which have the potential to contaminate both surface and groundwater. During the late 1980s and early 1990s the challenge of regenerating the area has been the responsibility of the Black Country Development Corporation. Central to plans for regeneration has been the improvement of the transport infrastructure within the area by the construction of the Black Country Spine Road, with the objective of opening up considerable areas of derelict land for development.

The route of an elevated section of the road was to pass across a derelict site located in the village of Great Bridge, near Tipton, to the eastern bank of the Oldbury arm of the Tame, the principal river draining the southern half of the Black Country. On investigation the site proved to include an infilled brick quarry or 'marl pit' containing a considerable thickness of made ground, much of which could be described as heavily contaminated. Contaminated groundwater located within the pit was thought to have the potential migrate in response to the construction of the road and to pollute the River Tame.

This paper details the steps taken to investigate of the nature of the contamination at the marl pit site, and to understand its hydrogeology and the subsequent design and implementation of the groundwater control measures necessary to ensure that potentially deleterious changes in surface and groundwater quality did not occur in response to the construction of the Black Country Spine Road embankment.

Background

Site history

In common with much of the Black Country, industrialization of the Great Bridge area dates back several centuries. Following limited mining of the Carboniferous Productive Coal Measures at depth beneath the site in the mid-eighteenth century, bedrock mudstones were extracted to provide raw materials for the manufacture of Staffordshire Blue Bricks. The

GORE, B. C. & CAMPBELL, I. M. 1998. Great Bridge Marl Pit: a case study in the prevention of contaminant migration. *In*: MATHER, J., BANKS, D., DUMPLETON, S. & FERMOR, M. (eds) *Groundwater Contaminants and their Migration*. Geological Society, London, Special Publications, **128**, 313–331.

resulting void, known locally as 'The Marl Pit', fell into disuse by the beginning of the twentieth century and subsequently became flooded.

For the next three-quarters of a century the pit was used for the unlicensed deposition of industrial waste materials. Although no detailed records exist of what was deposited within the pit during this period, phenolic liquors are known to have been dumped at the site in the 1960s. The Marl Pit was used as a licensed waste disposal site for the deposition of 'inert' industrial wastes from the mid-1970s and had become completely infilled by the late 1980s.

Geology

Made ground. The site was the subject of a comprehensive geotechnical and chemical investigation as part of the advance works for the Black Country Spine Road. During the investigation, more than 50 cable percussion holes were bored through the pit infill, demonstrating that the pit exceeded 2 ha in surface area and was over 30 m deep at its centre prior to abandonment (Fig. 1).

The materials infilling the pit were found to comprise clays and silts, ash sands and silts, gravels of slag, colliery spoil (including coal, mudstone and quartzite gravel), clinker and building rubble, together with poorly consolidated silt and with pockets of tar. The composition of the materials reflected the by-products and wastes derived from the former industrial land uses in the Great Bridge area, which are known to have included brick and tile works, collieries, iron and brass foundries, gasworks and slag works.

The distribution of the fill materials within the Marl Pit is naturally exceedingly complex. However, a simplified stratigraphy involving subdividing the infill into three layers (Fig. 2) was found to adequately describe the sequence of 'strata' within the infilled excavation. The stratigraphy consists of the following (in order of increasing depth):

- *Layer 1*: up to 6 m of clayey sands or sandy clays (this layer extended beyond the margins of the pit and mantles almost the entire immediately surrounding area).
- *Layer 2*: up to 23 m of slag gravel (which can be subdivided into upper and lower layers on

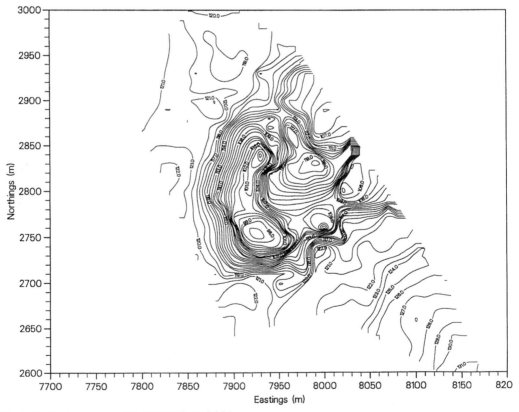

Fig. 1. Bedrock contours of the Marl Pit (mAOD).

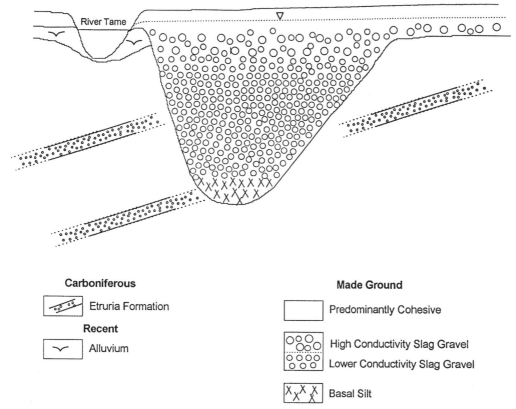

Fig. 2. Sketch geological cross-section of the Marl Pit area.

the basis of hydraulic conductivity, the upper layer being several metres thick).

- *Layer 3*: up to 9 m of poorly consolidated silt concentrated in the deepest parts of the pit.

Bedrock. The bedrock in the Great Bridge area is the Upper Carboniferous (Westphalian C), Etruria Formation, a sequence of non-calcareous mudstones and siltstones approximately 120 m thick, containing numerous coarse-grained sandstone and conglomerate lenses. These strata dip gently to the west beneath the site at around 7° (Fig. 2).

Hydrogeology

Made ground. The distribution of shallow groundwater within fill materials was found to be extremely variable. A number of high-level perched potentiometric surfaces were recorded in the predominantly cohesive Layer 1. Penetration of the saturated upper slag gravel Layer 2 (usually in the region of 4–6 m below ground level) was generally followed by a rise in water level of between 1 and 2 m; behaviour indicative of the presence of confining conditions caused by the presence of the predominantly cohesive upper layer of fill materials.

Estimates of the permeabilities of the upper slag gravel layer were calculated from a number of falling head permeability tests conducted during the site investigation. These indicated that the hydraulic conductivity of this layer was highly variable due to the presence of many alternative pathways for fluid flow (including voids left by insufficient compaction during placement). This zone of high conductivity within the upper parts of the slag gravel was around 2–4 m thick with a mean hydraulic conductivity of perhaps 2.9×10^{-4}–$1.2 \times 10^{-3} \, \mathrm{m \, s^{-1}}$.

The upper slag gravel layer was underlain by slag gravel and silt layers of lower hydraulic conductivities. These were investigated following the installation of a number of additional boreholes within the boundaries of the pit. Two pumping tests were conducted, the results of which indicated that groundwater at depth within the pit was also weakly confined and that the permeability of the slag layer exceeded that of the basal silt layer.

Best estimates of transmissivities obtained for these materials during the tests were:

- lower granular slag layer: $T = 12.5 \times 10^{-5}\,\text{m}^2\,\text{s}^{-1}$ ($K = 6.3 \times 10^{-6}\,\text{m}\,\text{s}^{-1}$ assuming a thickness of 20 m);
- basal silt layer: $T = 2.5 \times 10^{-5}\,\text{m}^2\,\text{s}^{-1}$ ($K = 2.5 \times 10^{-6}\,\text{m}\,\text{s}^{-1}$ assuming a thickness of 9 m).

Etruria Formation. The presence of confined, or semi-confined, groundwater was recorded whenever boreholes penetrated discrete sandstone units within the Etruria Formation. Although the potentiometric surface within individual sandstone units was variable, it was generally around 2 m higher than that in the made ground.

Investigation of the hydraulic properties of the Etruria Formation was restricted to a falling head test conducted on a single cable percussion borehole and a number of packer injection tests undertaken on a rotary cored hole. Both holes were located to the west of the site, on the opposite bank of the River Tame.

The sandstones within the Etruria Formation had an intergranular permeability of between 1.2×10^{-6} and $2.3 \times 10^{-6}\,\text{m}\,\text{s}^{-1}$ whilst the intervening mudstones exhibited a low, but measurable, secondary permeability (related to jointing) in the order of $3.5 \times 10^{-7}\,\text{m}\,\text{s}^{-1}$.

Groundwater flow model

Measurements of groundwater levels within the fill using standpipes indicated a gentle head gradient (around $1\frac{1}{2}°$) across the site declining from the southeast to the northwest. The potentiometric surface within the fill materials proved to be a gentle expression of the change in ground surface elevation over the site and was between 3 and 5 m below ground level; the presence of the pit having no discernible influence. The principal direction of groundwater flow was identified as being sub-parallel to the river in a north-northwesterly direction (Fig. 3) with the potentiometric surface in the extreme northwest of the site, close to the River Tame, being approximately 1.5 m above river level.

Monitoring of groundwater levels and contaminant concentrations in groundwater during the advance works for the contract provided some evidence for the presence of a secondary migration pathway which appeared to be associated with the presence of a shallow dry gully, or drainage ditch, in the south of the site (Fig. 3). The gully was some 15 m wide, 3–4 m deep and approximately 160 m long, running approximately E–W and terminating at the river at around 2.5 m above the water level.

The measured permeabilities of the pit infill implied that active recharge and significant throughflow would be restricted to the near-surface high conductivity slag gravel layer, and that this layer was likely to contribute a small amount of baseflow to the River Tame at the northwestern end of the site. The much lower hydraulic conductivity of the underlying slag gravel and basal silt layers implied that groundwater flow at depth within the pit was likely to be at much slower velocities. Groundwater within the deeper parts of the pit was therefore thought to effectively be contained by the surrounding Etruria Formation. The possibility of off-site groundwater migration from this region via the more permeable sandstone units within the Etruria Formation was restricted by their impersistance.

Contamination

Contaminants of concern

Chemical analyses of soil and water samples taken during the site investigation revealed the presence of extensive contamination of both fill materials and groundwater present within the pit (Tables 1 and 2). The principal potential risk posed by the contamination was the threat to the surface-water and groundwater resources in the vicinity of the site as a consequence of off-site contaminant migration. The chemical species most likely to be mobile were those found in elevated concentrations in both soil and water samples, i.e. phenolic hydrocarbons, arsenic, copper, lead, mercury, nickel, zinc, sulphates, sulphides and trichloromethane. Extremes of pH were also recorded during the investigation.

Of the contaminants of concern, the greatest potential for pollution of controlled waters was presented by the greatly elevated concentrations of phenols recorded in groundwater samples. The soil survey had shown that concentrations of phenolic substances increased with depth; a finding corroborated by analyses of groundwater taken from the upper 14 m of fill materials.

On the basis of the difference in density between phenolic compounds and water, it was therefore anticipated that the most highly contaminated groundwater would be found within the deepest parts of the pit. This hypothesis was

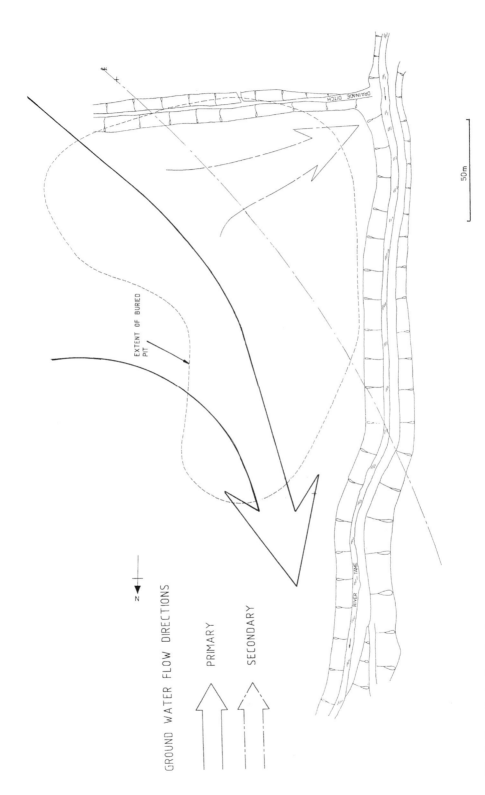

Fig. 3. Groundwater flow directions.

Table 1. *Summary of selected chemical analyses: made ground*

Determinand	Maximum	Minimum	Mean
pH	12.2	2.1	8.4
Electrical conductivity ($\mu s\,cm^{-1}$)	21 200	37	740
Phenol ($mg\,kg^{-1}\,C_6H_5OH$)	1 013	1	21
Toluene extractable matter (%)	13.3	0.01	0.13
Organic matter ($mg\,kg^{-1}$)	31.5	0.1	9.1
Arsenic ($mg\,kg^{-1}\,As$)	140	<2	8
Cadmium ($mg\,kg^{-1}\,Cd$)	610	<0.5	24
Chromium ($mg\,kg^{-1}\,Cr$)	698	<10	42
Copper (% Cu)	5.0	<0.001	0.019
Lead (% Pb)	3.1	<0.001	0.017
Nickel (% Ni)	3.8	<0.001	0.004
Zinc (% Zn)	2.0	<0.001	0.030

confirmed by chemical analysis of the groundwater extracted from the basal silt layer during test pumping in which a peak concentration of $8600\,mg\,l^{-1}$ phenol in water was recorded.

The risk to surface waters and shallow groundwater

The concentration of the heavily contaminated phenolic liquors in what was effectively a sump in deepest regions of the infilled excavation provided a good reason for supposing that the pit did not pose a significant risk to water quality in the River Tame. Migration of phenol from the sump was anticipated to be predominantly in response to concentration gradients (via diffusion) with only a limited role for advection because of the inferred low groundwater velocities within the fill materials at this depth. Phenol diffusing into the upper layers of the granular fill materials, where higher

groundwater flow velocities were expected, would be dispersed and diluted by advective transport mechanisms. The quantities of phenol subsequently migrating into the River Tame via fast-moving shallow groundwater was therefore thought to be very small, although measurable.

Although some minor seepages of phenol-contaminated water were visible along the banks of the River Tame adjacent to the Marl Pit, this supposition was supported by monitoring of water quality in the River Tame both upstream and downstream of the site. This provided no clear evidence of significant additional phenol contamination of the river which could be attributed to the presence of the Marl Pit (Table 3). In fact, the monitoring identified phenolic compounds as being present in river water upstream of the site, confirming that sources of phenol other than the Marl Pit may be found within the Black Country catchment of the River Tame.

Table 2. *Summary of selected chemical analyses: groundwater in made ground*

Determinand	Maximum	Minimum	Mean
pH	11.9	5.4	8.7
Electrical conductivity ($\mu s\,cm^{-1}$)	20 800	357	2 437
Chemical oxygen demand ($mg\,l^{-1}\,O$)	22 272	5	1 483
Biochemical oxygen demand ($mg\,l^{-1}\,O$)	11 000	5	505
Phenol ($mg\,l^{-1}\,C_6H_5OH$)	4 672	0.1	133.1
Total organic carbon ($mg\,l^{-1}$)	4 693	2	301
Total organo-chlorine compounds ($mg\,l^{-1}$)	15.5		
Arsenic ($mg\,l^{-1}\,As$)	3.00	<0.01	0.05
Cadmium ($mg\,l^{-1}\,Cd$)	0.23	<0.01	0.01
Chromium ($mg\,l^{-1}\,Cr$)	4.47	<0.02	0.05
Copper ($mg\,l^{-1}\,Cu$)	30.0	<0.02	0.24
Lead ($mg\,l^{-1}\,Pb$)	26.0	<0.02	0.21
Nickel ($mg\,l^{-1}\,Ni$)	3.01	<0.02	0.05
Zinc ($mg\,l^{-1}\,Zn$)	69.8	<0.022	0.8

Table 3. *Summary of water quality in the River Tame*

Determinand	Maximum	Minimum	Mean
Upstream			
pH	7.7	7.5	7.6
Electrical conductivity ($\mu s\,cm^{-1}$)	1436	1165	1287
Chemical oxygen demand ($mg\,l^{-1}\,O$)	79	17	41
Phenol ($mg\,l^{-1}\,C_6H_5OH$)	1.3	0.05	0.28
Downstream			
pH	7.7	7.4	7.5
Electrical conductivity ($\mu s\,cm^{-1}$)	11529	1166	2593
Chemical oxygen demand ($mg\,l^{-1}\,O$)	222	1.3	54
Phenol ($mg\,l^{-1}\,C_6H_5OH$)	0.1	0.05	0.06

The risk to groundwater within the Etruria Formation

The groundwater contained within the sandstones of the Etruria Formation, although not a major resource, is locally abstracted for industrial use, although there were no abstractions within the immediate vicinity of the site. The nearest abstraction point to the site, located approximately 300 m to the west of the Marl Pit on the opposite bank of the River Tame, was closed down some years before construction of the road commenced.

The risk to any abstractions further away was thought to be very low because of the lack of hydraulic continuity between individual sandstone beds and the potential for significant chemical and biological attenuation of phenolic contaminants as a result of long contaminant transport travel times (Darcy velocities within the Etruria Formation of between 3 and 25 m year^{-1} have been suggested).

Groundwater control measures

Implications of site development

The Black Country Spine Road was to pass directly over the centre of the pit in a southeast to northwest alignment on a raised embankment. The loading of the poorly consolidated fill materials within the pit by the embankment (especially the basal silt which was thought to be normally, or possibly under, consolidated) presented an obstacle to the construction of the road in the form of unacceptably large settlements. From a geotechnical perspective, the most effective solution to this problem was the preconsolidation of the material within the pit using a surcharge embankment, although numerous other options were considered.

Preloading involves the application of ground stresses equivalent to those likely to be imposed by the final construction in order to reduce the amount of long-term consolidation settlement affecting the completed works. Preloading can be achieved by construction of an earth embankment over the area to be loaded. Surcharging involves the addition of a temporary loading to the preload embankment to produce a total load greater than that which would be imposed by the final works. Consequently the surcharge will be capable of inducing the same degree of settlement as the preload embankment but in a significantly shorter period of time.

In the case of the Marl Pit, an embankment was designed to give a surcharging period of six months duration. After this time, excess material would be removed from the surcharge and the embankment reshaped to the final profile required for road construction. The surcharge embankment was built in stages to ensure pore-water pressures remained within tolerable limits and the stability of the embankment itself was not compromised. The total height of the surcharge embankment ranged from 4 m in the south to 12 m in the north; generally the top 4 m being the surcharge component.

By definition, surcharging of the Marl Pit would impose significant stresses on the pit infill which was thought likely to have a serious environmental impact on the vicinity of the site. The consolidation and dewatering of the basal silt, in particular, could mobilize large volumes of contaminated water with the potential to migrate off-site through near-surface fill materials and cause a serious deterioration in water quality within the River Tame. Clearly construction of the road embankment could not proceed without a mechanism for preventing such an occurrence in place.

Although various options were considered to limit or prevent contaminant mobilization (including obviating the need for a surcharge altogether by the total or partial excavation of materials within the marl pit and their replacement with clean-compacted material) the control of groundwater pressures and flows during surcharging by pumping represented the only feasible option given the geotechnical requirement for a surcharge embankment.

Design of groundwater control measures

A groundwater control scheme involving the extraction and treatment of phenol-contaminated groundwater from within the pit during the surcharging period was devised. The system aimed to prevent the migration of contaminated groundwater by dissipating excess hydraulic pressures generated by the surcharge whilst simultaneously generating an inward hydraulic gradient toward the pit and away from the river.

The National Rivers Authority, as the statutory regulator, was consulted at all stages with regard to the design and implementation of the proposed scheme. The NRA agreed that the approach to be adopted was the best practicable environmental option for the site, despite contaminated materials being left on-site after completion of the works. There was general agreement that there was a relatively low risk posed by the contamination before construction and that the control works would allow the *status quo* to be maintained. The use of a groundwater control system would also have the additional benefit of removing a significant volume of contaminated groundwater from the pit during its operation.

The control works as initially designed (Figs 4 and 5), involved the installation of a series of abstraction well points located along the up-gradient side of the surcharge embankment from the River Tame together with the construction of a physical barrier to contaminant migration. This involved the construction of a sheet pile cut-off wall between the surcharge embankment and River Tame along the principal line of groundwater migration, i.e. to the north-east of the surcharge embankment. Recognition of the possibility of a secondary migration pathway during the advance works led to a minor redesign of the works to incorporate abstraction points along the southwestern, down-gradient side of the surcharge embankment.

In total, 15 extraction wells were installed along the northeast side of the surcharge embankment and two to the southwest. Each well penetrated to the base of the pit infill and the upper 11 m of each well was cased off to prevent the entry of gases (including methane) from the unsaturated zone in the made ground. The wells were connected to a high-vacuum well point pump via a header manifold.

Although not directly related to the groundwater control measures, further protection to the River Tame was provided by construction of a temporary diversion channel (lined with high-density polyethylene) to allow construction of the toe of the surcharge embankment over the original course of the river.

Monitoring regime

Groundwater levels. To monitor the efficacy of the system, a total of 40 monitoring locations were used, subdivided into the following categories (Fig. 6):

• eight vibrating wire piezometers underneath the surcharge embankment (remotely monitored) to measure water pressure;
• 12 standpipe piezometers surrounding the surcharge embankment to measure water levels and sample;
• 18 standpipes located at a greater distance from the surcharge to measure water levels and sample from made ground and two standpipes to measure water levels and sample from Etruria Formation sandstones.

Using these instruments, changes in potentiometric level were measured prior to, during and following construction of the embankment.

Contaminant migration. To assess the effectiveness of the groundwater control system in preventing off-site contaminant migration, water samples were taken regularly from both the outer ring of 20 standpipes and the River Tame upstream and downstream of the site. These samples were analysed for the presence of the following potential contaminants:

• pH
• electrical conductivity
• ammonia
• chemical oxygen demand
• biochemical oxygen demand
• sulphide
• trichloromethane
• total phenol
• total zinc, cadmium, lead, nickel, copper, mercury and arsenic

Analyses collected over several months prior to commencement of surcharging were used to establish baseline conditions against which later readings could be compared (Table 4, Figs 12–14). Because of the length of turn-around

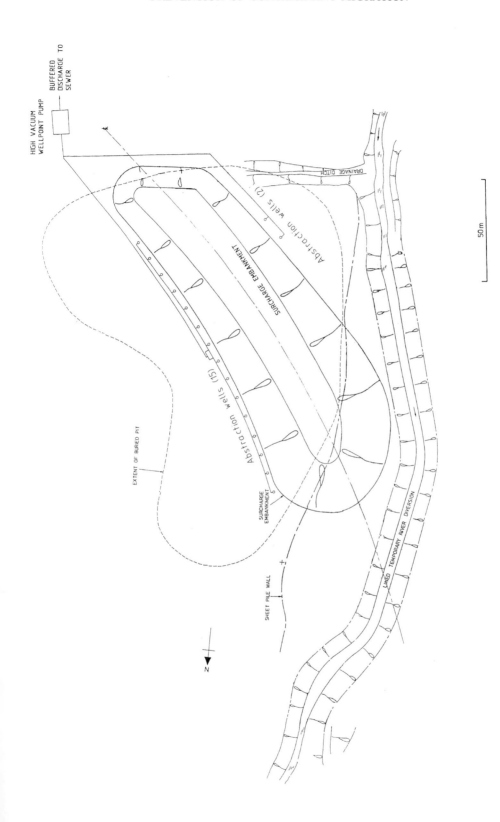

Fig. 4. Groundwater control measures: plan.

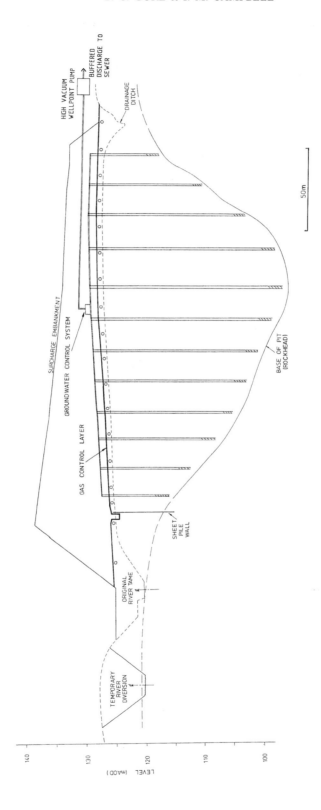

Fig. 5. Groundwater control measures: section (NNW–SSE).

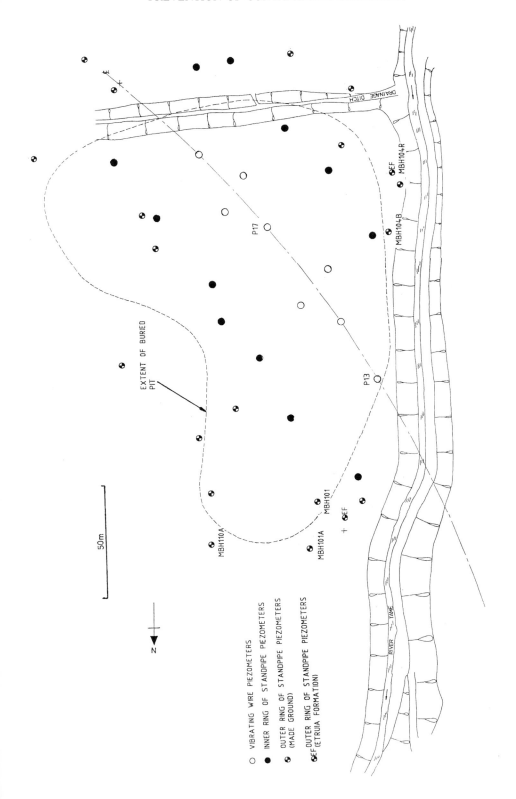

Fig. 6. Locations of groundwater monitoring instruments.

Table 4. Summary of baseline contaminant monitoring (peak values, all in milligrams per litre except where noted)

Determinand	MBH 101A	MBH 101	MBH 102R	MBH 102	MBH 104B	MBH 104R	MBH 105	MBH 106	MBH 107	MBH 108	MBH 109	MBH 110A	8SA15	Seepage	River Tame Upstream	River Tame Downstream
pH	7.8	–	7.9	–		–	7.3	8.5	–	7.8	9.3	11	8.8	8.2	–	–
EC (μS cm^{-1}, 20 °C)	3190	1260	1872	1057	2890	1614	2060	2330	1884	1593	643	928	1857	1870	1436	1448
COD	3705	221	147	113	792	5065	226	229	379	68	103	1084	216	384	79	59
BOD	3550						25	68		21.5	65	270	26	212		
Sulphide	12.2						0.6	0.04		0.4	2.4	1.6	<0.02	0.2		
Phenol	648	28.4					4.1	0.2		0.1	0.2	6.2	1.2	2.5		
Zinc	1.66		0.20	0.1	90.4	1873	0.86	0.08	0.20	15.0	0.71	2.09	0.09	0.05	0.6	0.05
Cadmium (μg l^{-1})	27						11	2		34	37	72	2	<20		
Lead	0.59						0.58	<0.1		3.05	0.17	0.66	0.1	<0.01		
Nickel	1.10						0.13	<0.02		0.51	7.23	1.35	2.89	4.11		
Copper	0.78						0.32	0.1		3.46	0.22	0.47	0.1	<0.02		
Mercury (μg l^{-1})	8.4						38	25		9.4	11	184	2.2	6		
Arsenic (μg l^{-1})	85						29	7		89	13	50	12	12		
Trichloromethane	<1						1	<1		<1	<1	1	<1	<1		
Ammonia	3.7						1.8	8.3		2.2	2.6	10.8	8.0	3.7		

required for chemical analyses, additional daily monitoring of indicators of water quality (pH, electrical conductivity) was taken at a number of key monitoring points located along the lines of postulated groundwater migration to provide an early warning of changes in water quality.

During the determination of baseline conditions, elevated concentrations of phenol in groundwater were noted not only in boreholes located both to the northeast of the surcharge embankment, along the postulated line of principal groundwater flow, but also to the west of the embankment. This finding led directly to the identification of the secondary migration pathway.

Water treatment and effluent disposal

Chemical analyses of contaminated water within the fill had shown that the phenolic component was biodegradable and that the whole was capable of treatment at the local water treatment works. A discharge consent (selected determinands from which are shown in Table 5) was negotiated with Severn Trent Water, allowing pumped groundwater to be disposed to trunk sewer (via foul water sewer) at a cost per unit volume based upon the volume and chemical composition of the effluent (principally chemical oxygen demand and suspended solids). To help in ensuring that the maximum contaminant concentrations acceptable to Severn Trent Water were not exceeded by short-term peaks, the effluent was buffered in on-surface in steel-holding tanks prior to discharge. The quality of water held in the storage tanks was monitored on a daily basis during pumping and the outflow

was continuously sampled for spot and 24 hour composite samples.

Performance

Operational phases

The operation of the groundwater control system passed through the following phases:

- Groundwater control pumping commenced   15 June 1993
- Start of surcharge construction   14 July 1993
- End of surcharge construction   30 September 1993
- Part of the pump system switched off   17 February 1994
- Entire pump system switched off   3 March 1994

These phases are shown in Figs 7–14 which illustrate changes groundwater level and concentrations of phenol in selected monitoring boreholes during the operation.

Groundwater pressures

Pre-surcharging. The groundwater control system was switched on one month before the commencement of surcharge construction for the following reasons:

(a) to ensure that the system was successfully controlling groundwater levels in the vicinity of the pit; and

(b) to allow immediate control of increases in groundwater level during embankment

Table 5. *Summary of discharge water quality, June/July 1993*

	Maximum	Minimum	Mean	Discharge consent
pH	10.3	9.6	10.1	6–12
Electrical conductivity ($\mu S\,cm^{-1}$)	5820	2370	4238	
Suspended solids ($mg\,l^{-1}$)	218	9	50	400
Ammonia ($mg\,l^{-1}\,N$)	12	0.3	8	50
Total sulphate ($mg\,l^{-1}$)	481	218	306	1000
Chemical oxygen demand ($mg\,l^{-1}$)	6190	1618	3736	20000
Phenol ($mg\,l^{-1}$)	1365	171	808	6000
Total cyanide ($mg\,l^{-1}$)	4.2	0.17	0.8	10
Zinc ($mg\,l^{-1}$)	0.38	0.02	0.09	5
Lead ($mg\,l^{-1}$)	0.12	0.005	0.03	5
Nickel ($mg\,l^{-1}$)	0.25	0.007	0.08	5
Iron ($mg\,l^{-1}$)	10.1	0.84	2.1	250
Chromium ($mg\,l^{-1}$)	0.2	0.02	0.05	5
Copper ($mg\,l^{-1}$)	0.29	0.015	0.07	5
Arsenic ($mg\,l^{-1}$)	0.378	0.005	0.143	2

construction by generating a substantial negative groundwater pressure regime prior to commencement of the works.

Following pump switch-on, the well point system, as designed, performed well; an inward hydraulic gradient being maintained (Figs 7–11).

During surcharging. Using the monitoring instruments installed under and around the surcharge embankment footprint, the drawdown resulting from the groundwater abstraction system and the increase and subsequent dissipation of pore-water pressures within the infill material beneath and around the surcharge embankment caused by embankment staged loading were monitored.

Following the decline in groundwater levels which immediately followed the switch-on of the groundwater control measures, two classes of response were noted in groundwater monitoring instruments with response zones in made ground. The first class was observed in vibrating wire piezometers located directly under the surcharge and in some standpipe piezometers in the immediate vicinity of the surcharge embankment. These instruments showed a stepped increase in groundwater level/pressure directly corresponding to staged loading of the surcharge embankment (Fig. 7). The magnitude of this build-up of groundwater pressures increased with the proximity of the monitoring instrument to the deepest areas of the pit (and the basal silt deposit) where peak pressures generated were

equal to, or exceeded, those recorded before pump switch-on (Fig. 8).

A stepped increase in groundwater levels was also observed in standpipe piezometers located away from the surcharge embankment along the direction of the primary migration pathway; behaviour indicating that excess groundwater pressures generated within the pit were being transmitted in this direction (Fig. 9).

The second class of response was observed in all remaining monitoring instruments located to the east and southwest of the surcharge embankment, at a greater distance from the groundwater control system and away from the direction of primary groundwater flow. In these standpipes there was no significant increase in groundwater pressures which could be attributed to surcharge construction; water levels being controlled solely by the operation of the groundwater control system (Fig. 10).

In a small number of boreholes responses different to those described above were observed. The two standpipes with response zones in Etruria Formation sandstones exhibited only a small increase in groundwater pressures attributable to surcharge construction and, in general, groundwater levels declined throughout the monitoring period (Fig. 11). Groundwater monitoring instruments to the south of the site, beyond the dry gully, did not respond to either the groundwater control or surcharging and were therefore thought to be located within a different hydrogeological system.

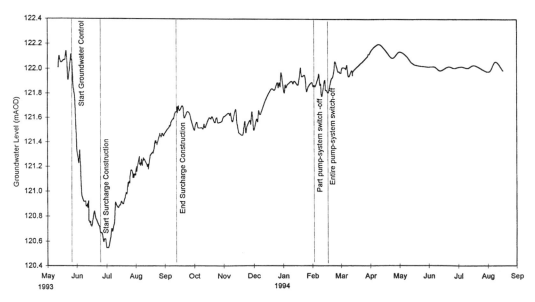

Fig. 7. Potentiometric level: vibrating wire piezometer P13.

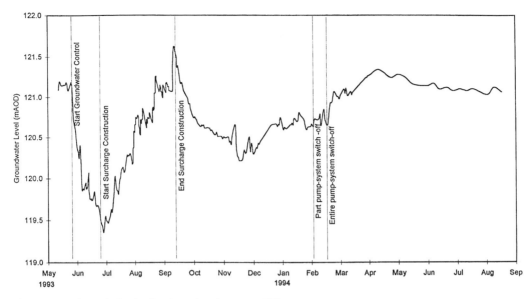

Fig. 8. Potentiometric level: vibrating wire piezometer P17.

Post-surcharging. In general, pore-water pressures beneath and around the surcharge embankment reached an equilibrium during a period of between two and six weeks after the end of surcharge construction (Figs 7 and 9). In those instruments in which the greatest overpressure had been observed, a decline in pressure to reach equilibrium occurred (Fig. 8). From this

it was surmised that primary consolidation due to surcharge construction had ceased. Part of the groundwater pumping system was switched off eight months after the start of surcharging and no noticeable increase in groundwater pressures or levels was noted subsequently (Figs 7–9). Based on this observation, the remainder of the system was shut down two weeks later.

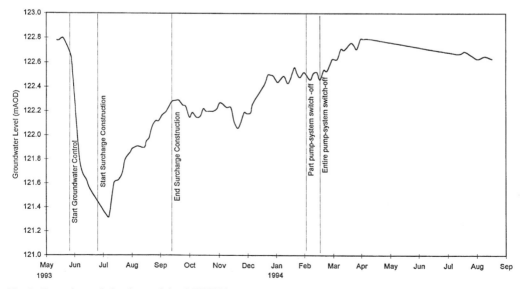

Fig. 9. Potentiometric level: standpipe MBH101.

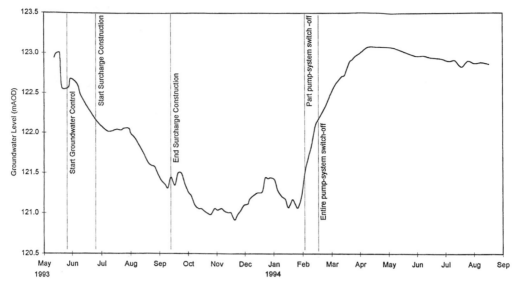

Fig. 10. Potentiometric level: standpipe MBH104B.

Over the following six to eight weeks, ground-water levels in all piezometers located within the boundary of the Marl Pit reverted (almost exactly) to those existing prior to groundwater lowering (Figs 7–10). Thus it could be reasonably assumed that conditions after pumping had ceased had reverted to those observed before groundwater control began and that the objective of maintaining the *status quo* had been achieved. Recovery of groundwater levels within the

Etruria Formation sandstones did not occur within the time frame of the study (Fig. 11).

Contaminant migration

During the determination of baseline conditions, only water samples from the outer ring of monitoring wells located along the postulated directions of groundwater flow (Fig. 4)

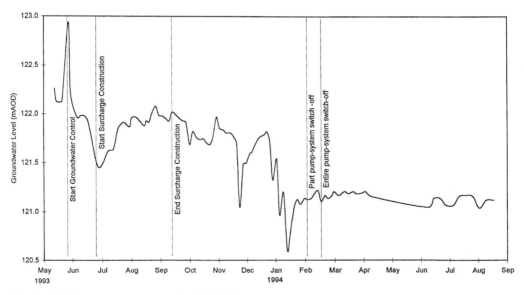

Fig. 11. Potentiometric level: standpipe MBH104R.

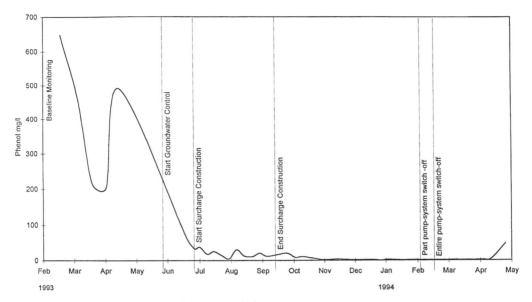

Fig. 12. Phenol concentrations: standpipe MBH101A.

unequivocally exhibited significant phenol contamination.

A very significant decrease in phenol concentrations was observed in a borehole located along the primary migration pathway over the period the pumps were on (Fig. 12). That this reduction was attributable to the operation of the groundwater control measures was shown by the rise in phenol concentrations which

followed pump switch-off (Fig. 12). In almost all other monitoring locations there was no significant change in the concentrations of phenol observed during groundwater control operations when compared with the variation observed during the baseline assessment period (Fig. 13). There was therefore no evidence that additional migration pathways had been created during the surcharging operation.

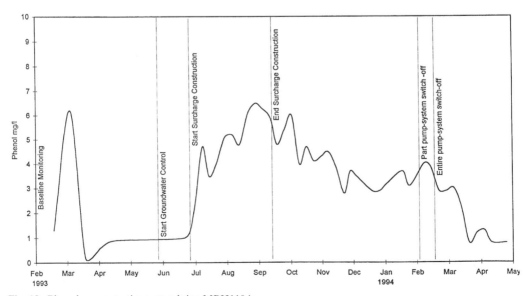

Fig. 13. Phenol concentrations: standpipe MBH110A.

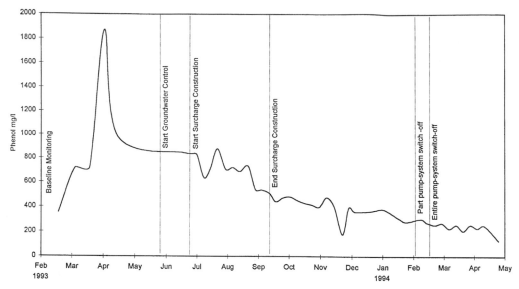

Fig. 14. Phenol concentrations: standpipe MBH104R.

Phenol concentrations in samples from the two boreholes with response zones within Etruria Formation sandstones steadily declined during the groundwater control operation (Fig. 14). This illustrates how the effectiveness of the groundwater control measures extended beyond reducing concentrations of phenol within the pit infill to removal of contaminated groundwater from plumes migrating off-site within permeable bedrock strata.

Quantity and quality of water abstracted

The estimated quantity of groundwater which would be expelled from the pit due to the surcharge was expected to be of the order of three million litres. However, due to the nature of the abstraction system devised, a greater volume of groundwater (up to 15 million litres) was actually abstracted. Discharge water quality was at its worst during June and July 1993 but did not exceed the discharge consent negotiated with Severn Trent Water (Table 5) and was therefore always capable of treatment.

Conclusions

This case study illustrates a practical response to minimizing the risk of additional surface water and groundwater contamination caused by contaminant migration in response to construction by the use of hydraulic control measures. The

location of the Marl Pit in an area of historic contaminative land-use such as the Black Country meant that although some off-site contaminant migration was occurring prior to road construction, the site could not be considered to be causing a serious pollution problem. The groundwater control measures were not designed to effect clean-up of contamination but were instigated to mitigate the negative environmental impact which could occur in response to the construction of the Black Country Spine Road embankment over the pit.

The results of analyses carried out both during the operational phase and subsequent to pump switch-off provided evidence that no additional migration of contaminated groundwater in response to surcharging occurred along the postulated groundwater flow and contaminant migration pathways. This was despite the transmission of groundwater pressures along the principal direction of groundwater flow. The scheme was therefore successful in achieving its stated objective of maintaining the *status quo* in respect of off-site contaminant migration.

In addition, the groundwater control measures appear to have had a considerable collateral benefit in removing contaminated groundwater not only from within the pit but also from the underlying Etruria Formation sandstones.

The primary reason for the success of the groundwater control operation was a sound conceptual understanding of the geology and hydrogeology of the area, including the identification of the principal contaminant migration pathways. The achievement of good hydrological control was aided by the geometry of the system, with much of the contaminated groundwater being contained within a sump in the deepest parts of the infilled excavation and therefore constrained by the surrounding low-permeability strata. The use of groundwater control in more open systems may present greater difficulties.

Pesticides in groundwater: some preliminary results from recent research in temperate and tropical environments

P. J. CHILTON, A. R. LAWRENCE & M. E. STUART

British Geological Survey, Maclean Building, Wallingford, Oxon OX10 8BB, UK

Abstract: The paper summarizes the preliminary results of field investigations of pesticide behaviour in the UK, Barbados and Sri Lanka. Field observations of groundwater pesticide concentrations in these studies range up to $100\,\mu g\,l^{-1}$, but are more commonly less than $5\,\mu g\,l^{-1}$. The risk of pesticide concentrations in pumped water reaching 10 or 100 times guideline values from normal agricultural use at recommended application rates is probably small. Simple laboratory determinations of degradation rates support earlier studies and general considerations of subsurface environmental conditions in suggesting that pesticides are likely to be more persistent in groundwater than in soils. Simple modelling of the saturated zone movement of pesticides towards wells indicates that concentrations in pump discharge are highly sensitive to aquifer porosity, pesticide half-life and extent of cultivated area. The greatest risk to groundwater from normal usage of pesticides in agriculture will occur where persistent compounds are applied over aquifers which are shallow, permeable and thin, and overlain by permeable soils.

During the past 15 years, nitrate has remained the most important issue in respect of groundwater pollution from agricultural practices. In contrast, the possibility of leaching of pesticides into the water environment has only been addressed more recently. The EC Drinking Water Directive set a stringent maximum admissible concentration for an individual pesticide in drinking water of $0.1\,\mu g\,l^{-1}$. In relation to analytical capabilities at the time, this was effectively a surrogate zero and, as for nitrate, compliance has produced significant treatment costs for the UK water industry.

In tropical regions, the growing demand for food will continue to result in increasing intensification of agriculture, often with irrigation and with heavy applications of agrochemicals. In many areas, such highly intensive crop production occurs on soils directly overlying shallow aquifers which are used for potable supply as well as for irrigation. There is increasing concern about the impact of the intensification of agriculture on groundwater quality, and it is important that the risks to groundwater quality posed by intensive agriculture and heavy use of agrochemicals be appraised, so that any necessary control measures can be introduced. This paper provides a brief review of the main factors which might determine pesticide migration to aquifers, supported by the preliminary results of field studies in the UK, Barbados and Sri Lanka.

Pesticide usage

The largest individual consumer of pesticides is the USA, followed by the countries of Western Europe. Developing countries together use only a small proportion of the total. Herbicides dominate pesticide use in temperate climates in Europe and North America, but insecticides are more commonly used elsewhere. Globally, pesticide use is concentrated on a small number of crops; more than 50% of the total is applied to wheat, maize, cotton and soya bean (Conway & Pretty 1991). In developing countries the highest applications are to plantation crops, although usage on vegetables is becoming more important. Total consumption of pesticides continues to increase; annual growth rates reached 12% in the 1960s but now are around 3–4% (Conway & Pretty 1991). In the UK, the most rapid growth has been associated with the increasing use of herbicides and fungicides in the cultivation of autumn-sown cereals. Increases in pesticide usage are now more rapid in many developing countries than in the developed world.

Observed occurrence of pesticides in groundwater

An increasing number of pesticides are being detected in groundwater in Europe and North America, as routine monitoring programmes are developed in response to tightening drinking water quality standards. Much less monitoring has been undertaken in tropical regions, largely because of the difficulty and cost of detecting pesticide residues at very low concentrations. All pesticide compounds pose a significant environmental health hazard since they are designed to be toxic and persistent. The EC Drinking Water Directive maximum admissible

CHILTON, P. J., LAWRENCE, A. R. & STUART, M. E. 1998. Pesticides in groundwater: some preliminary results from recent research in temperate and tropical environments. *In*: MATHER, J., BANKS, D., DUMPLETON, S. & FERMOR, M. (eds) *Groundwater Contaminants and their Migration*. Geological Society, London, Special Publications, **128**, 333–345.

concentration has been exceeded in some British water supply boreholes (Croll 1986; Lees & McVeigh 1988), although concentrations above $1 \mu g l^{-1}$ have not often been recorded. The pesticides most frequently encountered to date are herbicides: atrazine, simazine, mecoprop and isoproturon. The triazines are not extensively used agriculturally in the UK, and their occurrence results from their use as total weedkillers on railway tracks, highways, airfields and other paved areas.

The triazines, together with various soil insecticides, especially the carbamates and chloropropanes, have been found in concentrations exceeding $1 \mu g l^{-1}$ in shallow aquifers elsewhere in Europe and the USA (Wehtje *et al.* 1981; Zaki *et al.* 1982; Cohen *et al.* 1986; Jones *et al.* 1988). Most pesticide concentrations observed in these studies have been in the range $0.1-100 \mu g l^{-1}$, and it seems likely that concentrations significantly above this range can be attributed to local point-source contamination close to the well or borehole, rather than conventional agricultural use (Cohen 1990). More recently, the US Environmental Protection Agency has completed the first national survey of pesticides in private and public potable supply boreholes (EPA 1990). Atrazine and DCPA were the most commonly identified compounds, being found in 10% of community boreholes and 4% of rural domestic boreholes. However, none of the detections in the community water supplies were above the respective health advisory levels ($3 \mu g l^{-1}$ and $4 \mu g l^{-1}$ respectively), and only 0.6% of rural supplies exceeded these levels (EPA 1990).

Pesticide transport and behaviour

Pesticide leaching from the soil

The natural processes which govern the fate and transport of pesticides can be grouped into the following broad categories: leaching, volatilization, degradation, sorption and plant uptake (Fig. 1). Plant uptake is usually a small component. The mode of application and action of the pesticide are important factors in relation to soil leaching, since those targeted at plant roots and soil insects are more likely to be leached than those applied to the leaves. Volatile losses occur from the soil particles, from the plants and from soil moisture.

Pesticide compounds may degrade in the soil by microbial or chemical processes to produce metabolites and ultimately simple compounds such as ammonia and carbon dioxide. Soil half-lives for compounds in widespread use range from 10 days to several years, but for the most mobile pesticides they are normally less than 100 days. Insecticides used for soil treatment may persist in the soil for sufficient time to allow significant leaching to groundwater to occur. Moreover, some derivatives from partial oxidation or hydrolysis may be as toxic and mobile as the parent compound.

Most pesticide compounds have water solubilities in excess of $10 \, mg \, l^{-1}$ ($10\,000 \, \mu g \, l^{-1}$), and this is not a limiting factor in leaching from soils. The mobility of pesticides in solution in soil water will normally depend upon their affinity for organic matter and/or clay minerals. Pesticides that are strongly sorbed onto organic

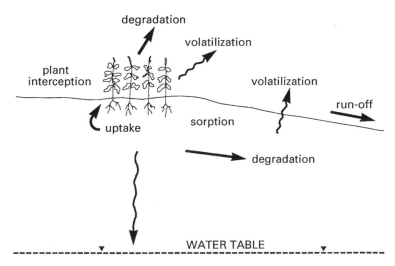

Fig. 1. Processes determining the fate of pesticides in the soil and unsaturated zone.

matter or clay particles are likely to be retained in the soil rather than being leached to groundwater. A possible exception to this might occur if strongly adsorbed pesticide residues were transported in a fissured formation on colloidal particles.

Transport in the unsaturated zone

Preliminary assessment can be made of the transport of pesticides from soils into groundwater systems, based on the physicochemical properties of the pesticides themselves and on knowledge of groundwater flow and aquifer properties. Following this approach, Foster et al. (1991) estimated the likely retardation in transport of pesticides with respect to a conservative, non-reactive solute, using the partition coefficients of each compound. Although the approach involves a number of important simplifying assumptions (Foster et al. 1991), it does show that transport through the matrix of many aquifer materials is likely to be slow. Except where the water-table is very shallow, pesticide residues from normal agricultural applications would be unlikely to reach the water-table in significant concentrations, if matrix transport was the only mechanism operating.

Pesticide compounds leached from permeable soils into the unsaturated zone enter an environment that contains less clay minerals and organic matter and has a greatly reduced indigenous microbial population. The attenuation processes that affect pesticides are likely, therefore, to be much less active beneath the soil zone. Thus the mobility and persistence of all pesticide compounds should be many times greater in the unsaturated zone than in a typical agricultural soil (Lawrence & Foster 1987).

Cavalier et al. (1991) carried out studies of the persistence of a number of pesticides in groundwater incubated at normal aquifer temperatures. Long lag periods were observed before any degradation occurred, and very long half-lives were measured (Table 1). The study was carried out on groundwater only, without aquifer material, and the half-lives may be unrealistically long. Even so, the results support the general expectation that persistence is likely to be significantly greater in aquifers than in soils.

The development of preferential flow in the unsaturated zone is of major importance in the consideration of pesticide transport into aquifers. This term is taken to include all forms of rapid downward movement in macropores and fissures, effectively by-passing the matrix. Preferential flow can be caused by numerous factors

Table 1. *Comparison of pesticide persistence in soil and groundwater*

Pesticide	Half-life, $T_{1/2}$ (days)		Ratio
	Soil	Groundwater	
Propanil	3	240	80
2,4-D	10	1200	120
Dichlorprop	10	900	90
Alachlor	20	1300	65
Metolachlor	40	800	20

Summarized from Cavalier et al. (1991).

(Foster et al. 1991) and is often associated with instability of downward flow in situations where a permeable formation is overlain by a somewhat less permeable soil horizon. Preferential flow could permit low to moderately persistent components to reach the water-table even where there are relatively thick unsaturated zones.

Given the complexity of pesticide transport in the soil and subsurface environment outlined above, and the fact that the behaviour of a particular compound may vary significantly depending upon conditions in the soil and aquifer e.g. moisture and organic carbon contents, pH, dissolved oxygen; (Chilton & West 1992), predicting pesticide residue concentrations in groundwater presents many difficulties and uncertainties.

Field investigations

For the study of pesticides, a general strategy similar to that followed for the earlier nitrate studies has been adopted (Chilton et al. 1993). Based on usage data, observed occurrence in groundwater and physicochemical properties, the compounds most likely to be leached to groundwater were identified. In the UK, sites were selected on the Chalk aquifer with good cropping records and history of usage of the selected compounds. Analytical methods, applicable to both water and solid material, were developed, together with sampling methods for the saturated and unsaturated zones of consolidated aquifers. Thus, centrifugation of crushed core material to remove the porewater (Kinniburgh & Miles 1983) has been successfully modified for pesticide analysis, even though larger volumes of water are required to obtain the necessary detection limits. Permanent observation boreholes have also been constructed to allow sampling of groundwater immediately beneath agricultural land to which the target pesticides have been applied in the process of normal cultivation practices.

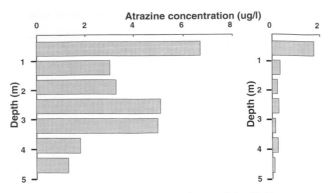

Fig. 2. Profiles of atrazine concentration in the unsaturated zone of the Chalk.

Collection and analysis of undisturbed samples from the unsaturated zone of the Chalk at three sites indicated hardly any residues of isoproturon above $0.5 \mu g \, kg^{-1}$ for the solid Chalk and $0.5 \mu g \, l^{-1}$ for the extracted porewater. Beneath fields to which atrazine had been applied in maize cultivation, concentrations of up to $6 \mu g \, l^{-1}$ were detected in porewaters. These were highest close to the surface, generally falling off rapidly with depth, but persisting to at least 5 m at two of the sites (Fig. 2). These findings are comparable to those of a similar study in the Granta catchment reported on by Clark *et al.* (1992). Regular sampling of groundwater from the uppermost part of the saturated zone has been undertaken at four of the sites, and reveals very low concentrations of both atrazine and isoproturon (Figs 3 and 4).

Barbados depends almost entirely on a relatively shallow, thin and highly permeable coral limestone aquifer for its water supplies. The island has a long history of sugarcane cultivation on large plantations. Two catchments providing 90% of the island's public water supplies were the subject of detailed assessments of pollution risk (Chilton *et al.* 1990), within which agricultural activities were an important component. Agricultural surveys established cropping regimes and fertilizer use, and identified the most important pesticides in regular use (Table 2). The choice of pesticides for study was based on considerations of mobility, solubility and persistence, which together provide some indication of susceptibility to leaching.

Fieldwork has comprised mainly regular sampling from existing public supply and irrigation wells. Atrazine was detected at concentrations usually in the range $0.5–3 \mu g \, l^{-1}$, throughout the year and in all wells sampled (Fig. 5). This suggests that atrazine may be more or less ubiquitous in the coral limestone aquifer of Barbados, and reflects its consistent usage over

many years (Chilton *et al.* 1994). There was a general correlation between atrazine concentration in groundwater and use of the compound around the sampled well. At the public supply sources (B to D in Fig. 5), concentrations remained relatively constant, reflecting uniform dilution in wells that are continuously pumped at substantial discharges. Atrazine concentration varied more markedly at irrigation wells (A in Fig. 5), which operate more intermittently and at lower discharges than public supply wells.

Although atrazine was widely detected in Barbados' groundwater, ametryn was very rarely detected. Both are widely used, and applied at about $4 \, kg \, ha^{-1} \, year^{-1}$ to sugarcane (Table 2). Further, as already stated, both compounds are regarded as relatively mobile and likely to be leached to groundwater. Perhaps degradation within the aquifer reduces ametryn to below detectable concentrations during migration through the unsaturated and saturated aquifer zones. The persistence of relatively similar compounds in groundwater may thus vary considerably, although the soil half-lives for both appear similar (Table 2). Neither diazinon nor dimethoate were detected in groundwater.

The field area in Sri Lanka is at Kalpitiya, a low-lying sand peninsula on the northwest coast. Coconut plantations are widespread and remain the principal crop, although in recent years there has been a significant expansion of horticulture. Irrigated horticulture involves triple cropping supported by heavy applications of fertilizer and pesticides. Soils are permeable, sandy and well-drained, and the underlying shallow sandy aquifer (Fig. 6) is also used for domestic supply. The soil insecticide carbofuran was selected for study, again on the basis of its properties (Cohen *et al.* 1986) and observed occurrence in groundwater elsewhere (Walker & Porter 1990). Fieldwork has been carried out at an experimental plot on an agricultural

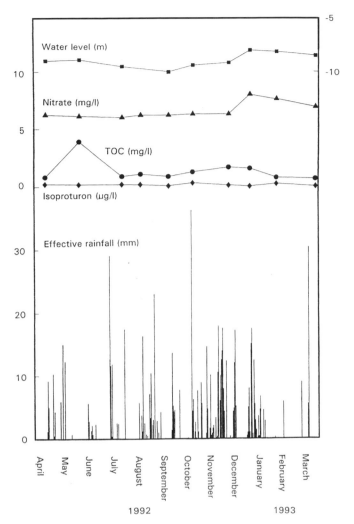

Fig. 3. Results of saturated zone monitoring in the Chalk beneath cereal cultivation.

research station. Eight 'mini-lysimeters' and twelve 50-mm-diameter piezometers completed at three depths (Fig. 7) were installed in a 0.2-ha field plot.

The carbofuran concentration in lysimeter drainage exhibited a similar pattern over three cropping seasons (Fig. 8). The pesticide was detected in the lysimeter drainage about five days after application and in the shallow piezometers six days later. The concentrations observed represent about 10% leaching of the first application and about 1% of the second application one month later. The following season the application was increased tenfold to enable the metabolites of carbofuran to be detected more easily, and the percentage leached to the water-table increased to 50% of the applied rate. The second application to this crop at the higher rate resulted in leaching losses similar to the previous seasons (Fig. 8). There are two possible explanations for the reduced leaching of second applications. Firstly, there could be a general increase in soil activity during the growing season as the applied irrigation water creates a more favourable environment for soil microbes, speeding up oxidation or cometabolic processes. Secondly, there may be enhanced degradation resulting from adaptation of the soil microbes to the presence of carbofuran. Clearly the rate of pesticide leaching to the water-table, even at the same site, can vary considerably and depends on the quantity applied and whether previous applications have been made recently.

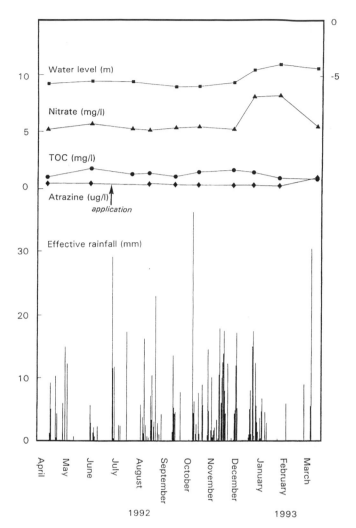

Fig. 4. Results of saturated zone monitoring in the Chalk beneath maize cultivation.

At the field site, the moist, aerobic conditions permitted oxidation of carbofuran to 3-hydroxy-carbofuran, which in turn was hydrolysed to the 3-hydroxy-7-phenol derivative. Both these compounds were detected in the lysimeter drainage but the total mass of metabolites detected corresponded to only about 5% of the parent compound. This suggests another metabolite, probably the 7-phenol derivative, was being produced but not detected.

Laboratory studies

Laboratory studies have focussed on the role played by microbiological activity in the degradation of pesticides in aquifer materials. Solid chalk and groundwater samples from several of the field sites were assessed for their indigenous microbial populations. Microcosm experiments have been carried out to provide a measure of pesticide persistence, by observing the disappearance of the original compound. Results appear to confirm earlier studies which indicated a period of microbial adaptation before degradation began, and confirmed that the most significant microbial activity is associated with the aquifer material rather than the groundwater. The experiments suggested that natural microbial activity in the Chalk is capable of degrading mecoprop. In microcosms most closely representing natural conditions, mecoprop was fully degraded within 60 days (Chilton *et al.* 1993).

Table 2. *Susceptibility to leaching of pesticides used in the Hampton Catchment*

Active ingredient	Use	Type	Acute oral toxicity[1]	Mobility class[2]	Solubility[3]	Soil half-life[3]	Total application (kg year^{-1})	Average application[4] (kg ha^{-1} year^{-1})
Asulam	H	TC	IV (4000+)	?	3	?	7293	2.83
Ametryn	H	T	III (3080)	4	1	4	6570	3.59
Atrazine	H	T	IV (2000)	?	5	4	6334	4.25
Methylarsonic acid	H	–	IV (1800)	?	1	?	6127	3.39
2,4-D Amine	H	Ph	II (700)	?	4	2	5996	3.38
Glyphosate salt	H	–	IV (4000)	2	1	3	4753	2.86
Paraquat	H	Py	II (150)	9	1	2	4542	3.74
Ionoxynil octanoate	H	–	II (110)	?	6	1/3	1651	0.75
2,D-Isooctyl	H	Ph	II (700)	?	6	?	1651	0.75
Diazinon	I	OP	II (300)	5	5	4	6240	15.7
Acephate	I	OP	II (900)	?	1	?	3724	19.7
Benomyl	F	C	V (10000+)	?	6	?	2247	8.11
Chlorothalonil	F	–	V (10000+)	?	6	?	1638	8.71

[1] WHO classification 1988–89 on LD50 (mg kg^{-1}) rates but adjusted to include dermal toxicity and other factors.
[2] Based on K_{oc} and K_{ow} partition coefficients (1 = most readily leached; 9 = unlikely to be leached (strongly sorbed)).
[3] From *Pesticide Manual*, 8th edition (1987).
[4] Average application rate for the plantations using the compound.
Use: H, herbicide; I, insecticide; F, fungicide.
Type: T, triazine; Ph, phenoxyl acid; OP, organophosphorus; TC, thiocarbamate; C, carbamate.
Solubility classes: 1 = >100 g l^{-1}; 2 = 10–100 g l^{-1}; 3 = 1–10 g l^{-1}; 4 = 0.1–1 g l^{-1}; 5 = 0.01–0.1 g l^{-1}; 6 = <0.01 g l^{-1}.
Soil half-life classes: 1 = <10 days; 2 = 10–30 days; 3 = 30–100 days; 4 = 100–300 days; 5 = >300 days.

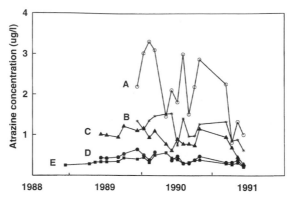

Fig. 5. Atrazine concentrations in groundwater from the coral limestone aquifer of Barbados.

In similar laboratory studies in Sri Lanka, carbofuran was metabolized within 40 days in incubations of water only and water with soil. In water only, the resulting carbofuran-7-phenol persisted at a constant concentration to the end of the incubation period, whereas in the soil/water mixture the metabolite had apparently completely disappeared after 60 days. The 7-phenol derivative may be lost from solution in the microcosm where soil is present by sorption onto the soil particles, by degradation by a different route utilizing carbon available in the soil as substrate, or by degradation by microbes which adhere to the soil surfaces. One or more of these processes may explain the absence of the 7-phenol metabolite in the lysimeter drainage and groundwater samples.

Parallel research in India showed that the degradation pathway for carbofuran in alluvial soils under paddy cultivation, where conditions are believed to be strongly anaerobic, was very different. There the main metabolite produced was carbofuran-7-phenol. This latter degradation

pathway has been observed elsewhere in paddy cultivated soils. It is important to establish the degradation pathway when assessing the risk to groundwater supplies firstly because the metabolites produced could be of very different toxicities to the parent compound and to each other, and secondly because the persistence of the different metabolites can vary. For example, the carbofuran-7-phenol metabolite appears more stable than carbofuran in groundwater.

Simple modelling

Once in the saturated zone, dilution and degradation are probably the most important processes attenuating pesticide concentrations. These processes can be very effective in reducing pesticide residues to acceptable concentrations. Attenuation will depend on the travel time to a well or borehole, the storage within the aquifer, the half-life of the pesticide and the percentage area of the catchment to the well that the

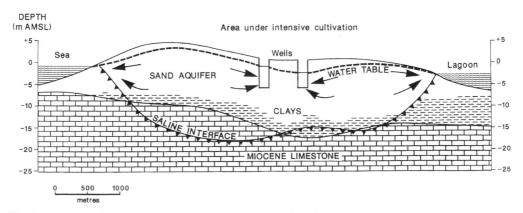

Fig. 6. Hydrogeological section of the Kalpitiya Peninsula, Sri Lanka.

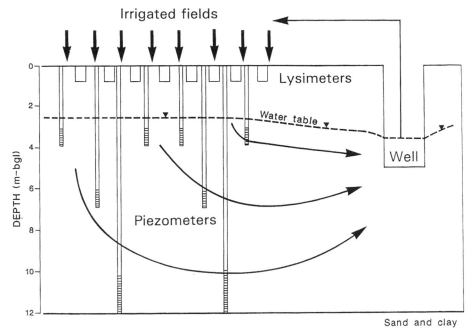

Fig. 7. Installation of small lysimeters and piezometers at the research plot.

pesticide is applied. The shallow, porous sand aquifer of the Kalpitiya Peninsula and the fissured coral limestone aquifer of Barbados represent two very different hydrogeological situations in this respect. A simple model has been developed to estimate possible maximum pesticide concentrations in groundwater and their duration in pumped wells, and to evaluate the inter-relationship of these factors (Lawrence & Barker 1993).

The model considers a single well with a circular catchment or capture zone, within which groundwater flows radially to the well (Fig. 9). Recharge is assumed to be uniform over the whole catchment and constant with time, and balances the abstraction from the well. Pesticide is applied at a constant rate over an area defined by two fixed radii (Fig. 9), and arrives at the water-table at a fixed concentration for a specified period of time. The fixed parameters employed in the model are given in Table 3. The output of the model is expressed as the concentration relative to the leached concentration at the water-table (Figs 10 and 11). The model was run for a range of porosities and pesticide half-lives, to assess the risk to water supplies from a typical range of compounds and to examine possible worst-case combinations of parameters. Although the model is based on the Sri Lanka situation, it was used to test the sensitivity of these parameters rather than to predict

precisely the likely pesticide concentrations in the well.

For compounds with half-lives less than about 36 days, residue concentrations in groundwater could be negligible by the time of a second application. For very persistent compounds, a 'build-up' of concentrations could occur, each application of the pesticide adding to the existing pesticide residues in the aquifer until water from the outer perimeter of the cultivated area has reached and been drawn from the well (Lawrence & Barker 1993).

The porosity of the aquifer affects both maximum concentration and the arrival and duration of the elevated concentrations. Increasing porosity increases dilution and therefore reduces pesticide concentration. Further, the time for water to move from a given radial distance to the well will increase as porosity increases. Conversely, low porosity would produce high peak concentrations and very rapid throughflow. An immediate implication is that frequent sampling, in relation to the timing of application, would be required to detect the peak concentrations in aquifers of low porosity. Routine, infrequent sampling could miss peaks or even miss positive detections completely.

The influence of extent of the cultivated area can also be investigated by varying the inner and outer radii (Fig. 9). Increasing the extent of the cultivated area will increase the maximum

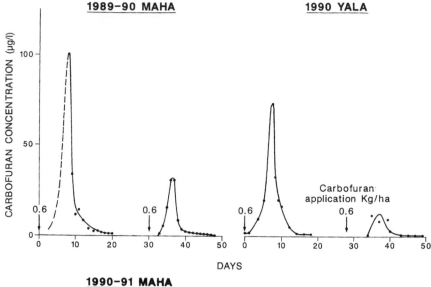

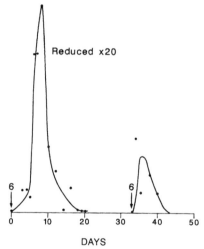

Fig. 8. Carbofuran in lysimster drainage over three cropping seasons in Sri Lanka.

and average pesticide concentrations at the well and the time required to reach maximum. Increasing the distance from the well to the inner boundary of cultivation will reduce peak concentrations and increase the first arrival time of pesticide residues in the well. This is an important consideration in controlling pesticide contamination of groundwater, as the influence of uncultivated protection areas around wells can be investigated. For pesticides with short half-lives, even relatively restricted uncultivated zones could have an important beneficial effect in reducing the likelihood of significant pesticide concentrations reaching wells in porous aquifers.

Conclusions

Both scientific investigation and routine monitoring of pesticides in groundwater present significant difficulties. It is not practicable for a water utility or regulatory authority to monitor for all of the large number of pesticides in regular use. This type of screening is prohibitively costly where potable supplies are based on a few high-discharge sources, and quite out of the question for large numbers of small, rural supplies. Choices have to be made, based on usage data, to focus on those compounds which are mobile and persistent and widely used. A strategy needs to be developed which focusses on the

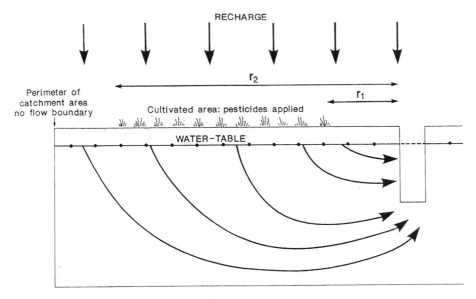

Fig. 9. Schematic section of the model construction.

high-risk pesticide compounds (i.e. those which are widely used, mobile, toxic and persistent) and on the most vulnerable aquifers.

Most of the published information on the persistence of pesticides refers to fertile, organic clay soils from temperate regions. Once beneath the soil, pesticides are likely to be more persistent because there is less organic matter and a lower level of microbial activity. The preliminary results from studies of pesticide persistence in groundwater suggest this is so, but there remains a great need for further information on pesticide persistence in groundwater and aquifers as opposed to soils, and for both soils and the deeper subsurface environment in tropical environments.

Degradation rates within an aquifer can vary significantly for relatively similar pesticide compounds, as suggested by the observations of atrazine and ametryn in the groundwater of Barbados. Further, the observation of different metabolites of carbofuran in the aerobic field

conditions of Sri Lanka and the anaerobic paddy cultivation in India, indicate that degradation pathways can vary significantly depending on the environmental conditions. Similarly, differences in metabolite detection between laboratory and field in Sri Lanka can be attributed to different environments. Finally, the enhanced degradation observed on second applications demonstrates the sensitivity of the subsurface microbial community to changing environments. It is important to establish degradation pathways and rates when assessing the risk of groundwater pollution, as metabolites may be of varying toxicity and persistence.

Maximum contaminant levels, based on toxicity considerations, are in the range 0.5–200 $\mu g\,l^{-1}$ for the most widely used compounds. Reported concentrations of pesticides in groundwater, those observed in the present studies and estimated by simple modelling are in the same

Table 3. *Fixed parameters used in the model*

Parameter	Value
Recharge rate	400 mm year^{-1}
Catchment area*	2.8×10^5 m^2
Aquifer thickness	10 m
Period of leaching to water-table (application)	14 days
Period between start of application	26 weeks

* Effective pumping rate = catchment area × recharge rate = 3.61 s^{-1}.

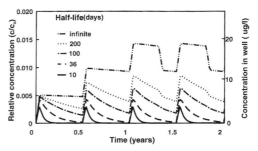

Fig. 10. Influence of pesticide persistence on residue concentrations in the pumped well.

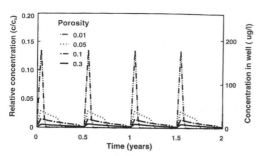

Fig. 11. Influence of aquifer porosity on pesticide concentrations for a half-life of 100 days.

general range, $0.1–100 \, \mu g \, l^{-1}$. The risks of pesticide concentrations in pumped groundwater reaching 10 or 100 times the guideline values as a result of normal agricultural use at recommended application rates are probably small. If higher concentrations are observed, they are likely to result from point rather than diffuse sources; that is, from spillages, poor disposal, soakaways, non-agricultural uses, etc.

The modelling has demonstrated that pesticide concentrations observed in pumped groundwater are highly sensitive to parameters such as aquifer porosity, pesticide half-life and extent of cultivated area. Pesticide residue concentrations can vary very rapidly with time, and one implication for monitoring is that, for the least porous aquifers, relatively frequent sampling intervals may be required if peak, or even positive, concentrations are not to be missed. It is difficult to predict pesticide concentrations unless these parameters are well-defined, but the model can be useful for indicating scenarios in which pesticide residues might pose a threat to drinking water supplies.

The greatest risk to groundwater supplies from pesticides will be where (a) aquifers are overlain by permeable soils and travel times to the water-table are short; (b) the aquifer is of low porosity and consequently dilution is small; and (c) the pesticide is relatively stable. Pollution of groundwater by pesticides is a potentially serious, and most certainly complex, environmental problem. It will require thorough investigation and careful monitoring to find a satisfactory solution and a reasonable balance of interests between the need for increased crop production and the requirement to maintain the quality of drinking water sources.

This paper is published by permission of the Director of the British Geological Survey, a component institute of the Natural Environment Research Council. The research programme in the UK was supported by funding from the National Rivers Authority under Contract No. 113, and that in Barbados, Sri Lanka and India by the UK Overseas Development Administration. The valuable participation of BGS colleagues and researchers in collaborating institutions in each country is gratefully ackowledged.

References

CAVALIER, T. C., LAVY, T. L. & MATTICE, J. D. 1991. Persistence of selected pesticides in ground-water samples. *Groundwater*, **29**(2), 225–231.

CHILTON, P. J. & WEST, J.M. 1992. Aquifers as environments for microbial activity. *In*: *Proceedings of the International Symposium on Environmental Aspects of Pesticide Microbiology*, Sigtuna, Sweden, August 1992, 293–304.

——, VLUGMAN, A. A. & FOSTER, S. S. D. 1990. A groundwater pollution risk assessment for public water supply sources in Barbados. *In*: *Tropical Hydrology and Caribbean Water Resources*. International Symposium, American Water Resources Association, 279–289.

——, STUART, M. E., GARDNER, S. J. *et al.* 1993. *Diffuse Pollution from Land-use Practices*. National Rivers Authority Project Record 113/10/ST.

——, LAWRENCE, A. R. & BARKER, J. A. 1994. Pesticides in groundwater: some preliminary observations on behaviour and transport in tropical environments. *In*: *Hydrological, Chemical and Biological Processes of Transformation and Transport of Contaminants in Aquatic Environments*. IAHS Publication 219, 51–66.

CLARK, L., GOMME, J., OAKES, D. B. *et al.* 1992. *Pesticides in Major Aquifers*. National Rivers Authority R & D Note 72.

COHEN, S. Z. 1990. Pesticides in groundwater: an overview. *In*: HUTSON, D. H. & ROBERTS, T. R. (eds) *Environmental Fate of Pesticides*. Wiley, New York, 13–25.

——, EIDEN, C. & LORBER, M. N. 1986. Monitoring groundwater for pesticides. *In*: GARNER, W. Y., HONEYCUTT, R. C. & NIGG, H. N. (eds) *Evaluation of Pesticides in Groundwater*. American Chemical Society, Washington DC, 170–196.

CONWAY, G. & PRETTY, J. 1991. *Unwelcome Harvest: Agriculture and Pollution*. Earthscan Publications, London.

CROLL, B. T. 1986. The effects of the agricultural use of herbicides on fresh water. *In*: *Proceedings of the WRc Conference on Effects of Land-Use Change on Fresh Water*, **13**, 201–209.

EPA 1990. *National survey of pesticides in drinking water wells. Phase 1 report*. US Environmental Protection Agency, EPA 570/9-90-015.

FOSTER, S. S. D., CHILTON, P. J. & STUART, M. E. 1991. Mechanisms of groundwater pollution by pesticides. *Journal of the Institution of Water and Environmental Management*, **5**(2), 186–193.

JONES, R. L., HORNSBY, A. G., RAO, P. S. C. & ANDERSON, M. P. 1988. Movement and degradation of aldicarb residues in the saturated zone under citrus groves on the Florida ridge. *Journal of Contaminant Hydrology*, **1**, 265–285.

KINNIBURGH, D. G. & MILES, D. L. 1983. Extraction and chemical analysis of interstitial water from soils and rocks. *Environmental and Scientific Technology*, **17**, 362–368.

LAWRENCE, A. R. & BARKER, J. A. 1993. *Simple groundwater model to predict pesticide concentration in pumped irrigation wells*. Technical Report, Hydrogeology Series, WD/92/49, British Geological Survey.

—— & FOSTER, S. S. D. 1987. *The pollution threat from agricultural pesticides and industrial solvents*. Hydrogeology Research Report No. 87/2, British Geological Survey.

LEES, A. & MCVEIGH, K. 1988. *An investigation of pesticide pollution in drinking water in England and Wales*. Friends of the Earth, London.

WALKER, M. J. & PORTER, K. S. 1990. Assessment of pesticides in upstate New York: results of a 1985–1987 sampling survey. *Groundwater Monitoring Review*, **30**, 116–126.

WEHTJE, F., LEAVITT, J. R. C., SPALDING, R. F., MIELKE, L. N. & SCHEPER, J. S. 1981. Atrazine contamination of groundwater in the Platte Valley of Nebraska from non-point sources. *Studies in Environmental Science*, **17**, 141–145.

ZAKI, M. H., MORAN, D. & HARRIS, D. 1982. Pesticides in groundwater: the aldicarb story in Suffolk County, New York. *American Journal of Public Health*, **72**, 1391–1395.

Investigation and management of pesticide pollution in the Lincolnshire Limestone aquifer in eastern England

J. SWEENEY[1], P. A. HART[2] & P. J. McCONVEY[1]

[1] *Environment Agency, Lincoln, Anglian Region, Northern Area, Waterside House,*
Waterside North, Lincoln LN2 5HA, UK
[2] *Environment Agency, Peterborough, UK*

Abstract: This case study describes pollution of the Lincolnshire Limestone aquifer, predominantly by the herbicide Mecoprop, to the north of Peterborough, Cambridgeshire. Key features of the case study are the occurrence of a series of former landfill sites located on the Lincolnshire Limestone aquifer outcrop, a nearby groundwater abstraction for public supply within the confined aquifer and the presence of a major geological fault and its associated control on groundwater movement. At the time of licensing, the fault was considered to hydraulically isolate the landfill sites from the abstraction.

A multiphase investigation, begun in 1990, was conducted by the National Rivers Authority and involved close liaison with the Cambridgeshire County Council, landowners, site operators and the Anglian Water Company. Investigations have focused on identifying and proving the source of pollution and determining its severity and extent. Remedial strategies and the implications of future change in hydraulic conditions have also been considered.

A medium-term risk management strategy for the area has been devised. This is in line with the UK Government's principles for contaminated land management.

This paper describes the study of a pollution incident in the Lincolnshire Limestone aquifer, north of Peterborough. In November 1990, the National Rivers Authority (NRA) became concerned about the concentration of the phenoxyacid herbicide Mecoprop in groundwater near the village of Etton (Fig. 1). Concentrations were found to vary from 1 to $3 \mu g l^{-1}$, and were rising. The UK Drinking Water Standard for individual pesticides is $0.1 \mu g l^{-1}$ (Fig. 2).

In response, the NRA conducted a search for suitable monitoring sites from which appropriate groundwater samples could be obtained. Some time previously, the NRA had requested the local waste regulatory authority to require the operators of three local landfill sites accepting putrescible waste (Swaddywell Pit, Ben Johnsons Pit and Ailsworth Road) to install a number of monitoring boreholes around their sites. Samples subsequently taken from these boreholes contained very high concentrations of Mecoprop (up to $828 \mu g l^{-1}$).

A fuller description of the case history is to be found in a series of reports held by the Environment Agency, Peterborough (NRA 1991*a*, *b*, 1992*a*, *b*, *c*, 1993).

Geological/hydrogeological setting

The Jurassic Lincolnshire Limestone is a major aquifer resource which is used extensively for public water supply. Groundwater movement within the limestone is predominantely by fissure flow. In this area (Fig. 3) the aquifer is recharged by rainfall in the outcrop area to the west around Stamford. Regional groundwater movement is down-dip, eastwards towards the Fens, where the aquifer becomes confined and where a number of important public water supply sources are situated. To the south of the village of Helpston lie a number of closed landfill sites, formerly limestone quarries. The landfill sites straddle the Marholm–Tinwell Fault, a major structural discontinuity which trends generally WNW–ESE through the area from Stamford to Peterborough. The fault upthrows the strata to the south and juxtaposes a Lincolnshire Limestone outcrop to the south against younger Jurassic strata to the north, where the confined Lincolnshire Limestone is at depth. Confining layers are the Upper Estuarine Series, Blisworth Clay, Kellaways Clay and Oxford Clay, which are interbedded with limestones and sandy limestones (Fig. 4). At times of average or high recharge under high water-table or average water-table conditions, groundwater levels within the confined zone of the aquifer can be artesian.

Description of investigations and findings

From 1990 to 1995 a major investigation was conducted by the NRA which required close liaison between the NRA, Cambridgeshire County Council, Anglian Water Services, the landfill site

SWEENEY, J., HART, P. A. & McCONVEY, P. J. 1998. Investigation and management of pesticide pollution in the Lincolnshire Limestone aquifer in eastern England. *In*: MATHER, J., BANKS, D., DUMPLETON, S. & FERMOR, M. (eds) *Groundwater Contaminants and their Migration*. Geological Society, London, Special Publications, **128**, 347–360.

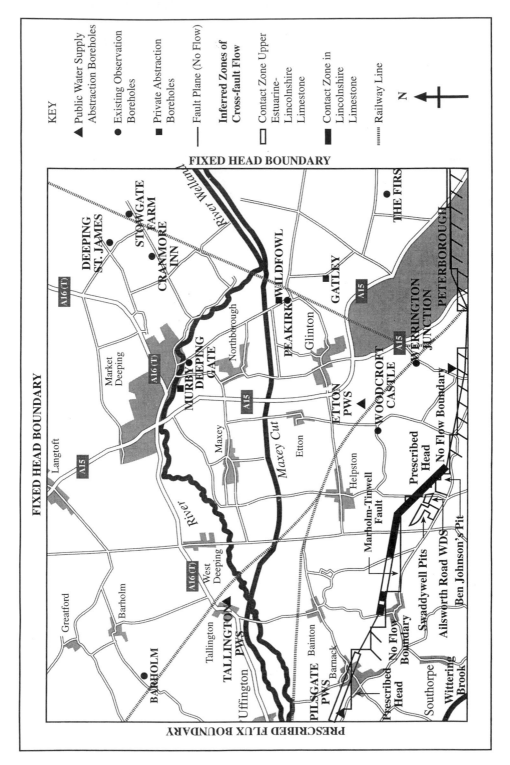

Fig. 1. Location of the study area and model domain.

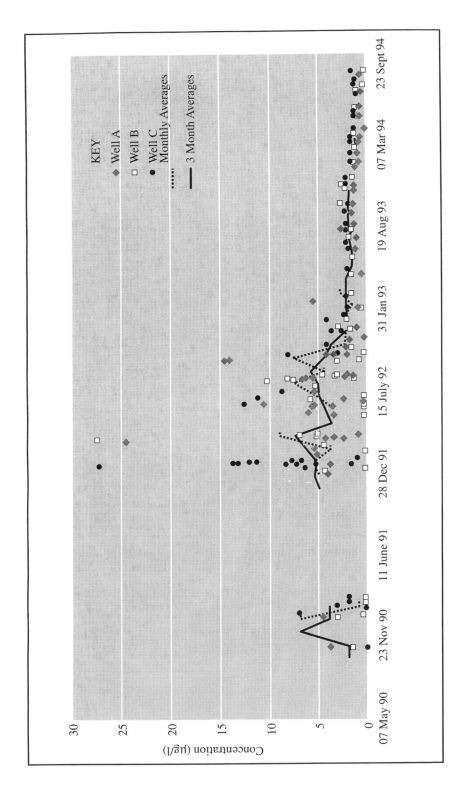

Fig. 2. Observed Mecoprop concentrations at Etton PWS.

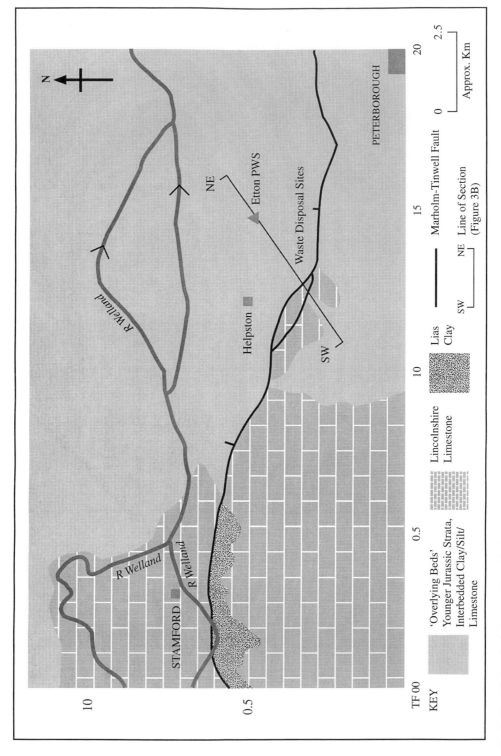

Fig. 3. Simplified geological map of investigation area.

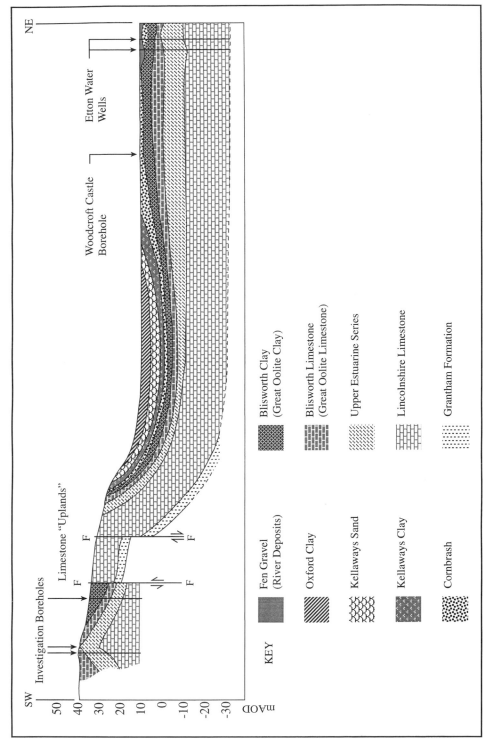

Fig. 4. Geological cross-section through Helpston waste disposal sites and Etton public water supply borehole.

operators and the landowners. With the 'source–pathway–receptor' approach in mind, investigations have been carried out in a number of phases as an understanding of the causes and effects of this pollution event has developed. Initial enquiries and monitoring established the following:

(a) Hydraulic transmission across the Marholm–Tinwell Fault occurs and is complex in nature. At the time of licensing of the three landfill sites in 1978, it was considered that

  – connections between the limestone blocks either side of the fault were limited, providing a barrier to flow from the unconfined aquifer into the confined aquifer;
  – the total length of fault across which leachate could migrate would be a relatively short distance of about 150 m;
  – any leachate migrating into the limestone would rapidly become diluted and attenuated in the aquifer.

(b) A significant amount of waste pesticide was deposited in two of the three landfill sites.

(c) There had been a complex history of waste deposit. Waste disposal site licences were issued to the site operators under the Control of Pollution Act 1974 (Part 1). Prior to implementation of the new Waste Management Licensing Regulations (Anon 1994), the site licences for all three sites were surrendered by the operators.

(d) Engineered containment measures were at best only very loosely specified.

Given that a significant amount of information needed to be acquired, collated and appraised, specialist consultants (Dames and Moore Ltd) were commissioned by the NRA to assist with the investigation. Their role was to review available data, specify cost, and carry out works to further evaluate the degree of pollution; following this, they were to identify and evaluate options for remediation.

Specific findings from the study of existing information were as follows:

(a) Contamination found in the Swaddywell Pit may have originated from the adjacent Ailsworth Road site. During the operational life of Ailsworth Road, seepages of leachate had been observed entering Swaddywell Pit from the Ailsworth Road site.

(b) Waste disposal records suggested that from one producer source alone, a total of approximately 40 tonnes of active agrochemical waste had been deposited in the landfill sites at Ailsworth Road and Ben Johnson's Pit.

(c) The likely existence of these waste deposits posed a major pollution risk to the Lincolnshire Limestone aquifer.

(d) The nearest public water supply at Etton was being polluted by the herbicide Mecoprop, and treatment was necessary to meet drinking water standards.

Obtaining conclusive proof that the landfill sites at Helpston were the source of pollution in the aquifer became fundamental. There being no other known source of Mecoprop contamination in the vicinity, seven monitoring boreholes were constructed close to the landfill sites, north of the Marholm–Tinwell Fault, to determine the degree of hydraulic connection across the fault.

Detailed analysis of geological logs and groundwater level measurements from these boreholes suggested the fault was acting as a partial barrier to the movement of leachate and groundwater into the confined part of the aquifer. However, chemical analysis of groundwater samples from these boreholes and from leachate wells within the landfill sites confirmed that some transmission of leachates across the Marholm–Tinwell Fault was occurring and established the landfill sites to be the source of the pollution.

Evaluating the complex interactions occurring within the aquifer and across the Marholm–Tinwell Fault, and attempting to determine the severity and extent of pollution and the mechanisms influencing plume movement became the next phase of investigation. Detailed field investigations using existing and additional boreholes, borehole geophysical logging, pump testing, tracers, water quality and water level monitoring were conducted during 1991/92. In parallel with this work, a groundwater/contaminant transport model was developed for the area using the TARGET (Transient Analyser of Reacting Groundwater and Effluent Transport) code.

Figures 5–7 depict some of the main findings of these investigations. To the north of the fault, several aspects became clear. A 'plume' was delineated in the aquifer heading in an east-northeasterly direction (Fig. 5). This decreased in severity from the fault zone at the landfill sites ($>1000 \mu g \, l^{-1}$) to the main part of an inferred plume front ($<100 \mu g \, l^{-1}$) west of Etton. The main controls over the distribution of contaminated groundwater included the flow rates across the fault, the prevailing hydraulic gradient and the rate and location of groundwater abstraction/discharge. Closer to the Etton abstraction, the effect of pumping is to attract

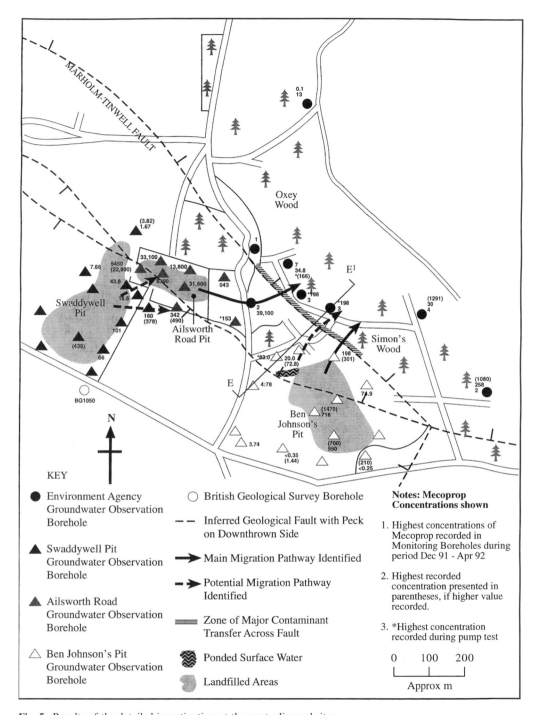

Fig. 5. Results of the detailed investigation at the waste disposal sites.

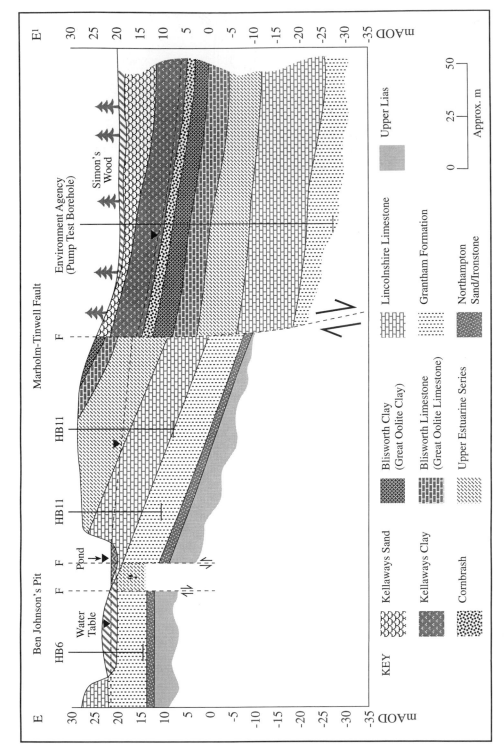

Fig. 6. Geological cross-section E–E'.

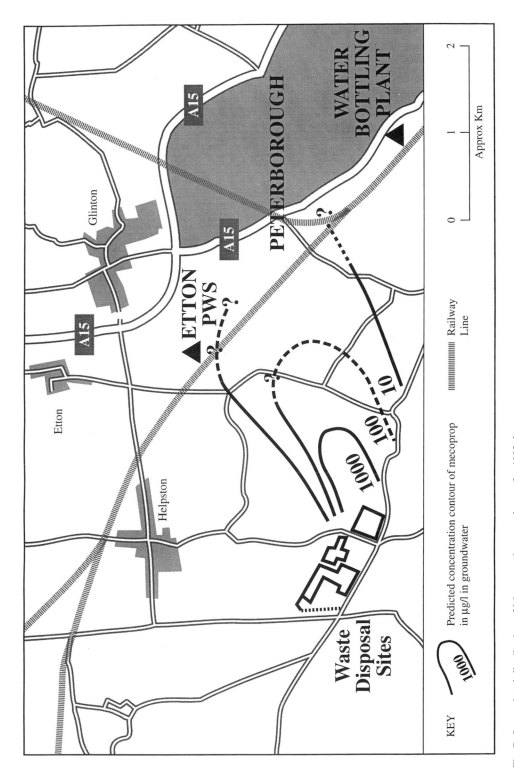

Fig. 7. Interpolated distribution of Mecoprop in groundwater for 1992 data.

part of the plume northeast towards the public water supply. No other polluting compounds were discovered in similar significant quantities to Mecoprop although elevated levels of ammonium and nickel were detected.

To the south of the fault, additional, detailed investigations confirmed the source of the pollution as being one or more of the landfill sites. At Ailsworth Road, Ben Johnson's Pit and Swaddywell Pit, Mecoprop was found in the groundwater at concentrations of $30\,000\,\mu g\,l^{-1}$, $100{-}500\,\mu g\,l^{-1}$, and $10{-}100\,\mu g\,l^{-1}$, respectively. The ability of the fault to transmit sufficient volumes of contaminant across was also confirmed.

Further work has been carried out by the British Geological Survey, specifically in the form of microcosm experiments, to evaluate the biogeochemical processes that influence the concentration of Mecoprop in the aquifer. This work suggests that redox conditions play a very important role in the degradation of this compound. Whilst under aerobic conditions biodegradation appears to be active, under anaerobic conditions little or no degradation appears to take place (Harrison *et al.* 1995).

Evaluation of remediation options

A number of options were identified which could bring groundwater quality within acceptable standards. Although a wide range of options were reviewed, the key criteria used to determine whether or not a particular set of options could be successful were as follows:

(1) maintenance of the continued abstraction for public water supply at Etton;

(2) Mecoprop concentrations in the Etton raw water not to exceed $10\,\mu g\,l^{-1}$, so that an acceptable degree of confidence is maintained in the overall treatment system, including the granular activated carbon treatment system installed by the Water Company at Etton to treat raw water.

Technical evaluation of the possible options concluded that the granular activated carbon provides the most robust groundwater treatment system, utilized at remedial well head locations; and isolation of landfill sites using a Bentonite grout curtain would be the most cost-effective means of cut-off, thus reducing their effect on the aquifer.

At that stage of the investigation, the contaminant transport model predictions indicated that combination of these two options would ensure that the concentrations of Mecoprop in groundwater arriving at Etton would remain below the $10\,\mu g\,l^{-1}$ threshold. However such a strategy would require long-term commitment in the form of groundwater treatment and landfill site remedial works. Additionally, those undertaking this work could incur very long-term liability for the sites. It could also be argued that isolation of the landfill sites would not provide a solution in the long term, and would only postpone the need for further action. On the other hand, to completely remove the waste body to a more appropriately designed disposal facility raised complex issues surrounding the Best Available Technique Not Entailing Excessive Cost (BATNEEC) and Best Practicable Environmental Option (BPEO) concepts. Additionally, significant technical, surface water pollution and health and safety concerns raised doubts about whether it would be possible to undertake such a massive waste relocation exercise.

Impact of groundwater level recovery

The investigation was begun when the region was experiencing a succession of below-average recharge years (1988–1992). Following the end of this 'drought' period, the hydrogeological regime of the area under investigation changed dramatically. In one year groundwater levels recovered from a record low to a record high. This recovery had important consequences for both model calibration and the scope of the investigation.

The contaminant transport model developed had been calibrated using historical records and could not be expected to perform satisfactorily under the extreme conditions now being experienced. Also, differences became apparent in water level predictions between the localized contaminant transport model and the existing regional resource flow model. These differences were reconciled to produce a single hybrid model, representing 'state of the art' for the area under investigation. However, predictions made by the hybrid model were still not consistent with observed conditions at key groundwater monitoring points. Thus the ability to evaluate the remedial options with accuracy was reduced. Further model development and additional observation boreholes were deemed to be required to improve model reliability to a satisfactory degree.

Throughout the previous phases of investigation, attention focused on groundwater pollution as surface water discharge did not occur due to the low groundwater levels being experienced. As a result, the risk and impact of contaminated groundwater being released to the surface water

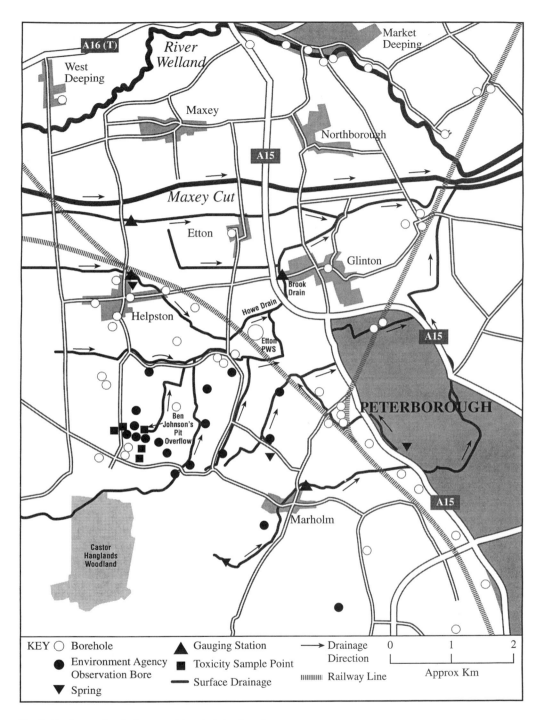

Fig. 8. Surface water drainage and borehole/spring sites.

environment had not been considered in great detail. In light of the increase in groundwater levels now being experienced, this scenario warranted further evaluation as to the potential implication of release of contaminated groundwater to the surface water system.

Toxicity of leachate

Recent investigations have sought to evaluate the environmental impacts of overflows from Ben Johnsons Pit, spring seepages and failure of disused boreholes and their subsequent uncontrolled discharge of water under artesian pressure ('wild boreholes'), on surface waters.

Key to being able to predict the environmental impact on surface waters are toxicity tests. These have been carried out using leachate from, and groundwater close to, the landfill sites. The toxicity of leachate to *Oncorhynchus mykiss* (rainbow trout), *Daphnia magna, Chlorella vulgaris* (algae) and *Lemna minor* (duckweed) were determined by the NRA with assistance from their Centre for Toxic and Persistent Substances. These organisms were selected as representatives of different levels of organization in the freshwater community.

Initial screening, using Mecoprop concentration as the pollution indicator, identified sites from which the most toxic samples could be obtained (Fig. 8). Where necessary repeat samples were taken; for example, repeat sampling confirmed Borehole HB13 to be the most toxic to *Chlorella* and gave the lowest corresponding Mecoprop concentration (the toxicity being due to the other leachate constituents) (Table 1).

The chemical composition of samples taken for toxicity testing, and their actual toxicity, varied widely. Such variance is to be expected since it is believed that liquid wastes were placed in trenches excavated in waste, with both liquid and solid wastes being highly variable in composition. In order to determine the worst-case environmental impact, the greatest toxicity values for each species were used (Table 1).

Tests indicated that the most toxic influences would be due to Mecoprop and ammonia. By determining leachate toxicity, as opposed to the toxicity of specific chemical compounds, possible synergistic effects were accounted for. Bearing this in mind, and also their high concentrations, high solubility and relatively rapid movement through the aquifer, Mecoprop and ammonia have been regarded as 'indicator' compounds in this case. Values for up to 79 chemical parameters were also determined, providing the basis to link the toxicity of a sample to any of the individual chemical compounds. The main benefit of this approach would be to reduce monitoring costs by using less costly chemical analysis to screen samples before conducting a comprehensive toxicity assessment.

This is demonstrated by results of Mecoprop and ammonia correlated with the proportion of leachate, and toxicity, in Table 2.

Determination of the concentration of Mecoprop and/or ammonia allows the leachate concentration to be estimated and thus the toxicity of any release of contaminated groundwater to the aquatic environment. It should be noted that a number of safety factors have been built in to this methodology:

- chemical composition is related to the most toxic leachate tested for each species;
- no allowance has been made for degradation of the leachate with time;
- it is assumed that leachate composition does not change as migration through the aquifer takes place, i.e. all contaminants become mobile.

Threshold values for levels of contaminants, based on uses and aquatic life, have been set at specific points in watercourses to ensure their protection. These threshold values have been established using toxicity data and results derived from the chemical analyses and if exceeded, would prompt remedial action. Comprehensive field investigations have identified all potential places of groundwater emergence in the area and surface flow routes to watercourses. At boreholes, artesian overflows are closely monitored, and will be restricted or prevented if chemical and/or toxicity threshold values are likely to be exceeded downstream.

The degree of urgency for borehole sealing work would be assessed by generating toxicity values from the indicator chemical results, and by taking into account dilution from baseline flow in the receiving watercourse.

All potential points of release eventually discharge into either the South Drain, Brook Drain or Folly River. Threshold values applied at key points on these watercourses could be used to ensure that watercourses downstream of these points are protected. The catchments area upstream of these points largely consists of field drainage systems which are frequently dry during summer months.

Medium-term management strategy

Until the calibration of the model is improved to a stage where long-term decisions can be safely based, a medium-term management strategy

Table 1. *Toxicity of leachates*

Species	LC_{50} (hours)	EC_{50} (hours)	Leachate (%)	Sample site
Rainbow trout	96	n/a	2.397	HA7
Chlorella vulgaris (green algae)	n/a	72	0.87	HB13
Daphnia magna	48	n/a	11.45	HA7
Lemna minor (duckweed)	n/a	168	1.4	HA7

LC_{50} = lethal concentration causing 50% mortality; EC_{50} = concentration causing an effect in 50% of the species.

has been devised. In addition, work commissioned by the Environment Agency (as successor to the NRA) to formulate methodologies for deriving clean-up standards for groundwater and soils, in association with further developments in the LANDSIM code and degradation studies, will assist in the decision making process regarding long-term strategies.

In determining the medium-term strategy the Environment Agency has taken into account the following:

(a) the suitable for use concept and cost–benefit approach proposed by the UK Government (Department of Environment and Welsh Office, 1994);
(b) the need to develop a sufficiently robust groundwater/contaminant transport model, capable of representing the complex hydrogeological setting around Helpston under a range of groundwater level conditions.

The following actions constitute the medium term management strategy:

• Intensive monitoring of surface and groundwaters in the affected area. Additional observation boreholes will be installed to monitor pollutant concentration and plume distribution.
• Development of improved groundwater/pollution modelling capability.
• Prevention/restriction of artesian release of groundwater, if surface water quality at key locations becomes unacceptable, as defined by toxicity assessment.

• Contingency plans made by those who may be affected by pollution will be supported by the EA, provided that the plans do not place at risk other users' rights and/or the environment.

Implementation of this medium-term strategy is proposed to protect the environment and to sustain the use of groundwater for public water supply.

Future strategy

It is clear that the severity and extent of this pollution event is influenced by a number of variable factors: groundwater levels, abstraction quantity and the rate at which pollution is introduced to the aquifer across the Marholm–Tinwell Fault. Each of these factors is largely unpredictable and beyond immediate control. Groundwater levels are dependent upon recharge and abstraction. Whilst abstraction is regulated by the EA, operation up to the licensed limits is at the discretion of the abstractor. The rate, concentration and point at which pollutant crosses the Marholm–Tinwell Fault currently cannot be controlled. The control of any of these factors would require significant capital and revenue expenditure.

Although several remedial management strategies appear to be feasible, they are technically complex. A rigorous evaluation of each strategy is essential to ensure that a technically robust and economical solution is found that is capable

Table 2. *Use of 'indicator' compounds for toxicity assessment*

Species	LC_{50} (Hr)	EC_{50} (Hr)	Mecoprop ($\mu g \, l^{-1}$)	Ammonia ($mg \, l^{-1}$)	Leachate (%)
Rainbow trout	96	n/a	456.49	12.83	2.39
Chlorella vulgaris (green algae)	n/a	72	21.05	1.14	0.87
Daphnia magna	48	n/a	2186.95	61.49	11.45
Lemna minor (duckweed)	n/a	168	269.3	7.5	1.4

of protecting current and future uses of this important groundwater resource. To achieve confidence in any long-term management option for the Helpston pollution incident, further improvements to the predictive model have been justified. Future commitment to this approach will be reviewed against the prospects for success after 1997.

To ensure that expenditure is effectively targeted, a robust long-term solution will be found that is capable of accommodating change. It would be unwise to recommend implementation of high-cost remedial action unless the robustness of the solution has been tested.

Bearing in mind the 'suitable-for-use' concept, the cost–benefit approach, and the fact that currently neither the environment nor public water supplies are being directly affected to an unacceptable degree, management of this pollution event by rendering pollution harmless through the dilute and disperse principle is an appropriate medium-term strategy.

Conclusions

It is concluded that by continued close monitoring of the landfill sites and by controlling the release of contaminated groundwater into the surface water system, the situation can be managed until it is possible to derive a costeffective, long-term strategy. In the meantime, unacceptable pollution of surface waters is being prevented and groundwater at Etton continues to be suitable, after treatment, for public water supply.

The authors are grateful to Environment Agency staff and Anglian Water Company staff who have contributed to the management of this incident through the Project Support Group and Project Steering Group. Dames and Moore Ltd, Manchester, have contributed a great deal of technical information to these groups. Assistance with the modelling was provided by K. Rushton, University of Birmingham.

The views expressed are those of the authors and not necessarily those of the Environment Agency.

References

ANON 1994. *The Waste Management Licensing Regulations 1994.* Statutory Instrument No. 1056. HMSO, London.
DEPARTMENT OF ENVIRONMENT AND WELSH OFFICE 1994. *Framework for contaminated land.* Document.
HARRISON, I. *et al.* 1995. *Fate of organic pollutants in groundwaters around landfill sites.* NERC, BGS Technical Report WE/95/50.
NRA 1991*a. Phase I: Preliminary assessment of groundwater contamination in the area around Helpston, Cambridgeshire.* Report to NRA. Dames & Moore.
——— 1991*b. Phase II: Installation of groundwater monitoring boreholes in the Helpston area of Cambridgeshire.* Report to the NRA. Dames & Moore.
——— 1992*a. Phase IIIA: Groundwater contamination investigation of the Helpston/Etton area.* Report to the NRA. Dames & Moore.
——— 1992*b. Phase IIIB: Groundwater contaminant transport model of the Helpston/Etton area,* Vol. 1. Final Report to the NRA, Anglian Region. Dames & Moore.
——— 1992*c. Phase IIIB: Groundwater contaminant transport model of the Helpston/Etton area,* Vol. 2. Final Report to the NRA, Anglian Region. Dames & Moore.
——— 1993. *Phase IV: Design of remedial works for groundwater contamination at Helpston/Etton, Cambridgeshire.* Feasibility Report for the NRA. Dames & Moore.

Index

Acephate, 339
acetates, 39–41, 260–1, 276, 279
acetone tank, 180, 188–9, 195–6
acid, extraction, 52–3, 269–79
    soils, 66, 68
    tar wastes, 5, 281–2, 283–311, 313
    waters, 23, 25–6, 66, 97–9, 101–18
adsorption, 11, 30–3, 63, 161–3, 174
advection, 88, 318
aerial transport, 20, 67, 255–6, 260, 265–7
aerobia, 203, 274–5, 338, 343
    iron, 109–14, 117–18
agriculture, 3–13, 64–74, 334–44
air sparging, 179–80
air stripping, 26, 33, 174–80, 196–200
airports, 25, 121–2, 147–57
Alachlor, 335
algae see Chlorella vulgaris
alkaline solutions, 53–4, 101–3, 260–3
alkanes, 131, 284
aluminium, 52–3, 57–60, 103, 110
    tar wastes, 284–5, 299–305
alunite, 271
amenity value, 95
americium, 266
Ametryn, 336, 339, 343
2,4-D amine, 335, 339
ammonium, 37, 49, 77–91, 103, 320–6
    pesticides, 334, 356, 358–9
    Siberia, 261, 269, 271, 275, 278–9
    sludge, 63–4, 69–74
    tar wastes, 284–5, 299–305
anaerobia, 151, 274–5, 340, 343
analysis, on site, 233–5, 237, 249–51
    soil gas, 149, 151
Anglian Water, 24–5, 65, 124–44, 160, 347
    and solvents, 201–14, 232, 251
anhydite, 111
anisotrophy, fractures, 293–302, 307–10
anthropogenic redistribution, 268–80
apatite, 268
appropriate person, 16
aragonite, 271
aromatics, 63, 161, 284–5
arsenic, 51–60, 63, 93–4, 103, 275
    Marl Pit, 316, 318, 320–6
    tar wastes, 284–5, 299–305
arsenpyrite, 93–4
asbestos, 20–1
ascorbate, 112
ash, pulverized, 51–60
assessment,
    tarry wastes, 285, 289, 309–10
assessment, 12, 16, 147–52, 156
asulam, 339
Atrazine, 5, 8, 334–44
attenuation, 5, 132–5, 265–80, 336–44
    qualities, 5, 11–12, 27
Attorney General's Reference No 1 of 1994, 15

bacteria, 203, 274–9, 308, 335–44
    and iron, 109, 199

Ballard v. Tomlinson case, 212
Barbados, 333, 336, 340–1, 343–4
barium, 52–4, 56–60, 271
Barlow, 54–60
barriers, fault, 266, 270–9, 353
BATNEEC (Best Available Technique Not Entailing
    Excessive Cost), 356
Belper, 283–210
Benomyl, 339
bentonite, 236, 284–5
benzene, 27–33, 131–48, 161, 169–76
    see also BTEX
benzole refining, 284
Bersbo mine, 112
beryllium, 271
beta activity, 261
bianchite, 109, 111
bicarbonate, 37, 78, 88, 272
    chlorinated solvents, 213, 231
    mine waters, 103, 107–8, 110
biodegradation, 11, 23, 27, 242, 308
    hydrocarbons, 151, 172, 174–80
    pesticides, 334–44, 356–60
    solvents, 181, 188–9, 203, 214
    Stamford, 131, 135, 142
biological treatment, 32, 174–80, 308–10
Birmingham, 10, 181
Black Country Spine Road, 313–31
Blisworth Clay, 347–51
BOD (biochemical oxygen demand), 63, 153
    tar wastes, 299–305, 318–20, 324–5
boreholes, 6–7, 36–41, 77–8
    Coventry, 232–43, 248–51
    hydrocarbons, 121–2, 126–44, 147–56
    Long Island, 160–1, 165–9
    Merrimack facility, 183–92
    pesticides, 350–60
    Sawston, 205–11, 219–27, 236–9
    Siberia, 255–63, 266–7, 271–9
    tar wastes, 289–310, 315–6
boron, 51–6, 59–60
boundary conditions, 193–5
BPEO (Best Practicable Environmental Option), 64,
    66, 356
brines, leach-mining, 268–79
BTEX (benzene, toluene, ethyl benzene and xylene),
    25–33, 129–40, 162–3, 169–80, 181
built-up area, remediation, 147–56
Bukinaii uranium, 265–76, 278–9
buried channels, 185–95, 219–20, 267, 276–9

CA (chloroethane), 188–9
cadmium, 51–3, 59–60, 63, 67
    acid mines, 97, 101–3, 107–8
    acid tars, 284–5, 299–305, 318, 320–6
caesium, 255, 260, 266, 271
calcite, 111, 268–79
calcium, 37–41, 52–60, 77, 84–5
    acid mines, 103, 107–8, 110
    acid tars, 284–5, 299–305
    Siberia, 260, 268–79
calcium bicarbonate, 177–8, 258

calcium carbonates, 260–3, 271–9
calcium–sodium sulphate–chloride, 268–9
CAM program, 260–3
Cambridge Water Company case, 201–15, 229–40, 243–51
  legal, 1, 6, 13, 20–1, 23–6, 181, 203, 217
carbamates, 334, 336–44
carbofuran and derivatives, 336–44
carbon, 266, 335
carbon adsorption, 197–200, 278
  see also PACT, GCA
carbon dioxide, 117, 151, 203, 274–9
carbon filtration, 25, 31–3, 197–200
carbon tetrachloride, 201
carbonate, 39–41, 104, 108–12
  Siberia, 258, 260, 271, 276
Carboniferous Coventry Sandstone Formation, 230–6, 242
Carnon valley, 23–6, 93–9
cassiterite, 93–4
cation exchange, 39–41, 91
cerium[144], 266
chalcopyrite, 93–4
Chalk, 8, 10–11, 24, 27, 49, 181, 335–6
  fissures, 75–80, 144, 220–7, 232
  Hertfordshire, 75, 80–1
  MTBE, 30
  putty, 75, 78, 80–1, 205, 211, 219–27, 232, 245
  Sawston, 201–5, 212, 217–20, 229–36, 243–51
Chalk Marl, 232, 243–4
Chelyabinsk, 253, 255, 265–7
Chernilshikov river, 256
Chlorella vulgaris, 358–9
chloride, 28, 32, 54, 57
  acid tars, 284–5, 299–305
  Hertfordshire, 75, 77–9
  mines, 82–91, 103, 107–8, 269–79
chlorinated solvents, 4–7, 13, 23–6, 181–252
  point sources, 217–27, 229–51
  Stamford, 123–32, 137–34
  USA, 165–80, 183–200
  see also TCA, TCE, PCB
chlorobenzenes, 63
chloropropanes, 334–44
Chlorothalonil, 339
chromium, 51–4, 56, 58–60, 63
  acid tars, 284–5, 299–305
  Marl Pit, 318, 325–6
  Siberia, 261, 271, 276
clays, 147–9, 160–1, 185, 189
  acid tars, 285–7, 314
  glacial, 36–7, 75–7
  radionuclides, 258–60, 276–8
Coal Measures, 283–311, 313–16
coals, 25–6, 49, 51–2, 57, 233–303
cobalt, 63, 101, 271, 284–5
COD (chemical oxygen demand), 299–305, 318–26
Codes of Practice, 7–8, 63, 69
coning, 94, 260
contaminant transport models, 144
contaminated land definition, 16
copiapite, 109
copper, 63, 82–91, 284–5, 299–305
  fuel ash, 52–3, 57–60

Marl Pit, 316, 318, 320–6
mines, 93–4, 97, 101, 108, 116, 271
coral limestone, 336, 340–1
cost effective, 7, 10–12, 212–14, 307–10
costs, 17–18, 179–80, 232, 240, 249–51
  pesticides remediation, 356–60
Coventry, 229–43, 245–51
Cretaceous, 165–80, 265–7, 276–9
  see also Chalk
cross-contamination, 232–43, 248–51
Cs[137], 255, 260
CSTR (completely stirred tank reaction), 112–14, 118
CTC (carbon tetrachloride), 201, 203, 220–7
curium, 266
CXTFIF program, 43–5
cyanide, 284–5, 325
cyclohexane extractable matter, 284–5

Daphnia magna, 358–9
data collection, 3–6, 10–13, 248–50, 262–3
DCA (dichloroethane), 188–9
DCE (dichloroethene) isomers, 188–9
DCPA, 334
Department of the Environment, N.I., 6, 8, 12
depth, 53–60, 97–9, 109–18, 211, 240–2
  and phenolic substances, 316–18
Derelict Land Grant, 283
detergents, radionuclides, 260–3, 276
dewatering, 25–6, 95–9, 296–307, 319–20
diazinon, 336, 339
1,1,-dichloroethene, 172, 189, 203
cis-1,2-dichloroethene, 188–9, 203
trans-1,2,-dichloroethene, 172, 188–9, 203
Dichlorprop, 335
diffusion, 33, 109, 318–9
diffusive exchange and transport, 242, 247–9
dilution, 12, 88, 88–91, 340–1, 360
dimethoate, 336
dioxins, 20, 63, 67
discharge, 6, 10–18, 95–9, 149–56, 330
  pesticides, 325, 330–1, 355–60
dispersion, 12–13, 38–48, 88–91, 144
dissolution, 43, 109–18, 151, 270–9
distribution coefficient, 35–48
distribution terminal, 121–2, 159–63
DNAPLs (dense, non-aqueous phase liquids), 11, 240–3, 307–8
  Chalk, 210–14, 217–28, 243–7
  Stamford, 129–31, 142
dolomite, 111
DPCA, 334
Drigg, 35–48
drinking quality, 8–13, 18–19, 24–5, 116
  and hydrocarbons, 129, 131
  inorganic contaminants, 52, 59–60, 79–91
  and solvents, 181, 203, 213–15, 240
DSC (Digested Sewage Cake), 63–4, 66, 69–74
dual porosity, 35–48, 203–5, 231, 237, 251
dual-phase organic solvent, 29–30, 32
duckweed see *Lemna minor*
DYN-EDM, 172
DYNFLOW program, 172, 187, 192–5
DYNPAL program, 172, 175, 178

DYNSYSTEM program, 172
DYNTRACK program, 170, 172, 187, 195–6

East Zinc Mine, 101–18
Eastern Counties Leather case, 181, 201–15, 229–40, 243–51
   legal, 6, 13, 20–1, 23–6, 181, 203, 217
EC (electrical conductivity), 285, 291, 303–4
   Marl Pit, 318–20, 324–6
EDB (dibromoethene), 165, 169–80
EDC (dichlorethene), 165, 169–80
Eh, 37, 77, 88–91, 108–12, 268–75
Environment Agency (EA), 123, 154, 334, 359–60
   regulatory, 1–13, 15–18, 20–1
Environmental Protective Agency (EPA), 203, 214–15, 334
epidote, 268
episodic contamination, 142–4
Esturine series, 127–8, 347–60
ethene, 203
ethyl benzene, 31, 131, 170, 172
   see also BTEX
Etruria formation, 315–16, 319, 325–6, 328, 330–1
Etton, 347–60
European Commission Communication 96 (59), 15–21
European Directive on the Protection of Groundwater from Dangerous Substances, 6, 8–11, 67, 73, 203, 212–15
European standards, drinking water, 203, 334–5
   spillage, 116, 129, 131, 143–4
   regulation, 8, 24, 333–4
   waste water, 64, 73

fatty acids, 84–5
fault zones, 103, 266, 269, 347–60
feldspars, 258, 268, 270, 276, 278
ferric hydrates, 57–60, 119–12, 150–3, 271–5
ferrihydrite, 109–18
ferroaluminium–magnesium brines, 269–70
filters, 199
finite-element grid, 192–5
fisheries, 95, 116
fissure flow systems, 27, 88–91, 203–5, 219–27, 229–47
fissures, 123–44, 128, 335, 347–60
flow, 37–47, 77, 107, 127, 211
FLOWPATH program, 144
fluoride, 82–91
flushing, 101, 112–18, 131–2
fly ash, 20, 49
forestry, 64–6, 68
foreseeability, 6, 20–1, 23–6, 211–14
Fracture Index (FI), 299, 301
fractures, 94, 106–7, 132, 143–4, 258
   Carboniferous, 231, 245–7, 281–310, 291–302
Fuller's Earth, 284, 287, 293
fungicides, 334, 339
furans, 20, 63

galena, 93–4
gamma activity, 261
gamma-logging, 126, 236, 352
gaseous discharge, LRW injection, 278
GCA (granular activated carbon), 33, 161–3, 174–5, 178–80, 356

germanium, 103
gibbsite, 271
glacial sands, 35–48, 75–80, 127–8, 219
   USA, 165–80, 185–95
glyphosate salt, 339
goethite, 110–112
goslarite, 109, 111
GPZ (Groundwater Protection Zones), 7, 9–13, 15, 76
Graham v. Re-Chem International Ltd., 15, 20
gravel, 148–50, 159–63, 185–92
Great Bridge, 313–31
green algae *see Chlorella vulgaris*
greenockite, 111
Guideline concentrations, 16, 24–5, 28, 33, 203
guidelines required, 144
Gwash River, 123–44
Gwenap Planning District, 93–9
GWMN (groundwater monitoring network), Tomsk, 257, 261–3, 265–7, 275–80
gypsum, 57–60, 111, 270–1

haematite, 111, 271
half life, 334, 336, 338, 341–44
halotrichite group, 271
Hampton catchment, 339
Heathrow, 147–56
Helpston, 5, 128, 347–60
herbicides, 5, 8, 334, 339
Hertfordshire PWS1, 75–91
heterogenic aquifers, 35–48, 49, 75–91, 200, 293–310
   hydrocarbons, 123–44, 165–70, 179–80
historical responsibility, 6–7, 11–13
Hutching and Harding Ltd., 201
hydration, 109–18, 337–8
hydraulic barrier, Krasnoyarsk, 266
hydraulic conductivity, 41–7, 94–5, 194–200, 231–2
   Lincolnshire Limestone, 132, 134
   Siberia, 266–70, 271–5, 276–9
   tar wastes, 293–310, 315–16
hydraulic gradient, 293–302, 316, 352–6
   and hydrocarbons, 128–9, 134–44, 147–52, 167–8, 320
   Merrimack, 185–96
   mines, 104–7, 276
   USA vertical, 170, 172, 175, 190–2
   vertical, 201–5, 231–2, 240–2, 291–307
hydrocarbon pollution, 5–6, 121–80, 281–311
hydrocarbons, C_{16}–C_{23}, 284
hydrochemical controlled front, 267–79
hydrogen, 278
hydrogen chloride, 67
hydrogoethite, 271
hydromicas, 258, 278
hydrophilic and -phobic MTBE, 30–3

incineration, 20, 64–7, 73–4, 308–10
injection, LRW, 255–63, 265–7, 275–9
insecticides, 334, 336–44
insurance, 6
Integrated Pollution Control (IPC), 7
investigation details, 232–40, 286–91, 335–40, 347–56
   Chalk, 77–81, 219–227
   USA, 168–71, 183–7
iodine, 35–48, 131, 266

ion exchange, 88, 174
Ionoxynil octanoate, 339
iron, 37, 51–3, 82–91, 325
  acid mines, 49, 90–1, 93–9, 101, 107–18
  leach-mining, 269–79
  oxides, 51, 57–60
  precipitation, 150–3, 174, 199
  tar wastes, 284–5, 299–305
iron–aluminium co-precipitation, 271–5
2,D-Isoctyl, 339
isoprotuon, 334, 336–7

jarosite, 109, 111, 271

Kalpitiya, 333, 336–44
kaolinite, 36, 258, 276, 278
kaolinite-hydromica, 268, 270
Kara Sea, 255–6
Karakatinsky artesian basin, 266–75
karsification, 123, 128, 142–4, 281
Kelloways Clay, 127–8, 347–56
kerosene, 147–56
knowingly permitting, 6–7, 13, 15–17, 20–1, 23–6,
    213–4
Krasnoyarsk-26, 253, 255, 265–7
krypton[85], 266
Kysykl-Kum mining area, 267–76, 278–9

lagoons, 51, 53–4, 59, 75–8, 181
  acid mines, 97–9, 101, 117–18
  acid tars, 284–6, 294–5, 306–7
  sludge, 64–6, 68–70
  slurry, 75, 77, 80–1, 91
landfill, 3–5, 12–13, 83, 90–1, 313–31
  acid tars, 284–5, 293, 307–10
  Hertfordshire, 64–6, 73–5
  and Mecoprop, 347, 352–60
  movement, 159–60, 163, 307–8
  Stamford quarry, 123–6, 142–3
lathanium, 271
LDS (Liquid Digested Sludge), 63, 70
leachate, 51–60, 82–91, 334–40
  Mecoprop, 347, 352, 358–60
leaching, mines, 95, 109–14, 267–75
  nitrogen, 63–4, 67–74, 75–91
  tracer, 37–47
  unsaturated zone Stamford, 134–5
lead, 10, 63, 67, 174, 271
  acid mine, 93–4, 101, 103, 116
  acid tar, 284–5, 299–305
  fuel ash, 51–4, 56, 58–60
  Marl Pit, 316, 318–26
leakage, 147, 165
legal case law, 4–13, 15–21, 23–6, 203,
    211–15
Lemna minor, 358–9
liability, 1, 6, 15–21, 23–6, 212–14
Lias, 127–8, 159, 347–60
lime, 64, 97–9, 104–6, 117–18, 287
limits, European, 64, 116, 143
  and MTBE, 29–33, 129–31
  pesticides, 4–5, 358–60
Lincolnshire Limestone, 123–45, 159, 347–60
lithium, 54, 56, 59, 271

LNAPLs (light, non-aqueous phase liquids), 10–11,
    287–8, 307–8
  spillage, 121–2, 131–9, 142–4, 165–80
local authority powers, 7–13, 16–18
longitudinal dispersivity, 37–48
LRW injection, 265–268, 275–9
lubricating oils waste, 281–310
LUS (Liquid Undigested Sludge), 63–4
lysimeters, 337–8, 340–4

magnesium, 37–9, 52–5, 58–60, 78–91, 299–305
  and mines, 103, 107
  Siberia, 260–3, 269–79
maintenance problems, 197–200
mallardite, 110
manganese, 37, 82–91, 174, 261, 284–5
  and mines, 101, 103, 107–12, 116–18, 269–79
  and PFA, 52–3, 59–60
Margerson and Hancock v. J. W. Roberts Ltd, 20
Marholm–Tinwell Fault, 125–8, 347–60
Marl Pit, 313–31
mass balance, 88–91
Mayak, 253, 255, 265–7
MC (methylene chloride), 165, 169–80
Meaford, 54–60
Mecoprop, 333–4, 338, 347–60
melanterite, 109–11
mercury, 63, 284–5, 316, 320, 324–6
metabolites, pesticides, 336–44
metals, 5, 25–6, 49, 93–9, 101–20, 199
  acid tars, 284–6, 299–305
  inorganic sources, 49, 51–60, 63–75
  radionuclides, 255, 260, 263–76
methane, 151, 175, 278
methodologies, 11, 144, 183, 247–9
  point sources, 232–40, 245–50
Methylarsonic acid, 339
methylene chloride, 172
2-methyoxy-2-methylpropane see MTBE (methyl
    tertiary butyl ether)
Metolachlor, 335
micas, 36, 258, 268–80, 276
Mickley seam, 285, 286
microbial acitivity, 49, 69, 274–9
migration, acid tars, 223–7, 287–310
  LRW, 261, 266, 276–9
mineral oil, 203, 265–80, 284–5
mineralogical controlled front, 267–79
mines, 93–9, 101–20, 267–75, 283–310
MINTEQ2 program, 110–12
modelling, 9, 12–13, 27, 340, 356–60
  dispersion rates, 35–47, 144
  heterogenic aquifers, 225–7, 262–3
  Long Island, 172–80
  Merrimack, 187, 192–6
  mine discharge, 98–9, 110–16
  radionuclide migration, 260–3
  sorption, 35–47
  SPZ, 144
  WATEQ4F program, 57
MODFLOW/MODPATH programs, 144
molybdenum, 51–4, 56–60, 63, 275, 284–5
monitoring, 24, 77–81, 229–52, 308
  hydrocarbons, 121–2, 155–6, 178–80

Marl Pit, 320–31
Merrimack valley, 185–92, 196–200
mines, 91, 95–9, 104–18
pesticides, 336–44, 347–60
Stamford, 135–44
Tomsk-7, 259–63, 265–7, 275–80
montmorillionite, 258, 276, 278
monzanite, 268
Mount Wellington mine, 93–6
MTBE (methyl tertiary butyl ether), 23, 25–6, 27–34, 282
spillage, 123–45, 162–3, 169–80
mullite, 51

N-availability, 63, 67–8
Nangiles mine, 95–9, 117
napthalene, 134–5
nasturan, 268
National Rivers Authority (NRA), 3, 5–12, 15–18, 213, 320
Cambridgeshire, 123–45, 154, 358–60
pesticides, 347, 352, 358–60
Wheal Jane, 93, 95–9
Nene, 159–63
neodynium, 277
neutralization, pH barriers, 270–9
newberryite, 271
nickel, 63, 284–5, 299–305, 356
acid mine, 101, 103, 108, 116
fuel ash, 56, 58–60, 512–4
Marl Pit, 316, 318, 320–6
Siberia, 261, 271, 276
niobium, 266
nitrate, 77, 82–91, 103, 151, 260
agriculture, 3–4, 9–10, 13, 49
breakdown, 63, 67–74, 276
PVA, 54–5, 59
radionuclides, 260–3, 268–79
Nitrate Sensitive Areas (NSA), 7, 9–10, 18, 68–9
Nitrate Vulnerable Areas (NVA), 7, 9–10, 18
nitric acids, 67, 276–9
nitrite, 77–91, 82–91, 270–9
nitrogen, 63–75, 274–5, 278
NLF, 291
Northampton Sands, 127–8, 159, 354
NSO (polar resins), 284–5
Nuratinskaya, 269–76
Nuratinsky, South mountains, 268

Ob river, 255–8, 259, 276
Obsky artesian basin, 276–7
oils, mineral, 284–5, 304
on-site analysis, 237, 245
Oncorhynchus mykiss, 358–9
Oolite series, 127, 144, 347–56
ore precipitation, radionuclides, 267–79
organochlorines, 23–6, 33, 63, 69, 334–44
organophosphates, 339
otavite, 110–11
oxalates, 112, 276
Oxford Clay, 127–8, 347–51
oxidation, 108–18, 270–9, 334, 337–8
Long Island, 174–5, 178–80
oxidation–reduction fronts, 267–79

oxides, 52–3, 199
oxygen, 64–6, 108, 151, 278, 335
oxyhydrates, 33, 41, 150–2
Ozersk see Mayak

PACT (powered activated carbon treatment), 174
paddy, 340, 343
PAHs (Polycyclic Aromatic Hydrocarbons), 63, 161, 284–5
Pakefield, 5
pancake, 147–56, 161–3, 165–80
Paraquat, 339
passive remediation, 153, 161–3
pathogens, 63, 67, 69
pathway defined, 16
pathways, 27, 240–50, 281–2, 284–310
Lincolnshire Limestone, 132–44, 352–60
Marl Pit, 315–6, 329
Merrimack valley, 185–96
mines, 94, 104–6, 116–18, 293–310
PCBs, 63, 67, 284–5
PCE (perchloroethylene or tetrachloroethene), 4, 23–6, 260
Chalk, 201–14, 218–27, 229–30, 243–7
Merrimack, 188–9
Stamford, 129–31, 137–44
pesticides, 4–6, 12, 63, 281–2, 333–45, 347–360
petroleum spills, airport, 121–2, 147–57
distribution terminal, 121–2, 165–80
MTBE, 25–6, 27, 29–33, 162–3
service station, 21, 121–2, 159–63
storage depot, 121–2, 123–44
PFA (pulverized fuel ash), 51–61, 67
pH, 25–6, 36–7, 52–4, 84–5, 174
acid tar waste, 285, 291, 300–4
leach-mining, 269–79
Marl Pit, 316–25
mine water, 95–9, 101, 107–18
and pesticides, 335
phenols, 63, 131, 340
acid tars, 281–2, 299–305, 313–31
phenoxyl acid, 339, 347–60
phosphate, 63, 67, 269–79
phthalate esters, 63
piezometers, 31–2, 234, 250–1, 261, 320
pesticides, 335–8, 340–5
piezometric heads, 190–5, 204, 258–63, 292–302
planning, 7–8
plug failure, 95, 117–18
plume, 5–6, 54, 97–9, 195, 352–60
hydrocarbons, 135–8, 147–56, 161–3
Marl Pit, 330
modelled, 5–6, 11–12
PWS1, 82–91
radioactive, 253–8, 261–3
Sawston, 5–11, 20, 243–51
plutonium[239], 260–3, 274, 278–9
point sources, 5–10, 75–91, 217–28, 229–52
pesticides, 344, 356–60
polishing pond, 117–18
polygons, Siberian LRW, 256–63, 266
pore space, 11, 35–48, 205, 221
radionuclides, 270–1, 275–9
porewater, 3, 11, 32, 251, 335–8

Chalk, 205–11, 227, 236–40, 243–7
and minerals, 51–60, 90
porosity, Coal Measures, 283–310
    leach-mining, 270–1, 275–9
    and pesticides, 341–4
    tracer tests, 36, 43–7
potassium, 37, 39, 52–5, 58–60, 77–91
    mines, 103, 109, 269–79
praesodynium, 266
precipitation, 43, 104, 109–18, 267–79
propane, 175
Propanil, 335
pump, 326–31, 352, 356
pump and treat, 10, 33, 80, 181
    Merrimack factories, 190–200
    spillage, 152–6, 161–3, 172–80
pyrite, 51, 93–4, 103, 109–14

quality, 4, 8–13, 18–21, 24–6, 27, 129
    Marl Pit, 318–20, 324–5, 330
    mine discharge, 95–9, 107–8
    monitoring, 3, 5–13, 240–50, 300–3
    of plume remediation, 11–12, 81–91, 129
    Siberia, 258–63
quartz, 36, 51, 110–11, 258, 268, 270, 278

radionuclides, 35–48, 253, 255–80
    discharge, 253, 255, 257, 260–3, 266, 269–79
radium, 271
rainbow trout see Oncorhynchus mykiss
recharge, 88–91, 93–5, 104–18, 258
    Merrimack, 189, 193, 197–200
    Stamford, 128, 132, 162–3, 347
recovery pit, 162–3
recycling sludge, 63–75
redistribution of elements, 268–80
redox, 43, 51, 274–9, 356
reduction, 85, 109–18, 174
remediation, 6–7, 15–16, 307–10, 358–60
    hydrocarbons, 32–3, 147–56, 161–3, 172–80
    Merrimack factories, 196–200
    methodologies, 11, 75–91, 152–6, 307–10
    mine water, 99, 101–18
Remediation notices, 17–18
resins, tarry, 284–5
responsibility, 3, 6–8, 11–13
retardation, 31–2, 35–48, 142, 265–79
rhodium[106], 266
rhodochrosite, 110–11, 118, 271
risk assessment, 8–13, 16, 25–6, 169–72, 214–15
River Purification Boards, 8–9
Rock Quality Designation (RQD), 299, 301
römerite, 109
rubidium, 58, 266
ruthenium[106], 260, 266
Rylands v. Fletcher, 13, 20–1, 211–13

sampling, 219–24, 232–40, 248–51, 289–91
    Heathrow, 150–2
    injection LRW, 276
    mines, 97, 101–3, 108
    pesticides, 336–42
    PFA, 54–60
    standards, 24, 36, 107, 214, 245

sandy aquifers, 31–2, 257–61, 265–7, 276, 336–44
saturation indexes, 111
Sawston, 201–15, 229–50, 243–251
    see also Eastern Counties Leather Co.
Scottish Environmental Protection Agency (SEPA), 6,
    8–9
selenium, 51–4, 56–60, 63, 275, 284–5
service station spillage, 121–44
settling ponds, 53–4, 255–7
settling tanks, 117–18, 208
Seversk, 253, 255–63, 265
sewage, 9, 49, 63–74, 212
Siberia, 253–64, 265–80
siderite, 103, 110–12, 118, 271
silica, 52–4, 57–60, 103, 107–10
    Siberia, 37, 261–3, 271–9
silicate phase, 41, 109–12
silicic acid, 276–9
silver, 63, 101, 103
Simazine, 5, 334
site personnel, 232
site specific work, 9, 11–13, 151–6
skimming pumps, 33, 152–3
slag, 314–16
sludge, treatment, 63–79, 117, 174
slurry, 7, 49, 75–8
smithsonite, 110–11
Smithtown Clay, 165–70, 172, 175
sodium, 37, 39, 77–8, 84–91
    and mines, 103, 107–8, 269–79
    nitrate brines from LRW, 276–9
    tracer test, 52–5, 57–60
soil conditioner, 11, 64, 68
soil gas, Chalk, 219, 223–7, 232
    hydrocarbons, 123, 135, 151, 161, 168–80
    Sawston, 205–11, 237–40, 243–51
    solvents, 183–90, 196–80, 232, 240–9
soil-water partition coefficient, 25, 27, 29–34, 135–44,
    172
solubility, hydrocarbons, 131–8, 147, 152–3, 172
    minerals, 5, 51–61, 109–118, 270–9
    MTBE, 29–33, 30–1, 123–44
    pesticides, 334–44, 358–60
    solvents, 123–44, 172, 201–3, 227, 240–51
solution mining, 267–79
solvent usage surveys, 28–33, 63, 227
sorption, 11, 52–3, 242, 248, 334–5
    kinetics, 35–48, 88–91
    radionuclides, 268–9, 278–9
source defined, 16
Source Protection Zones, 144
Soviet Union, 123, 253, 255–64, 265–80
sphalerite, 103, 109
sphene, 268
spillage, petroleum, 25, 121–80, 211–14
Sri Lanka, 334, 336–44
Sr[85], 255, 260
Staffordshire Blue Bricks, 313
Stamford, 123–45, 347–60
storage, 183–5, 196–200, 211–14
    sludge, 63–4, 68–70
    tanks, 7–8, 203, 255–6
strategies, 12–13, 144, 199, 248–50, 308–10
strontium, 35–47, 53–4, 56–60, 271

strontium[90], 255, 260, 266
sulphate, 39, 75, 77, 84–6, 198
  acid mines, 101, 103, 107–9, 117–18
  PFA, 54–5, 57–60
  radionuclides, 260–3, 267–79
  reduction, 117–118, 274–8
  tar wastes, 284–5, 291, 299–305, 316
sulphide, 58, 284–5
  Marl Pit, 316, 320, 324–5
  mines, 84–5, 93, 103, 109–18, 276
  oxidation, 117–18, 278–9
sulphur, 52–3, 57, 111
sulphuric acid leaching, 268–79, 281, 284
Superfund, 213–14
surcharging, 319–31
surface reservoir, LRW, 253–6, 266–7
surfactants, 63

tailings dam, 97–9, 101
Tame river, 313, 316–20, 318–19
tar, 281–310, 314–31
target defined, 16
TARGET program, 352–60
TBME (Tertiary Butyl methyl Ether) see MTBE
TCA (1,1,1,-trichloroethane), 23, 155–6, 172
  Chalk, 201–14, 220–7
  Coventry, 240–7
  Merrimack facility, 188–9, 196
  Stamford, 20, 129–31, 137–44
1,1,2,-TCA (1,1,2-trichloroethane), 131
TCE (trichloroethene), 23–6, 155–6
  Chalk, 201–11, 218–27
  Coventry, 229–30, 240–7
  Sawston, 201–11, 229–30, 243–50
  Stamford, 129–31, 136–44
  USA, 165–80, 188–9, 196–200
TeCE, 244
technelium[99], 266
technogenic leaching solutions, 271–9
temperature, 30, 261, 281, 288
  mines, 97, 101–3, 278
Tertiary Butyl Methyl Ether see MTBE
tetrachloroethene see PCE (perchloroethylene)
thermal drying sludge, 64, 66
thermal oxidation, 174–5, 178–80
Thiobacillus ferrooxidans, 109
thiocarbamate, 339
thiocyanate, 284–5
thorium, 271
timing migration, 5–6, 39–41, 54–60, 144, 211
  Siberia, 261, 266, 275, 278–9
tin, 93–4
titanites, 268
titanium, 271
TOC (Total Organic Carbon), 37, 84–91, 337–8
  acid tars, 285, 299–305, 318
toluene, 29–33, 196
  extractable matter, 284–5, 318
  spillage, 131, 138–44, 169–70, 172
  see also BTEX
Tom river, 256–63, 266, 276–7
Tomsk-7, 253, 255–64, 265–6, 275–9
tourmaline, 268
toxicity standards, 29–33, 339, 356–60

tracer tests, 35–48, 75–7, 90–1, 353
  chlorinated solvents, 205–11, 237, 242
  hydrocarbons, 32–5, 144
  tar wastes, 282, 296
transmissivity, Carboniferous, 231, 283–310
  Chalk, 81, 205–11, 232
  Marl Pit, 316
  Merrimack deep aquifer, 189–90
  mine voids, 97–9, 293–305
  Siberian aquifers, 258, 276
  Stamford, 128, 134, 276
Triassic Sandstones, 4, 10–13, 30, 181
triazines, 334, 339
trichloroethylene, 4, 6
1,1,2-trichlorofluorethane, 23
trichloromethane, 316, 320, 324–5
tritium, 43, 90–1, 260, 266
tungsten, 93–4

Upper Esturine Series, 347–51
Urals, 266
uraninite, 253, 255–6, 268–79
uranium deposits, 267–76, 278–9
USA liability, 213–14
USC (Undigested Sewage Cake), 63
Uzbekistan, 253, 267–76

vanadium, 51–3, 58–60, 271, 275, 284–5
VC (vinyl chloride), 188–9, 203
vertical migration, 181, 301–10
  chlorinated solvents, 190–2, 201–11
  hydrocarbons, 167, 172–3, 175
  point sources, 231, 240, 245, 250
vitrifaction, 67, 308
VOC (volatile organic compounds), 174–80, 308
  monitoring, 165, 168–72, 183–90, 195–200
voids as pathways, 95–9, 293–305, 315
  acid mines, 104–6, 112–18
volatility, 23, 34, 70, 278, 334
  ammonium, 69–74
  fuel ash minerals, 51, 53
  MTBE, 25, 33
  solvents, 201, 219–20, 232–40, 247–51
  Stamford, 132–5, 165–80
vulnerability mapping, 9

walkovers, 232, 245, 248
Wansford, 159–63
waste water, 7–8, 57, 64, 152–6, 161–3, 196–80
WATEQ4F program, 57
water discharge, 93–9, 104–18, 358–60
Water Protection Zones, 8–10, 76–8
Water Resources regime, 6–13, 15–21, 25–6
water table, 128–34, 142–5, 147–56, 172
weathering, 51–60, 94, 128
  Merrimack facility, 187, 189, 192
  tar, 284, 298, 305
  see also Chalk, putty
Welland River, 123–44, 134
West Zinc Mine, 101–18
Wheal Jane, 25–6, 93–9, 117
WHPA program, 144
willemite, 111
wolframite, 93–4

Works Notices, 16–18

xenon[131] and [133], 266
m-xylene, 29–33, 131, 170, 172
p-xylene, 32, 131, 170, 172
o-xylene, 131, 170, 172
xylene, 134, 140
    see also BTEX

Yenisey river, 255–6

yttrium[90], 266, 271

Zhelenogrosk see Krasnoyarsk-26
zinc, 51–3, 58–60, 67, 271
    acid tar, 284–5, 299–305
    inorganic sources, 82–5, 90–1
    Marl Pit, 316, 318, 320, 324–6
    mines, 97, 101–18
zircon, 268
zirconium[65], 266